THE GRASS CROP

The physiological basis of production

SERIES EDITOR
E. H. Roberts
Professor of Crop Production, Department of Agriculture and Horticulture, University of Reading

THE GRASS CROP

The physiological basis of production

Edited by
Michael B. Jones
and
Alec Lazenby

London   New York
CHAPMAN AND HALL

First published in 1988 by
Chapman and Hall Ltd
11 New Fetter Lane, London EC4P 4EE
Published in the USA by
Chapman and Hall
29 West 35th Street, New York NY 10001

© 1988 Chapman and Hall Ltd

Printed in Great Britain by
St Edmundsbury Press Ltd
Bury St Edmunds, Suffolk

ISBN 0 412 24560 4

All rights reserved. No part of this book may be reprinted, or reproduced or utilized in any form or by any electronic, mechanical or other means, now known or hereafter invented, including photocopying and recording, or in any information storage and retrieval system, without permission in writing from the publisher.

British Library Cataloguing in Publication Data

Jones, M. B.
  The Grass Crop.
  1. Grasses
  I. Title   II. Lazenby, Alec
  633.2′02′0912   SB197

ISBN 0-412-24560-4

Library of Congress Cataloging-in-Publication Data

The Grass Crop.

  Includes bibliographies and index.
  1. Grasses.   2. Forage plants.   I. Jones, M. B.
(Michael B.), 1946–   .   II. Lazenby, Alec.
SB197.G75   1988   633.2   87-24245

ISBN 0-412-24560-4

Contents

Preface	ix
Contributors	xi
Commonly used units and abbreviations	xiii

1 Introduction – the history of improved grasslands   1
E. L. Leafe
1.1 The world's grasslands   1
1.2 The Graminae   3
1.3 Tropical and temperate grasses – C_3 and C_4 pathway   4
1.4 Improved grasslands   7
1.5 Recent developments in understanding the physiology of the grass crop   14
1.6 Potential production of pastures   17
References   20

2 The grass plant – its form and function   25
M. J. Robson, G. J. A. Ryle and Jane Woledge
2.1 Introduction   25
2.2 The seed and seedling establishment   26
2.3 The vegetative shoot   30
2.4 The reproductive shoot   35
2.5 Tiller production   37
2.6 Biomass production   39
2.7 Leaf photosynthesis   42
2.8 Meristematic activity and the utilization of assimilates   50
2.9 Respiration   55
2.10 The plant and the community   62
References   73

3 The regrowth of grass swards   85
Alison Davies
3.1 Introduction   85
3.2 Regrowth in vegetative plants   87

3.3	Changes in sward structure during regrowth	94
3.4	Regrowth in different defoliation systems	101
3.5	The regrowth of reproductive swards	107
3.6	The influence of the animal on sward regrowth	113
3.7	Conclusions	117
	References	117

4 The effects of season and management on the growth of grass swards 129
A. J. Parsons

4.1	Introduction	129
4.2	Seasonal changes in the environment	130
4.3	Seasonal production	134
4.4	Management	148
4.5	Physiological basis for optimizing production from grassland	164
	References	169

5 Mineral nutrients and the soil environment 179
D. W. Jeffrey

5.1	Introduction	179
5.2	The role of associated organisms	181
5.3	The continuum between soil and plant	183
5.4	A view of the essentiality of elements – defining essentiality in physiological and ecological terms	184
5.5	Seasonal changes in nutrient ion deployment within the plant	186
5.6	Macronutrient cycles	190
5.7	The competitive status of grasses and their macronutrient supply	195
5.8	Acclimation of grass plants to growth-limiting soil characteristics	197
	References	201

6 Water relations 205
M. B. Jones

6.1	Introduction	205
6.2	The hydrological cycle	206
6.3	Development and effects of water stress	217
6.4	Implications for yield	234
6.5	Conclusion	236
	References	236

Contents

7	**Physiological models of grass growth**	**243**
	J. E. Sheehy and I. R. Johnson	
	7.1 Introduction	243
	7.2 Model terminology	244
	7.3 Requirements for crop physiological models	244
	7.4 Light interception and photosynthesis	246
	7.5 Respiration	251
	7.6 Partition of assimilates	254
	7.7 Rate of utilization	260
	7.8 Transformation of asimilate into plant tissue	266
	7.9 Regrowth	268
	7.10 Death of tissue	268
	7.11 Concluding remarks	269
	Acknowledgements	269
	References	269
	Definition of symbols	272
8	**The effects of pests and diseases on grasses**	**277**
	R. T. Plumb	
	8.1 Introduction	277
	8.2 Insect pests	278
	8.3 Nematodes	283
	8.4 Fungi	284
	8.5 Bacteria	296
	8.6 Viruses	296
	8.7 Concluding remarks	302
	References	302
9	**The grass crop in perspective: selection, plant performance and animal production**	**311**
	Alec Lazenby	
	9.1 Introduction	311
	9.2 Selection	312
	9.3 Growth and DM yields	317
	9.4 Quality	326
	9.5 Utilization	328
	9.6 Meeting the nutritional requirements of the animal	330
	9.7 Systems based on grass	332
	9.8 Postscript	353
	References	354
	Index	**361**

Preface

Grass is a very important world crop. In some countries, for example the UK, Australia and New Zealand, animal products from grassland make a greater contribution to the value of agricultural production than does any other crop. Yet research being undertaken to further our understanding of the factors affecting the growth and productivity of grasslands has trailed in the shadow of the determined efforts made to improve our knowledge of cereals and, to a somewhat lesser extent, legumes. However, in spite of its low profile, grassland research has resulted in considerable advances in our knowledge in the last 20 years, and we feel that this book provides a timely opportunity to bring together some of this work in a review of what is primarily the ecophysiology of the temperate grass crop. Unlike other crops grown for their grain or vegatative parts, grass and grassland products are used almost entirely for the feeding of ruminant animals; the interaction of the sward and the animal thus adds an extra dimension to investigations of the productivity of grassland.

No one author could adequately encompass the breadth of work covered in the book. Acknowledged experts have therefore been selected as contributors to provide an up-to-date review of their own specialized areas. Whilst multi-author texts can cause problems of lack of uniformity of approach, each contributor has been made aware of the contents of the other chapters in an attempt both to provide continuity and to prevent glaring overlaps.

In Chapter 1 Ted Leafe reviews the world distribution of natural and managed grassland, and provides an historical perspective in which he examines the importance of grasslands in Britain during the evolution of current farming systems. He also provides a review of some of the more important landmarks in grassland research since the Second World War. The next chapter, by Mike Robson, George Ryle and Jane Woledge, forms the foundation for the rest of the book. The physiology of the individual grass plant is considered in relation to its structure, as subsequently is the community as a model seedling sward in a controlled environment. In Chapter 3 Alison Davies moves further in the direction of the community of plants which form the grass sward, and reviews perhaps the most characteristic feature of the grass crop – its regrowth following defoliation by either cutting or grazing. Tony Parsons further develops this theme of the plant community in its field environment in Chapter 4. In particular, he reviews the dual influences of seasonal changes in climate and harvesting management by cutting or grazing, on the growth and productivity of grass swards. The influences, on grass growth, of two particular environmental constraints are examined in the next two chapters. In Chapter 5 David Jeffrey examines the influences of nutrient availability on grasses, and emphasizes the important role of the rhizosphere in plant–soil interactions, whilst Mike Jones in Chapter 6 discusses the effect of water stress on growth and the possibilities for adaptation in the field environment. Much of the physiological information discussed in previous chapters is brought together in Chapter 7. Here John Sheehy and Ian Johnson demonstrate how mechanistic models can provide a powerful tool for understanding the complex of interactions between the grass plant and its environment. The growth of all crops can be severely restricted by pest and disease attack, and in Chapter 8 Roger Plumb reviews the wide range of those found on temperate grasses, and how they reduce yield. Finally, in Chapter 9 Alec Lazenby provides an overview of the grass crop, laying particular emphasis on its ultimate fate, namely its utilization by the ruminant animal.

The book provides a useful reference source for those seeking to understand or unravel the influences of both the environment and management on grass productivity. It should therefore be of interest to researchers in all areas of crop productivity, to those studying crop physiology and agricultural botany, and to advanced students in agriculture. We also venture to hope that the book might make a small contribution to the improved management of grasslands for animal production.

Michael B. Jones
Alec Lazenby

Contributors

A. Davies
AFRC Institute for Grassland and Animal Production
Welsh Plant Breeding Station
Plas Gogerddan, Aberystwyth
Dyfed SY23 3EB, UK

D. W. Jeffrey
Department of Botany
Trinity College, University of Dublin
Dublin 2, Ireland

I. R. Johnson
Department of Agronomy and Soil Science
The University of New England
Armidale, NSW, Australia

M. B. Jones
Department of Botany
Trinity College, University of Dublin
Dublin 2, Ireland

A. Lazenby
University of Tasmania
GPO Box 252C, Hobart
Tasmania 7001, Australia

E. L. Leafe
3 Manor Barns, Middle Street
Ilmington, Shipston on Stour
Warwickshire CV36 4LS, UK

A. J. Parsons
AFRC Institute for Grassland and Animal Production
Hurley Research Station, Hurley
Berkshire SL6 5LR, UK

R. T. Plumb
Crop Protection Division
AFRC Institute of Arable Crops Research
Rothamsted Experimental Station
Harpenden AL5 2JQ, UK

M. J. Robson
AFRC Institute for Grassland and Animal Production
Hurley Research Station, Hurley
Berkshire SL6 5LR, UK

G. J. A. Ryle
AFRC Institute for Grassland and Animal Production
Hurley Research Station, Hurley
Berkshire SL6 5LR, UK

J. E. Sheehy
AFRC Institute for Grassland and Animal Production
Hurley Research Station, Hurley
Berkshire SL6 5LR, UK

J. Woledge
AFRC Institute for Grassland and Animal Production
Hurley Research Station, Hurley
Berkshire SL6 5LR, UK

Commonly used units and abbreviations

a.i.	Active ingredients
CGR	Crop growth rate
C_4 pathway	C_4 dicarboxylic acid pathway
C_3 plants	Plants fixing atmospheric CO_2 directly through the RPP-pathway
C_4 plants	Plants having the C_4 dicarboxylic acid pathway
DMD	Dry matter digestibility
DSE	Dry sheep equivalents
D-value	Proportion of digestible organic matter in dry matter
E	Evapotranspiration
E_A	Net assimilation rate or unit leaf rate
F_A	Leaf area ratio
g	Leaf conductance to water vapour
g	Gram
GJ	Gigajoule (joules $\times 10^9$)
h	Hour
ha	Hectare ($10\,000\,m^2$)
I	Irradiance
IRGA	Infrared gas analysis
J	Joule
kg	Kilogram
LAI or L	Leaf area index
ME	Metabolizable energy
MJ	Megajoule (joules $\times 10^6$)
P	Photosynthesis (single leaf or canopy)
P_m	Light-saturated photosynthesis or photosynthetic capacity
P_n	Net photosynthesis (single leaf or canopy)
P_g	Gross photosynthesis (single leaf or canopy)
PAR	Photosynthetically Active Radiation (400–700 nm)
Q_{10}	Temperature coefficient related to thermochemical reactions at a temperature separation of 10 °C

R	Respiration
RuBP	Ribulose 1,5-bisphosphate
RPP-pathway	Reductive pentose phosphate pathway
R_w	Relative growth rate
s.e.	Standard error
S/R	Stocking rate
t	Tonne (10^6 g)
UME	Utilized metabolizable energy
W	Dry weight of plant biomass
W	Watt (1 J s^{-1})
WSC	Water-soluble carbohydrate
WUE	Water use efficiency
α	Photochemical efficiency
ψ_a	Atmospheric water potential
ψ_l	Leaf water potential
ψ_r	Root water potential
ψ_s	Soil water potential
ψ_t	Turgor potential
ψ_π	Osmotic potential

CHAPTER 1
Introduction –
the history of improved grasslands

E. L. Leafe

1.1 THE WORLD'S GRASSLANDS

Girdling the Earth north and south of the equator are the world's grasslands. Their distribution is determined by climate, soil and topography, but their existence is due also to biotic factors, principally grazing animals, and fire. Man, also, has extended the world's grasslands by clearance of forests for cropping and for grazing his domestic animals (Fig. 1.1). Indeed, cultivated grasslands, established in regions whose natural climax is forest, are the most intensively used and, agriculturally, the most productive. The evolution and ecology of the grasslands are complex and controversial (Moore, 1964) and man's activities have so modified the world's grasslands that, today it is often difficult to distinguish the natural from the man-made. The casual observer viewing the brown top clad foothills of New Zealand's South Island may be surprised to learn that the species (*Agrostis tenuis*) only arrived in New Zealand with the European settler! There is disagreement as to whether the natural grasslands represent a climax community and about the extent to which the distribution of grasslands is determined by climate as against soil and biotic factors (Moore, 1964). Limited rainfall, or at least the occurrence of a dry season, especially when reinforced by other factors, particularly fire and grazing, appears to be a common factor preventing the formation of a forest canopy.

In all, the grasslands account for almost one-quarter of the world's cover of vegetation (Shantz, 1954). The natural grasslands – the Prairies of North America, the Pampas and Savannah of South America, the Steppes of Asia, the Veldt of Africa and the grasslands of Australasia are perhaps the best known – range over tropical and temperate regions with widely varying amounts and patterns of rainfall (Moore, 1964). There is great diversity of species, but the Graminae dominate and give their name to the ecosystem. Surprisingly, of the many thousands of species within the Graminae, only about 40 are extensively cultivated in pastures (Hartley and Williams, 1956) and the

Figure 1.1 □ The major grazing areas of the world. ▦, grasses prominent in the natural communities which can be grazed with little or no modification, ■, grasses not prominent in the natural communities which have been modified to produce grasslands. (Adapted from Moore, 1964).

The Graminae

cultivated species are derived from woodland or forest margin habitats rather than great natural grasslands (Hartley, 1964).

1.2 THE GRAMINAE

Of all the families of plants, indisputably we owe most to the Graminae. They provide, directly, the major part of our carbohydrate diet and energy through our consumption of the grains of rice, maize, wheat, oats and barley sorghum, and the refined products of sugar cane. Indirectly, animals that graze pastures, or are fed the fresh or conserved products of grassland and other graminaceous forages, provide us with a major source of protein and fat.

The structural features of the grasses allows them to withstand both grazing and fire, and to survive periods of adverse climatic conditions such as drought and extremes of heat and cold. Growing points, at least in the vegetative plant, are near to the ground, inaccessible to grazing animals and well protected by enveloping leaves. Basal and intercalary leaf meristems enable the partially defoliated leaf to continue growth. Growth of tillers from axillary buds of compressed basal internodes provides limitless opportunities for regrowth, and the rhizomatous and stoloniferous habits are often well developed (see Chapter 3). Grasses range from extremely persistent, long-lived perennials to short-lived annuals completing their life cycle, opportunistically, between the

TABLE 1.1 *The classification of the Gramineae according to Prat (1960)*

Subfamily:		FESTUCOIDEAE
	Tribes:	Festuceae, Hordeae, Agrostideae, Aveneae, Phalarideae Stipeae, Monermeae and the unplaced genus *Beckmannia*
Subfamily:		PANICOIDEAE
	Tribes:	Paniceae, Andropogoneae, Maydeae, and also the 'petites tribus' Arthropogoneae, Boivinelleae, Isachneae, Melinidieae and a number of unplaced genera
Subfamily:		CHLORIDOIDEAE
	Tribes:	Chlorideae, Zoysieae, Eragrosteae, Pappophoreae, Sporoboleae and some unplaced genera
Subfamily:		BAMBUSOIDEAE
	Tribes:	Bambuseae, Arundinarieae, Dendrocalameae, Melocaneae and others not listed
Subfamily:		ORYZOIDEAE-OLYROIDEAE (PHAROIDEAE)
	Tribes:	Oryzeae, Olyreae, and three unplaced genera: *Orthoclada*, *Pariana* and *Streptochaeta*
Subfamily:		PHRAGMITIFORMES
	Tribes:	Arundineae, Danthonieae, Arundinelleae, Aristideae, and fifteen unplaced genera including *Distichlis*, *Ehrharta*, *Elytrophorus*, *Micraira*, *Microlaena*, *Tetrarrhena* and *Uniola*

onset of rains and the dry season. Similarly, their adaptive morphology, such as the reduction of leaf area under drought stress, and their physiological mechanisms such as the storage of reserve carbohydrate, and stomatal and osmotic regulation (Pollock *et al.*, 1980; Jones *et al.*, 1980; Begg and Turner, 1976) (see also Chapter 6) fit them admirably to their niche.

The Graminae is one of the largest plant families, containing more than 600 genera, and Burbidge (1964) describes their systematics as '....not particularly satisfactory'. Of the sub-families (Prat, 1960), the Festucoideae contains all the cultivated grasses of the temperate region, while the Panacoideae and Chloridoideae contain most of the tropical grasses (Table 1.1). The important grasses of temperate agriculture belong to two tribes – the Festuceae and the Agrostideae – and Spedding and Dieckmahns (1972) list *Lolium*, *Phleum*, *Dactylis* and *Festuca* as the genera of most value to British agriculture. The important cultivated grasses of the tropics and sub-tropics are found mainly within the genera *Panicum*, *Pennisetum*, *Andropogon*, *Digitaria*, *Chloris*, *Cynodon*, *Cenchrus*, *Setaria*, *Paspalum*, *Eragrostis* and *Sorghum*, while the two C_3 genera *Phalaris* and *Bromus* are important in regions of high summer temperature and low rainfall. Other genera, such as *Bouteloua*, *Brachiaria* and *Themeda*, are dominant components of some natural grasslands.

1.3 TROPICAL AND TEMPERATE GRASSES – C_3 AND C_4 PATHWAY

The observation by Kortschack *et al.* (1965) and Hatch and Slack (1966) that the biochemical pathway for carbon assimilation is different in some tropical and temperate Graminae generated intense interest in the distinction between the so-called C_4 and C_3 grasses. In the C_4 species 4-carbon acids form the primary stable product of carboxylation, whereas in C_3 species phosphoglyceric acid, a 3-carbon compound, is the earliest detectable product. Furthermore, it was shown that two distinct carboxylating enzymes were involved; ribulose bisphosphate carboxylase oxygenase (RuBP carboxylase/oxygenase) in the initial step of the Calvin cycle in the C_3 species, and an additional carboxylating enzyme, phosphoenolpyruvate carboxylase (PEP carboxylase), in the C_4 species. From an agronomic point of view, the important character is the very much higher rate of carbon assimilation per unit leaf area in tropical C_4 species such as sugar cane, maize, sorghum, *Cynodon* and most *Panicums* compared with temperate C_3 species in genera such as *Lolium*, *Dactylis*, *Agrostis* and *Festuca*. This has important consequences for potential yield and water use efficiency, which is discussed in Chapter 6. Also, C_4 species generally have a higher temperature optimum for photosynthesis and, unlike C_3 species, their leaves do not become light-saturated for photosynthesis at high light intensities.

Associated with the different biochemical pathways in C_4 and C_3 species are other features. First, the possession by C_4 species of a prominent and well-developed bundle sheath, containing specialized chloroplasts, which was found to be the site of rapid fixation of CO_2 (Hatch, 1976); the so-called Kranz anatomy. Second, the long-observed differences between species in CO_2 compensation point; the equilibrium CO_2 concentration reached when a species is placed in a closed, illuminated assimilation chamber (Moss, 1962). C_4 species have a compensation concentration around $5\,\mu l\,l^{-1}$ CO_2, whereas C_3 species have a compensation concentration around $50-100\,\mu l\,l^{-1}$. Third, it was found that C_3 species have a high rate of photorespiration in normal air, while photorespiration is absent or present only at very low level in C_4 species.

The unravelling of the biochemistry of the C_4 and C_3 species and the interrelating of the features of the so-called 'C_4 syndrome' (C_4 pathway, Kranz anatomy and chloroplast dimorphism, low photorespiration and CO_2 compensation point) is one of the most fascinating stories of modern plant biochemistry and physiology (Edwards and Walker, 1983). Briefly, the current view is as follows (Fig. 1.2). In C_4 species there is, first, a carboxylation phase, catalysed by PEP carboxylase. This occurs in the mesophyll cells, using the energy provided by the mesophyll chloroplasts, and is common to all C_4 species. The C_4 acids so formed, primarily aspartate or malate, are transported from the mesophyll to the bundle sheath cells where they are decarboxylated, the pathway varying somewhat between species. Within the bundle sheath cells the CO_2 arising from the decarboxylation of the C_4 acids is joined with ribulose bisphosphate by the reductive pentose phosphate pathway (the RPP pathway,

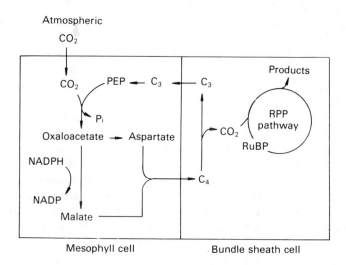

Figure 1.2 The C_4 pathway of photosynthesis, showing the division between mesophyll and bundle sheath cells. NADP is nicotinamide-adenine dinucleotide phosphate oxidized, NADPH is nicotinamide-adenine dinucleotide phosphate reduced, PEP is phosphoenolpyruvate, P_i is inorganic phosphate, RuBP is ribulose bisphosphate.

or Calvin cycle). This reaction, catalysed by RuBP carboxylase/oxygenase which is arguably the most important enzyme in nature and certainly the most abundant in plant kingdom (Edwards and Walker, 1983), produces two molecules of phosphoglycerate which are then reduced to triose phosphate using solar energy trapped by the bundle sheath cell chloroplasts. From triose phosphate the familiar products of photosynthesis, sucrose and starch, as well as other short- or long-term storage products are formed. Sucrose is the principal mobile metabolite, and is transported to meristematic regions of the plant where it is used to provide both energy and the building blocks for growth. Thus, in a sense, C_4 photosynthesis in the mesophyll cells acts as a pump, delivering high concentrations of CO_2 to the bundle sheath cells, where it is incorporated by the Calvin cycle, thus supporting the high rates of carbon fixation characteristic of C_4 plants compared with C_3 plants, which possess only the Calvin cycle.

One feature of the Calvin cycle is that the key enzyme, RuBP carboxylase/oxygenase, has an affinity not only for CO_2 but also for oxygen. Indeed, oxygen competes with CO_2 and the enzyme functions partly as an oxygenase. The result is that a proportion of CO_2 receptor, ribulose bisphosphate, is oxygenated and split to yield one molecule of phosphoglycerate and one of phosphoglycolate. The latter then enters a sequence of reactions leading to the loss of CO_2. This is the process of photorespiration. In the bundle sheaths of C_4 plants the concentration of CO_2 is high and hence the photorespiratory pathway is low or absent, at least at normal CO_2/O_2 ratios. In C_3 plants the concentration of CO_2 at the chloroplast is lower, and significant oxygenation of the CO_2 receptor takes place. At low oxygen concentrations in air this is reduced, whereas at high concentrations it is increased.

The consensus is that the C_4 characteristic is a derived condition since it is polyphyletic, and that C_4 species arose from C_3 progenitors (Woolhouse, 1983). It is common to refer to C_4 species as the tropical grasses and C_3 species as temperate grasses. The distinction is not, however, so clear-cut (Edwards and Walker, 1983). The genus *Panicum* contains both C_3 and C_4 species, as well as species which are considered to be intermediate in characteristics between C_3 and C_4 (Bouton and Brown, 1981), although this has been challenged (Woolhouse, 1983). All of the true temperate grasses possess the C_3 pathway, but *Spartina townsendii* (*sensu lato*), which arose from hybridization between a native and an American species and is widely distributed in coastal areas of England, possesses the C_4 syndrome (Long, 1983; Woolhouse, 1983).

Pastures containing both C_3 and C_4 species exist in regions where there is a suitable seasonal pattern of climate. The C_3 species produce their yield in the cool moist season, or in spring and autumn, while the C_4 species dominate in the hot dry season (Tieszen and Detling, 1983). Examples are *Bouteloua gracilis* (C_4) and *Agropyron smithii* (C_3) from the mixed grasslands of North American Prairies and *Danthonia* species (C_3), *Themeda australis* (C_4) grassland of Australia. The possibility, in suitable climatic regions, of

Improved grasslands

establishing sown pastures containing both C_3 and C_4 species is an attractive one and has been tried, for example, in the Sydney basin of Australia, and in New Zealand, where the *Paspalum*/perennial ryegrass/white clover pasture is said to have the highest potential for production of any (Levy, 1970). In practice, however, the management of such mixed pastures presents difficulties because one or other species tends to become dominant.

1.4 IMPROVED GRASSLANDS

1.4.1 'Natural' grasslands

Although the distinction between 'natural' and 'artificial' grasslands is clear enough when viewed from the extreme ends of the spectrum, there is by no means an obvious separation between the classes. Thus, man has considerably affected the natural grasslands by the deliberate use of fire, by exerting influence on the populations of grazing wild animals, by the introduction of other species of animals, both wild and domesticated, and by other activities. Sometimes these effects have been disastrous (Moore, 1964; Edwards, 1981). The European settler introduced domestic animals into the drier natural grasslands of Australia, Africa and America, with little conception of the precarious and vulnerable nature of these ecosystems compared with the robust man-made grasslands of Europe. The consequences are well documented and, although in the last 50 years a more enlightened attitude has developed towards the management, preservation, and restoration of the natural grasslands (Edwards, 1981), destruction and degradation still continues, especially when drought leads to an imbalance between the needs of the grazing animal and the amount of forage grown, as for example in the Sahel region of Africa today.

Improvement of grassland takes many forms. Many natural grasslands have been improved by the simple expedient of stock control and controlled grazing (Moore, 1962), and this has often led to a change in botanical composition without the deliberate introduction of new species. More recently the productivity of natural pastures has been raised dramatically by the casual or deliberate introduction of legumes, notably in Australia, and this has been linked with the recognition and correction of major and trace element deficiencies, particularly phosphorus and molybdenum. However, not all natural grasslands are on soil of low mineral status: large tracts of the Steppes, the Prairies and the African grasslands are on soils of high fertility and favourable rainfall, and they have become major centres of arable cropping.

A further stage of improvement is reached when new grass species are introduced deliberately, often with an accompanying legume, with or without the destruction of the indigenous grasses. Large areas of natural grassland, particularly in regions where climate and rainfall are favourable, have been improved in this way. Such improved grasslands converge with, and are often indistinguishable from, the grassland derived from forest by man's activity. New Zealand's perennial ryegrass/white clover pastures, for example, are of dual origin (Levy, 1970; Daly, 1972).

1.4.2 Cultivated grasslands – development and improvement of grassland in Britain

Within the cultivated grasslands, improvement usually means the upgrading of pastures of low fertility or ones that have been neglected, but the processes of improvement are similar in both cases. These were areas derived from forest, initially for arable cropping in most cases, and generally the climate was favourable for grass growth. Their establishment, in Europe at least, dates back many centuries. According to the vicissitudes of economics, war and politics, these grasslands have served for much arable cropping as well (Davies, 1960) and, of course, alternate arable cropping and grassland is still common in many areas. Much of the grassland of central England occupies land that was once cultivated, as the pattern of ridge and furrow so clearly shows.

The history of the development and improvement of the cultivated grasslands begins with the history of civilization and the transition from a nomadic life, based on hunting and fishing and the collection of food from the wild, to a settled life in which crops were cultivated and animals domesticated. This transition appears to have occurred in many places in the world over a long period but began more than 10 000 years ago. The evidence of earliest developments is sparse and depends much on the scant remains of bones of animals discarded by early man. Only from the last few thousand years does a more detailed picture emerge and a record of the history of grassland begin to appear. It is in the countries bordering the Mediterranean and in Europe that the best historical evidence exists of the development of grassland. It is beyond the scope of this chapter to trace the development of grassland throughout Europe and the Mediterranean, let alone the world. However, the development and improvement of grassland in Britain, as well as being of interest in its own right, illustrates the stages in the development and intensification of grassland in a temperate climate.

The bones of domesticated as well as wild animals found in association with the dwellings of the neolithic inhabitants of Britain show that they were a settled people, but we have little evidence that their animals were grazed on other than wild vegetation. The Celts, who followed the stone-age people, left evidence, especially in the West Country, of enclosed grazing areas as well as arable land (Hoskins, 1955), and to this extent they may qualify as the first grassland improvers. Enclosure may, however, have had the purpose of protection against marauding animals rather than containment of stock, and improvement – if it occurred – was incidental. Later in this period forest clearance and improved implements enabled more-fertile and heavier soils to be cultivated and it is likely that, as the fertility of the forest soils was depleted by grain growing, land 'tumbled down' to pasture. The Roman period saw the building of towns and roads which survive to this day, and the continued clearance of forest to feed the population, which it is estimated rose to between 0.5 and 1.5 million, with perhaps 750 000 acres (300 000 ha) under cultivation (Hoskins, 1955). Britain became a major grain-producing country and although there is little evidence of deliberate cultivation of pastures, the Romans began the drainage of marshes for grazing.

The Saxons, who followed the Romans, were notable agriculturalists who vigorously continued clearing the forests by felling and burning, establishing new arable land and growing wheat, oats, barley and rye (Harvey, 1955). The agricultural achievements of the early Saxons, and the many villages they founded, have left their mark on Britain to this day. The recorded history is flimsy, at least until the spread of the monasteries after the middle of the first century AD, but it is clear that within the Saxon village, feed for animals was provided for by the grazing of common land, from grain and its by-products and by the foraging of man and animals. Since only a nucleus of animals could be maintained over the winter, a large proportion were killed in the early autumn, thus providing salt beef for the Saxon winter diet. Little deliberate improvement of grassland appears to have taken place. As graphically described by Harvey (1955), 'To the Saxon, grass was an arbitrary gift of Nature which they could use but not increase: only in later ages did man master the secret of sowing and cultivating grass as a crop'.

The Norman Conquest (1066) saw great social and political changes in Britain. Much land passed to the nobility and thence to larger landowners. Despite the creation of the Royal Forests, woodland and the 'waste' continued to be cleared to add to the open fields and common grazing of the villages. The two-field system sometimes gave way to a three-field system (Orwin and Orwin, 1967) which included peas and vetches. Meadows, established on land less suitable for cultivation, formed part of the field system. They were allocated in strips (doles) in the same way as cultivated land, and were closed-off to animals until after the hay crop was taken for winter feed. However, the number of animals which could be overwintered was still limited by the scarcity and unpredictability of winter feed.

This was a period of great expansion of grazing land for sheep. England's burgeoning prosperity owed much to wool; the great monastic houses and feudal magnates grazed immense flocks of sheep over their granges in the North, the limestone downs, the Cotswolds and the drained marshes. By the end of the 14th century it is estimated that there were 2.5 million people in England and 8 million sheep (Hoskins, 1955).

Medieval grassland was of two kinds. First, the meadows and common grazing land which were an integral part of the tillage system of the open fields and where grazing rights were enjoyed by holders of arable holdings and others in the village. Second, there were large tracts of grassland; mainly sheep-walks associated with the monasteries, the aristocracy and, as time went by, increasing numbers of yeoman farmers.

The Tudor period (1485–1603) is notable in English agriculture for the 'enclosures' of the open fields and common grazings, which led to the system of farming we know today. Much has been written of the hardship caused, but there is little doubt that, without enclosure, agriculture could not develop and achieve economies of scale to feed the now growing population. There was little incentive to improve husbandry in the communal system where the fruits of one man's innovation (or indolence) were shared by all the village (Prothero, 1888). Opposition to enclosure was strongest in central England where the open field system was predominant and, although the pace of enclosures ebbed and flowed over the next 300 years, the process continued until much land was converted to pasture for cattle and sheep, with consequent human depopulation. The replacement of tillage by pasture on the heavy soils of the Midlands is vividly described by Hoskins (1955): 'By the end of the 16th century there was, if not a continuous belt of grassland on the Liassic uplands of Northamptonshire and Leicestershire, at least something very near it, and tens of thousands of cattle and sheep grazed over what had been the arable lands of Medieval peasantry'. Such is the origin of the famous Midland 'fattening pastures' and the reason for the characteristic ridge and furrow of the grassland of the region. Initially the pastures were enormous, reflecting the size of the open fields, but stock management and the need of shelter demanded subdivision, so hedges were planted and the green patchwork of the English landscape began to take shape.

By the beginning of the 18th century it is estimated that about half of the open fields and the associated common grazings had been enclosed. There was a further surge under the Hanoverians, especially George II and George III, and by 1850 enclosure of the open fields was virtually complete. Alongside the enclosures, agricultural methods were also changing. The rotations which had reigned fundamentally unchanged since Saxon times began to give way to different methods. The dissolution of the monasteries (1536–1539) had led to a change in land ownership but had also removed a source of learning from which agriculture had previously benefitted. In the 17th century their place was taken by the landed gentry who occupied their leisure in farming and of whom Prothero (1888) says '… in their retirement conferred greater benefits on the well-being of England than they had ever done by their political activities'.

The agricultural improvements which took place from the beginning of the 17th century were, in the first place, improvements to arable agriculture. Improvement to grassland proper lagged behind and seemed to arise more haphazardly. The value of perennial ryegrass was recognized, particularly after seed had been imported from the continent, but the quality appeared variable and the practice of collecting seed from the hay crop probably led to the selection of less desirable forms (Russell, 1966). Perennial ryegrass was indigenous to England, and became an important component of 'permanent'

Improved grasslands

pastures established after the enclosures. It was also a component of the pastures long established on the drained marshes. However, the production of seed from the desirable leafy forms of perennial ryegrass derived from these sources belongs to a later age.

The first systematic account of the English grasses and their value to agriculture appears to be that of Stillingfleet (1759), who castigated English farmers for their neglect of English grasses (Russell, 1966). However, it is recorded that the famous Thomas Coke of Norfolk and the Duke of Bedford employed children to collect seed of desirable grasses to establish improved pastures. The native grasses which Stillingfleet recommended included many that we no longer regard as desirable, e.g. meadow foxtail (*Alopecurus pratensis*), crested dogstail (*Cynosurus cristatus*), annual meadow grass (*Poa annua*) and sweet vernal (*Anthoxanthum odoratum*), reflecting what was then present in the best 'natural' pastures. It was from among the 'artificial' grasses that, later, the most valuable species were to emerge; the ryegrasses (*Lolium perenne* and *L. multiflorum*), cocksfoot (*Dactylis glomerata*), which was imported from Virginia in 1763, and timothy (*Phleum pratense*), named after Timothy Hansen of Maryland from where it was exported. Though these grasses were present in the English flora, the stimulus for their development appears to have been the importation of improved strains from abroad. Looking back, one is struck by the slow and halting way in which grassland husbandry developed in England and by the fact that in agriculture, mainland Europe, and the Low Countries in particular, were much in advance.

During the closing years of the 18th century and the early 19th century there was further progress in identifying the best species and defining suitable mixtures. The value of the ryegrasses was recognized although Curtis (1798) states enigmatically 'Ray [Rye] grass still continues to be the only grass whose seed can be purchased for the purpose of laying down meadow and pasture land; and how inadequate that grass is for such a purpose is known by every farmer'. Possibly this was to promote the sale of his own proprietry seed mixtures! Nevertheless, the early history of ryegrass as a sown pasture species was far from auspicious.

The next major account of pasture grasses is by Sinclair in 1816 (see Russell, 1966) by which time both timothy and cocksfoot, as well as ryegrass, were recommended constituents of pastures and the recommended seeds mixtures began to acquire a modern look. Much experimentation with these mixtures was carried out by Sinclair on the Duke of Bedford's estate at Woburn, recipes for long-term grass and for alternative husbandry were recommended, and advice on management was given. By the 1840s, Lawes had shown the value of treating bones or rock phosphate with acid to produce 'super' phosphate. The arable experiments at Rothamsted were laid out in 1852 and these were followed, in 1856, by the Park Grass experiments which examined the effects of manurial treatments on the productivity and composition of centuries-old pasture. Agricultural depression in the late 19th century led to much 'grassing down' of arable land. Enlightened agriculturists deplored the poor state of much of this grassland and work by Somerville and Gilchrist at Cockle Park, in the north of England, showed how it could be improved by better management and resowing with selected seed mixtures.

Although in the early years of the present century agriculture was beginning to benefit from scientific advances in chemistry and biology, it remained relatively depressed until the outbreak of World War I. At the beginning of the war import of food was unhindered, and there were no serious food shortages, but by 1917 the spectre of hunger was real and there began the policy of ploughing up grassland for arable crops, principally wheat. As on another occasion, 40 years later, the fertility built up in grassland soil served Britain well, and serious hunger was averted. A few years before the war, Stapledon who is probably the best known of British grassland pioneers, had begun his work. In particular, he recognized the importance of species and varieties which had been demonstrated by earlier pioneers. He was also imbued with a sense of urgency to select and breed better grasses and legumes for British farmers, and in particular to provide better varieties for the re-establishment of grassland ploughed up during the war, than the dubious commercial strains then commonly available. The Welsh Plant Breeding Station (WPBS) was founded in 1919 with Stapledon as first Director, and there began an illustrious period of grassland development in Britain.

Authoritative accounts of the contribution made by Stapledon, Jenkin and their colleagues at the WPBS have been given by Russell (1966) and Lazenby (1981). Stapledon's was an ecologist's view of grassland. He recognized the importance of breeding, but above all he stressed the importance of interactions between species and the interaction between plant and the grazing animal. The principal breeding contribution came from his colleagues, Jenkin in particular, whose S-varieties of perennial ryegrass dominated British grassland for almost half a century. It was recognized that the most valuable source of genetic material for selecting and breeding lay among the centuries-old and highly productive grazing pastures of the Midlands and the marshes. Such ecotypes were leafy, persistent, productive and responsive to high fertility. Careful selection from these and other sources, controlled hybridization and rigorous control of multiplication and seed production transformed the quality of herbage seed and, together with improvements in husbandry, due to the efforts of such men as Martin Jones and William Davies, this was a period of great significance in the development of British grassland.

Stapledon was an indefatigable advocate of ley farming (Stapledon and Davies, 1948). He believed that only by regular ploughing and reseeding could grassland of quality be achieved and maintained. Undoubtedly he was influenced by his experience on the Welsh hills, and by his experience of ploughing and subsequent re-establishment of grassland after the war. He was also attracted philosophically by the biological and ecological 'fitness' of a system of agriculture in which the fertility built up under the ley was exploited

Improved grasslands

by succeeding arable crops in a cycle that was largely self-sustaining and capable of high productivity in the hands of skilled farmers. It is sad that despite Stapledon's missionary zeal, because of the intensity and high degree of specialization in modern farming, mechanization and the availability of cheap fertilizers and pesticides, interest in ley farming has waned. Also, in recent times it has become clear that on the all grass or predominantly grassland farm, frequent resowing is not necessary to maintain productivity and that under skilled management (drainage, stocking and fertilizer) highly productive swards composed of desirable species can be maintained almost indefinitely. Nevertheless, the awareness of grassland that Stapledon created, and the way many people were inspired, had a lasting influence on grassland in Britain and many other parts of the world.

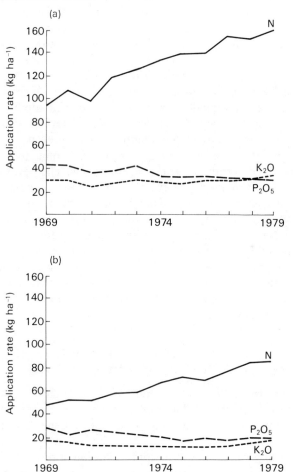

Figure 1.3 Fertilizer use on grassland in England and Wales, 1969–1979. (a) Leys; (b) permanent-grass. (From Church and Leech, 1981.)

The post-war period has been well-documented by Lazenby (1981). The features that stand out are, first, the intensification of grassland production, attendant upon the availability of relatively cheap and subsidized nitrogenous fertilizer. The increase in application rate has been dramatic (Fig. 1.3). Research to define the requirements for other major and minor nutrients was also necessary, as was research to develop systems of utilization in grazing and conservation. Second, mechanization has played a part in the intensification of grassland. Resowing can be carried out with greater ease and reliability, crops for conservation can be cut quickly and to time, and silage making has overtaken haymaking as a means of conserving feed for winter because if its greater productivity and reliability. Third, the further simplification of the swards sown and the increasing dominance of ryegrasses, both Italian and perennial, in intensively managed pastures. Single-species and single-cultivar swards have become commonplace and grassland farming has taken on an aspect of arable farming–that of monoculture. Fourth, specialization in farming has progressively separated arable from grassland farming and, while mixed farming still holds an important position between the intensive arable regions of the East and the all-grass farming of the West, ley farming in the manner advocated by Stapledon has become rare. Unhappily, white clover, which had played such an important role in grassland development, diminished in importance and disappeared altogether from many intensively used pastures.

Fifth, considerable advances have been made in defining forage quality and understanding the digestion of forages in ruminants. The development of the *in vitro* digestibility method (Tilley and Terry, 1963) improved knowledge of ruminant digestion and the calculation of feeding requirements (Anon., 1980) have played an important part in the improvement of animal production from grassland. Finally, stability of demand for grassland products has been achieved by overt and covert intervention and control of prices and imports by government and the European Economic Community. The farming community may find legislation and statutory control irksome, but undoubtedly this has been important in halting the depressingly familiar cycle of prosperity and depression in British agriculture.

1.5 RECENT DEVELOPMENTS IN UNDERSTANDING THE PHYSIOLOGY OF THE GRASS CROP

Later chapters give a detailed account of the state of knowledge of the physiology of the grass crop. What is presented here is a brief summary, albeit a selective and personal one, of some of the developments over the past 40 years.

In the post-war years there was a rapid expansion of research. In particular, the development of instruments for measurement, recording and control enabled the growth and yield of crops to be studied more quantitatively and incisively than before. It is interesting to trace the stages in this development.

Recent developments

Thus, the traditional and oldest measure of crop performance is final harvest yield. This is of prime interest to the farmer, of course, but is uninformative as a basis for understanding why crops yield well or badly, why yields vary from place to place and year to year, and why some managements are more successful than others.

A series of intermediate harvests as the crop grows is more informative, and when applied to grass crops it is found that their increase in weight of dry matter follows a logistic curve reminiscent of the growth of a population of microorganisms. A major step forward was the application of the technique of growth analysis, developed from the work of F. G. Gregory in 1917, which provided a basis for comparing growth rates (relative growth rate) and related the pattern of dry matter change obtained by sequential harvesting to certain physical characteristics of the crop, particularly the area of leaves and their duration (see Chapter 2). The crucially important concept of leaf area index was developed, and a quantitative understanding of crop growth began to emerge. The technique of growth analysis was applied to grasses with great skill by Robson and Jewiss (1968) and others, and this was an important step in the quantification of grass growth. However, there were drawbacks, and the success of the early attempts to understand grass growth in the field and to resolve the importance of individual environmental and management factors was limited. In retrospect it is easy to see why. Only limited characterization of the environment was possible, and the newer physiological techniques developed in the laboratory had not reached a stage where their use in the field could be contemplated. The principal difficulty was the discrepancy in timescales between the primary physiological processes, such as photosynthesis and respiration, which responded almost instantaneously to changes in the environment, and the outcome in terms of changes in dry weight, which could be measured with adequate precision only on a timescale of several days or a week or more. The concept of 'net assimilation rate', derived from changes in dry weight and the leaf area, did not give the insights into the physiology of the crop that were necessary to understand its response to the climatic environment and to management.

Nevertheless, important advances were being made in understanding the environmental control of growth and development of grasses (Cooper, 1968), and the development of the controlled environment cabinet played a central role in this (Evans, 1963). Building on earlier work in the cereals, the influence of photoperiod and vernalization on the flowering of the main economic grasses was elucidated. The response of the individual plant to the main environmental variables, of which light and temperature were the most important, and its response to defoliation were evaluated (Milthorpe and Ivins, 1966). The development of the infrared gas analyser (IRGA) enabled the gas exchange of individual leaves and plants to be measured and the processes of photosynthesis and respiration to be measured directly and on an appropriate timescale.

One advance which was vitally important, but which depended much less upon instrumentation, was the description of the developmental morphology of grasses, and particularly of patterns of tillering and tiller life history (Langer, 1956).The perennation of grasses by the process of tillering is their most characteristic feature and one which sets them apart from nearly all other crop plants. Agronomically, tillering, and the way it is affected by management and the environment, is immensely important. The understanding of tillering and its control is therefore fundamentally important to understanding the physiology of the grasses (see Chapter 3).

Thus, in grassland research, the 1950s and 1960s were a period of intense activity to define the response of the individual plant to its environment and to manipulation, especially by defoliation. This laid the essential foundation for understanding the growth of grass and mixed grass/legume swards in the field. However, as with other crops, there were additional difficulties in the field. First, the community of plants and the resultant crop canopy required the development of new concepts and techniques of measurement. The marrying together of the concept of canopy architecture and light interception (Monsi and Saeki, 1953) with aspects of the earlier method of growth analysis and with characterization of the photosynthetic attributes of leaves (Rabinovitch, 1951) led to the development of mathematical models of crop canopies and predictive calculation of crop photosynthesis and growth (de Wit, 1965; Monteith, 1965; Loomis, 1983). Such models provided the conceptual and integrative framework for the physiologist (see Chapter 7). Second, in the field, environmental parameters fluctuate both systematically and randomly, sometimes rapidly and sometimes slowly. The development of sensors to measure, data-loggers to record and computers to collate and analyse these data, was crucial to the task of understanding crop growth in the field (Woodward and Sheehy, 1983).

By the 1960s the IRGA and associated equipment and techniques had developed to a point where it was possible to make reliable continuous (or at least long-term) measurements of the gas exchange of crop canopies in the field, and hence obtain a direct and near-instantaneous measurement of crop photosynthesis, respiration and, in some cases, transpiration. Two approaches were available; the so-called aerodynamic method in which the flux of CO_2 is calculated from the CO_2 concentration profiles above the crop and a 'transfer coefficient', and enclosure methods in which a small portion of the crop is enclosed in a transparent cover. Both methods have been used in grassland; for example, the aerodynamic method was used to study natural grassland ecosystems in the Canadian Prairie (Ripley and Redman, 1976) and the enclosure method to study the intensively managed grass crops in the UK and subseqently in the rest of Europe and Australia (Leafe *et al* ., 1974).

The combination of new techniques, new concepts and expanding knowledge of the physiology of the individual plant, led to rapid progress in the 1970s in understanding the growth of the grass crop in the field. Some of the

Potential production of pastures

long-standing questions surrounding the growth of the crop were answered and, in particular, the reasons for the characteristic seasonal pattern of production of the temperate grasses were discovered (Leafe *et al.*, 1974). The underlying physiology of response to the major environmental variables and the management began to be elucidated, and the past decade has seen a powerful armoury of physiological, physical and mathematical methods brought to bear on the daunting complexities of the grass sward harvested by grazing animals (see Chapter 4).

Among the many parallel advances in plant physiology during this period which have contributed to our understanding of grassland, several others should be mentioned. The discovery and elucidation of the C_3 and C_4 pathways in photosynthesis has already been mentioned. The quantification of respiratory losses of CO_2 and the provision of a rationale to account for them (McCree, 1970; Penning de Vries, 1974), though controversial, has enabled the carbon balance of crops to be better understood and predicted. The pioneer work by Donald and Black (1958) was one of the earliest attempts to quantify crop growth by consideration of the carbon balance, and deserves highest recognition even though the concept of critical, or optimum, leaf area proved to be an oversimplification. There has also been major progress in understanding the water relations of plants and crops (see Chapter 6) and their response to mineral nutrients, particularly nitrogen (see Chapter 5). It must be recognized that, in practice, nitrogen and water supply have an overwhelming influence on grassland production (Loomis, 1983), and this has led to debate on the relative contribution to grassland production of plant breeding, intensification of management, particularly increased nitrogen application (Alberda, 1971) and improved utilization (Lazenby, 1981).

Finally, we have now entered an era where physiological insights and criteria are used increasingly in selection and breeding. Identification of genotypes of temperate grasses having slow rates of mature tissue respiration has led to the development of higher-yielding lines (Wilson, 1975) and there can be little doubt that as the techniques of genetic manipulation are applied to the development of improved grasses, knowledge of their biochemical and physiological processes will become increasingly important. Similarly, the refinement of management practices, including cutting, grazing, mineral nutrient application and irrigation, to achieve the optimum economic yield in a way that is environmentally acceptable, will increasingly depend on understanding the processes by which the plant and crop respond to management and their environment.

1.6 POTENTIAL PRODUCTION OF PASTURES

The calculation of the potential yield of crops is complex and, as Stern (1975) points out, the concept is an elusive one. Furthermore, there are few situations in which it is economic to attempt to produce the maximum yield regardless of the cost of inputs. Nevertheless, the concept has value in providing a baseline against which the performance of actual crops may be judged, and for the insight it gives into the processes and factors influencing yield. When comparing the actual and potential yields of crops it is important to define clearly whether the 'yield' referred to is the harvestable yield, the above-ground yield or the total above- and below-ground yield (i.e. plant biomass). Similarly, it is important to define the period of assesment, i.e. annual yield, average daily yield, maximum daily yield, etc., and to define whether the yields are from whole fields subject to normal agricultural practice, from experimental plots or from container-grown crops.

Basically, two approaches have been used to explore the potential yield of grass crops. In one approach swards have been grown under conditions as near to ideal as possible: with water and mineral nutrients non-limiting, with pests, diseases and weeds controlled and, generally, a relatively long period between successive defoliations. Using this approach Cooper (1968) has reported an annual harvestable yield of $29\,t\,ha^{-1}$ from container-grown swards of the C_3 grass *Lolium perenne*. From experimental field plots Alberda (1968, 1971) has reported harvestable yields of $22\,t\,ha^{-1}\,year^{-1}$ and yields of around $20\,t\,ha^{-1}\,year^{-1}$ appear to be not uncommon from experimental plots of temperate grasses in northern Europe when there are liberal inputs of nutrients and water. Lower yields of around $15\,t\,ha^{-1}\,year^{-1}$ have been obtained from what is considered to be the best agricultural practice in the UK, while an average of $11.1\,t\,ha^{-1}\,year^{-1}$ has been obtained from an infrequent cutting regime in the recent series of grassland manuring trials in the UK (Morrison, 1980). The average for agricultural practice will, of course, be much lower than this, but it must be remembered that most pastures receive much less than the optimum fertilizer application.

These figures are from swards havested by cutting. Grazed swards present additional conceptual and practical problems. Recently, good progress has been made in resolving these difficulties but there appears to be no reliable figure yet for the potential production of grazed swards (Johnson and Parsons, 1985) (see Chapter 4).

A second approach to the determination of potential yield is theoretical calculations using models based on the known photosynthetic and respiratory characteristics of the species in question, a knowledge of assimilate utilization, the structure of its canopy and data on the climatic environment. Generally, the effects are discounted of factors which, in theory at least, it is possible to correct (such as drought, mineral deficiency, pests and diseases). Alberda (1977), using the model of de Wit (1965), gives a figure for the total annual dry matter production for the grass crop in northern Europe of $45\,t\,ha^{-1}\,year^{-1}$. If an harvest index of 60% is assumed (Alberda, 1968) this gives a harvestable yield of $27\,t\,ha^{-1}\,year^{-1}$, which is close to the yields already achieved in practice in experimental plots and container-grown swards.

Leafe (1978) made an analysis of the production of perennial ryegrass in the

Potential production of pastures

UK based on combination of measured data and theoretical calculations. Use was made of extensive data on the carbon exchange of ryegrass canopies in the field, and on patterns of assimilate distribution determined in the growth room and field. This analysis sheds interesting light on the influence on yield of management, climatic factors, and the innate physiology of the plant. For example, if during the normal growing season in the UK (April–October), every day were cloudless and the canopy maintained full light interception and exhibited its maximum photosynthetic capacity, the theoretical above-ground dry matter yield would be $45.0\,t\,ha^{-1}\,year^{-1}$. However, based on a 10-year average of actual irradiance, yield is reduced to $30.5\,t\,ha^{-1}\,year^{-1}$. This is reduced to $25\,t\,ha^{-1}\,year^{-1}$ if allowance is made for those periods when light interception is incomplete following defoliation (five cuts in the growing season), and to $19.5\,t\,ha^{-1}\,year^{-1}$ if account is taken of reduced leaf photosynthesis and hence canopy photosynthesis in the vegetative regrowths following the first cut (Woledge and Leafe, 1976). The yield predicted from this calculation based on the carbon balance of the crop is therefore close to yields obtained in practice from crops amply supplied with mineral nutrients and water.

These actual and predicted yields for a temperate (C_3) grass growing in northern Europe may be contrasted with yields from C_4 species. For example, an annual harvestable yield of $85.0\,t\,ha^{-1}\,year^{-1}$ has been reported for napier grass (*Pennisetum purpureum*) growing in experimental plots with ample nutrients and water, and in a tropical environment with considerably higher radiation than northern Europe and little seasonal variation in temperature (Vincente-Chandler et al., 1959). An informative comparison of C_3 and C_4 species is made by Loomis (1983), who shows the clear superiority of C_4 species (among them Napier grass, sugar cane, forage sorghum, maize and bermuda grass) over C_3 species, including perennial ryegrass, at low latitudes and thus high irradiance and temperature, but a trend towards the reverse at high latitudes. The analysis also demonstrates the great importance of the length of the growing season in determining the annual yield of both C_3 and C_4 species.

Differences in the response of temperate (C_3) and tropical (C_4) species to temperature, radiation and water stress may be of adaptive significance, although Woolhouse (1983) urges great caution in interpretation. They do, however, have a strong bearing on potential yields. Thus, C_4 species generally grow faster at their optimum (higher) temperature than C_3 species, but may grow more slowly at lower temperatures and may be killed by low temperature (Ludlow, 1976). Associated with this are the temperature optima for photosynthesis. In general, C_4 grasses have a higher optimum temperature for photosynthesis and a much higher rate of photosynthesis at that optimum than C_3 species (Tieszen and Detling, 1983). Because of the greater affinity for CO_2 of PEP carboxylase than RuBP carboxylase/oxygenase, similar rates of CO_2 uptake may be maintained at smaller stomatal opening in C_4 species than C_3 species, thus reducing transpiration. Thus, water use efficiency in carbon fixation is higher in C_4 than in C_3 species, and this influences their potential productivity under water stress (see Chapter 6).

The concept of potential production cannot, perhaps, be applied very usefully to the natural grasslands since these are usually severely limited by nutrient deficiency, water shortage and other factors, rather than the biological potential of the plants present. However, the *actual* production of some natural grassland systems is of interest, and Tieszen and Detling (1983), in their review of the productivity of grassland and tundra, present data for six North American systems based on data from Sims and Singh (1978). These range from an annual above-ground dry-matter production of less than $1 \, t \, ha^{-1} \, year^{-1}$ for shrub steppe to $3.46 \, t \, ha^{-1} \, year^{-1}$ for tall-grass prairie, with an average for the six systems they examined of $2.23 \, t \, ha^{-1} \, year^{-1}$.

The large differences between the yield of cultivated and natural grasslands underlines the point made by Loomis (1983) in his authoritative review of the productivity of agriculture systems. He points out that shortage of mineral nutrients, and nitrogen in particular, is the most general cause of yield limitation in the absence of water stress. In grassland, both natural and cultivated, as in arable agriculture, it is mineral deficiency and water stress which generally limits yield rather than limitations of photosynthetic capacity (see Capters 5 and 6). Nevertheless, in some intensive systems of agriculture, where nutrients and water are supplied in abundance, the point may now have been reached where photosynthetic capacity is the limitation to increasing productivity. Furthermore, the adverse effects of mineral deficiency and water stress are mediated to a considerable extent through the photosynthetic process. Alleviation of the effects of stress on photosynthesis by breeding, genetic manipulation and by management is, thus, a major research goal.

REFERENCES

Alberda, Th. (1968) Dry matter production and light interception of crop surfaces. IV. Maximum herbage production as compared with predicted values. *Neth. J. Agric. Sci.*, **16**, 142–53.

Alberda, Th. (1971) Potential production of grassland, in *Potential Crop Production* (eds P. F. Wareing and J. P. Cooper), Heinemann, London, pp. 159–71.

Alberda, Th. (1977) Possibilities of dry matter production from forage plants under different climatic conditions. *Proc. 13th Int. Grassland Congr., Leipzig*, pp. 61–69.

Anon. (1980) *The Nutrient Requirements of Ruminant Livestock,* The Agricultural Research Council, Commonwealth Agricultural Bureaux, Slough.

Begg, J. E. and Turner, N. C. (1976) Crop water deficits. *Adv. Agron.*, **28**, 161–216.

Bouton, J. H. and Brown, R. H. (1981) Differences in photosynthetic types in the Laxa group of the *Panicum* genus: cytogenetics and reproduction, in *Proc. 14th Int. Grassland Congr., Lexington, Kentucky,* pp. 176–8.

Burbidge, N. T. (1964) Grass systematics, in *Grasses and Grassland* (ed. C. Barnard), Macmillan, London, pp. 13–28.

References

Church, B. M. and Leech, P. K. (1981) *Fertilizer Use on Farm Crops in England and Wales, 1980,* Ministry of Agriculture, Fisheries and Food, London.

Cooper, J. P. (1968) Energy and nutrient conversion in a simulated sward. *Report of the Welsh Plant Breeding Station, 1967,* pp. 10–11.

Curtis, W. (1798) Practical observations on the British grasses best adapted to laying down on improving meadows and pastures, 3rd edn, Couchman and Fry, London.

Daly, G. T. (1972) The grasslands of New Zealand, in *Pastures and Pasture Plants* (ed. R. H. M. Langer), A. H. and A. W. Reed, Wellington, New Zealand, pp. 1–39.

Davies, W. (1960) *The Grass Crop – Its Development, Use and Maintenance,* E. and F. N. Spon, London.

Donald, C. M. and Black, J. N. (1958) The significance of leaf area in pasture growth. *Herbage Abstr,* **28,** 1–6.

Edwards, G. and Walker, D. (1983) C_3, C_4: *Mechanisms, and Cellular and Environmental Regulation, of Photosynthesis,* Blackwell Scientific Publications, Oxford.

Edwards, P. J. (1981) Multiple use of grassland resources, in *Proc. 14th Int. Grassland Congr., Lexington, Kentucky,* pp. 64–69.

Evans, L. T. (ed.) (1963) *Environmental Control of Plant Growth,* Academic Press, New York. (Proceedings of a Symposium held at Canberra, Australia, August 1962.)

Hartley, W. (1964) The distribution of the grasses, in *Grasses and Grasslands* (ed. C. Barnard), Macmillan, London, pp. 29–46.

Hartley, W. and Williams, R. J. (1956) Centres of distribution of cultivated pasture grasses and their significance for plant introduction, in *Proc. 7th Int. Grassland Congr,* pp. 190–201.

Harvey, N. (1955) *The Farming Kingdom,* Turnstile Press, London.

Hatch, M. D. (1976) The C_4 pathway of photosynthesis: mechanism and function, in CO_2 *Metabolism and Plant Productivity* (eds R. H. Burris and C. C. Black), University Park Press, Baltimore, pp. 59–81.

Hatch, M. D. and Slack, C. R. (1966) Photosynthesis by sugarcane leaves. A new carboxylation reaction and the pathway of sugar formation. *Biochem. J.,* **101,** 103–11.

Hoskins, W. G. (1955) *The Making of the English Landscape,* Hodder and Stoughton, London.

Johnson, I. R. and Parsons, A. J. (1985) Use of a model to analyse the effects of continuous grazing managements and seasonal patterns of grass production. *Grass Forage Sci.,* **40,** 449–58.

Jones, M. B., Leafe, E. L. and Stiles, W. (1980) Water stress in field-grown perennial ryegrass. II. Its effect on leaf water status, stomatal resistance and leaf morphology. *Ann. Appl. Biol.,* **96,** 103–10.

Kortschack, H. P., Hartt, C. E. and Burr, G. O. (1965) Carbon dioxide fixation in sugar cane leaves. *Pl. Physiol.,* **40,** 209–13.

Langer, R. H. M. (1956) Growth and nutrition of timothy (*Phleum pratense*) I. The life history of individual tillers. *Ann. Appl. Biol.,* **44,** 166–87.

Lazenby, A. (1981) British grasslands; past, present and future. *Grass Forage Sci.,* **36,** 243–66.

Leafe, E. L. (1978) Physiological, environmental and management factors of importance to maximum yield of the grass crop, in *Maximizing Yields of Crops* (eds J. K. R. Gasser and B. Wilkinson), ARC Symposium Proceedings, HMSO, London.

Leafe, E. L. Stiles, W. and Dickenson, S. E. (1974) Physiological processes influencing the pattern of productivity of the intensively managed grass sward. *Proc. XII Int. Grassland Congr., Moscow*, pp. 442–57.

Levy, E. B. (1970) Pasture plants in relation to their environment, in *Grasslands of New Zealand*, 3rd edn (ed. A. R. Shearer), Government Printer, Wellington, New Zealand, pp. 108–37.

Long, S. P. (1983) C_4 photosynthesis at low temperatures. *Plant, Cell Envir.*, **6**, 345–63.

Loomis, R. S. (1983) Productivity of agricultural systems, in *Encyclopedia of Plant physiology, New Series, Vol. 12D* (eds O. L. Lange, P. S. Nobel, C. B. Osmond and H. Ziegler), Springer-Verlag, Berlin, pp. 152–71.

Ludlow, M. M. (1976) Ecophysiology of C_4 grasses, in *Water and Plant Life* (eds O. L. Lange, L. Kappen and E.-D. Schulze), Springer-Verlag, Berlin, p.p. 364–86.

McCree, K. J. (1970) An equation for the rate of respiration of white clover plants grown under controlled conditions, in *Prediction and Measurement of Photosynthetic Productivity* (ed. I. Setlik), PUDOC, Wageningen, pp. 221–9.

Milthorpe, F. L. and Ivins, J. D. (eds) (1966) *The Growth of Cereals and Grasses*, Butterworths, London. (Proc. 12th Easter School in Agricultural Sciences, University of Nottingham, 1965.)

Monsi, M. and Saeki, T. (1953) Über den Lichtfaktor in den Pflanzengesellschaften. *Jap. J. Bot.*, **14**, 22–52.

Monteith, J. L. (1965) Light distribution and photosynthesis in field crops. *Ann. Bot.*, **29**, 17–37.

Moore, C. W. E. (1964) Distribution of grasslands, in *Grasses and Grasslands* (ed. C. Barnard), Macmillan, London, pp. 182–205.

Moore, R. M. (1962) Effects of the sheep industry on Australian vegetation, in *The Simple Fleece: Studies in the Australian Wool Industry* (ed. A. Barnard), Melbourne University Press in association with the Australian National University, pp. 170–83.

Morrison, J. (1980) The influence of climate and soil on the yield of grass and its response to fertiliser nitrogen, in *The Role of Nitrogen in Intensive Grassland Production* (eds W. H. Pruins and G. H. Arnold), PUDOC, Wageningen, pp. 51–7.

Moss, D. N. (1962) The limiting carbon dioxide concentration for photosynthesis. *Nature*, **193**, 587.

Orwin, C. S. and Orwin, C. S. (1967) *The Open Fields*, 3rd edn, The Clarendon Press, Oxford.

Penning de Vries, F. W. T. (1974) Substrate utilisation and respiration in relation to growth and maintenance in higher plants. *Neth. J. Agric. Sci.*, **22**, 40–4.

Pollock, C. J., Riley, G. J. P., Stoddart, J. L. and Thomas, H. (1980) The biochemical basis of plant response to temperature limitations. *Report of Welsh Plant Breeding Station for 1978*, pp. 227–46.

Prat, H. (1960) Vers une classification naturelle des Graminées. *Bull. Soc. Bot. Fr.*, **107**, 32–79.

Prothero, R. E. (1888) *The Pioneers and Progress of English Farming*, Longmans and Green, London.

References

Rabinovitch, E. I. (1951) *Photosynthesis and Related Processes,* Wiley/Interscience, New York.

Ripley, E. A. and Redman, R. E. (1976) Grasslands, in *Vegetation and the Atmosphere* (Part II, *Case Studies* (ed. J. L. Monteith)), Academic Press, London.

Robson, M. J. and Jewiss, O. R. (1968) A comparison of British and North African varieties of tall fescue (*Festuca arundinacea*). III. Effects of light, temperature and daylength on relative growth rate and its components. *J. Appl. Ecol.,* **5,** 191–204.

Russell, E. J. (1966) *A History of Agricultural Science in Great Britain,* George Allen and Unwin, London.

Shantz, H. L. (1954) The place of grasslands in the earth's cover of vegetation. *Ecology,* **35,** 142–5.

Sims, P. L. and Singh, J. S. (1978) The structure and function of ten western North American grasslands. III. Net primary production, turnover and efficiencies of energy capture and water use. *J. Ecol.,* **66,** 573–97.

Spedding, C. R. W. and Dieckmahns, E. C. (eds) (1972) *Grasses and Legumes in British Agriculture,* Commonwealth Agricultural Bureaux, Farnham Royal, Buckinghamshire, U.K.

Stapledon, R. G. and Davies, W. (1948) *Ley Farming,* Faber and Faber, London.

Stern, W. (1975) Actual and potential photosynthesis production, in *Photosynthesis and Productivity in Different Environments* (ed. J. P. Cooper), Cambridge University Press, Cambridge, pp. 661–72.

Stillingfleet, B. (1759) *Miscellaneous Tracts Relating to Natural History, Husbandry, Physick and, in Addition, Calendar of Flora,* 3rd edn, Baker and Leigh, London.

Tieszen, L. L. and Detling, J. K. (1983) Productivity of grassland and tundra, in *Encyclopedia of Plant Physiology,* (New Series, Vol. 12D) (eds O. L. Lange, P. S. Nobel, C. B. Osmond and H. Ziegler), Springer-Verlag, Berlin, pp. 173–203.

Tilley, J. M. A. and Terry, R. A. (1963) A two-stage technique for *in vitro* digestion of forage crops. *J. Br. Grassland Soc.,* **18,** 104–11.

Vincente-Chandler, J., Silva, S. and Figarella, J. (1959) The effect of nitrogen fertilisation and frequency of cutting on the yield and composition of three tropical grasses. *Agron. J.,* **51,** 202–6.

Wilson, D. (1975) Variation in leaf respiration in relation to growth and photosynthesis of *Lolium. Ann. Appl. Biol.,* **80,** 323–38.

Wit, C. T. de (1965) *Photosynthesis of Leaf Canopies,* Agricultural Research Reports 663, IBS, Wageningen.

Woledge, J. and Leafe, E. L. (1976) Single leaf and canopy photosynthesis in a ryegrass sward. *Ann. Bot.,* **40,** 773–83.

Woodward, F. I. and Sheehy, J. E. (1983) *Principles and Measurements in Environmental Biology,* Butterworths, London.

Woolhouse, H. W. (1983) The effects of stress on photosynthesis, in *Effects of Stress on Photosynthesis* (eds R. Marcelle, H. Clijsters and M. van Poucke), Martinus Nijhoff/Dr W. Junk, The Hague, pp. 1–28.

CHAPTER 2
The grass plant – its form and function

M. J. Robson, G. J. A. Ryle and Jane Woledge

2.1 INTRODUCTION

The form of a plant reflects its function. The morphology of grasses is a product of their genetic make-up and the environment they experience, a key component of which is the grazing animal. Forage grasses have evolved to withstand periodic defoliation. Agriculturally important north-temperate grasses remain vegetative throughout most of the year, with the growing points, from which new leaves are produced, held at or near ground level on unelongated stems. Thus, when the leaves are harvested, whether by cutting or grazing, the majority of the growing points escape. Any leaves which are not harvested, entirely or in part, senesce and die. Turnover is rapid. At the height of the growing season, a typical grass shoot may bear three live leaves and produce a new one every 7–10 days (Alberda and Sibma, 1968). The entire leaf canopy can be replaced within as little as 3–4 weeks.

It is not only the leaves that die and are replaced, grass shoots, or tillers as they are known, are also relatively short-lived. Most die within 12 months of being produced, some after only a week or two, while a few persist into a second year (Langer, 1956; Robson, 1968). Some are shaded by larger neighbours, some are weakened by cutting or uprooted by grazing, some are attacked by pests or diseases and some become reproductive. The crop survives through the continuous production of new vegetative tillers from axillary buds on older surviving shoots.

Although the perenniality of our grasslands is dependent on vegetative regeneration, the seasonal cycle of the crop is dominated by flowering. In spring, following a precise sequence of environmental conditions, the shoot apex of many mature established tillers switches from the production of leaf initials to that of floral parts, and the stem begins to elongate. The transition from vegetative to reproductive growth is closely associated with a number of changes in the basic physiology of the plant (Pollock and Jones, 1979; Parsons and Robson, 1980, 1981a,b) which, in turn, play a major role in bringing about the high and sustained rates of dry matter production characteristic of the spring crop (Parsons and Robson, 1982). Flowering also profoundly influences the seasonal patterns of tiller production and death. Not only does flowering seal the fate of the reproductive tiller, in that its ability to produce leaves is terminated, but the dense shade created at the base of the crop leads to the death of many smaller, vegetative tillers. Thus, flowering leads to a high tiller mortality in spring followed by a peak production of replacement vegetative tillers in early summer (Langer, 1956; Robson, 1968) (see Chapter 3).

In newly established grass swards the individual plant is the basic unit and the way in which its form and function – its morphology and physiology – respond to environment, management and the increasingly influential presence of neighbours are crucial to its success and that of the crop. However, in older swards, and at least half UK grassland is more than 20 years old, the individual plant – the product of a single seed – loses its identity. Many plants die, while those that survive and increase in size fragment into smaller units as a result of tiller death and the lateral invasion by new tillers of adjacent gaps in the sward. The basic unit becomes the tiller, or any group of tillers that retain functional organic connections.

In this chapter we start with the seed, its germination and the establishment of the seedling. We continue with the growth and development of the young plant, and its response to the conditions it experiences. We end with a consideration of how individual plants combine and interact to produce a crop community, and how biomass production and tissue turnover lead to the achievement of a ceiling yield of harvestable dry matter. *Lolium perenne* (perennial ryegrass), the most widely used of the north-temperate forage grasses, it taken as our standard, although reference is made to other species where appropriate, and when data are available.

2.2 THE SEED AND SEEDLING ESTABLISHMENT

2.2.1 The seed

The grass seed begins its development as a single fertilized ovule within the ovary of a flower, or floret. As well as the ovary, topped by its twin stigmas, each floret contains three stamens and two very small scales or lodicules, all enclosed within two large bracts known as the lemma and palea (Fig. 2.1a). A number of florets make up a spikelet (Fig. 2.1b) and a number of spikelets the inflorescence, whose type varies with species (Fig. 2.1c).

The great bulk of the seed is starchy endosperm, the function of which is to sustain the embryo during germination and early seedling growth. The embryo consists of a primary shoot or plumule within a protective sheath, the coleoptile, and a primary root or radicle also within a sheath, the coleorhiza. These are attached by a short mesocotyl to a flat shield-like structure, the scutellum, one face of which abuts the endosperm (Fig. 2.2a).

As the seed matures its coat or testa fuses with the ovary wall or pericarp to

The seed and seedling establishment

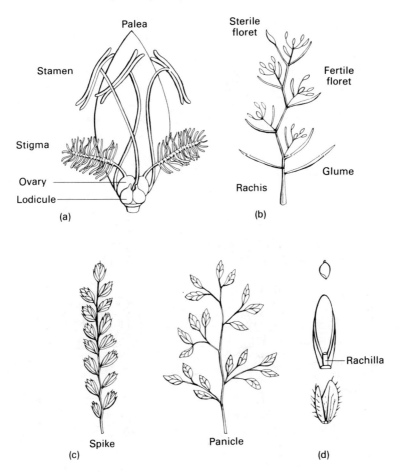

Figure 2.1 The grass inflorescence (a) Floret with lemma removed to expose reproductive organs. (b) Diagrammatic spikelet showing two basal glumes, five fertile florets and one sterile (terminal) floret. (c) Types of inflorescence: spike and panicle. (d) Types of 'seed' (from the top): naked grain of *Phleum pratense*, enclosed grain of *Lolium perenne* and an entire spikelet of *Holcus lanatus*. (After Armstrong, 1948; Langer, 1972.)

form a dry, one-seeded, indehiscent fruit known technically as a caryopsis, but commonly referred to in the *Gramineae* as a grain. As sold by the seed merchant, and sown by the farmer, the grain may be naked, e.g. in *Phleum pratense* (timothy), or enclosed by the persistent lemma and palea and attached to a portion of the spikelet axis, or rachilla, as in *Lolium perenne*. In some agriculturally less important grasses, the 'seed' may be the entire spikelet (Fig. 2.1d).

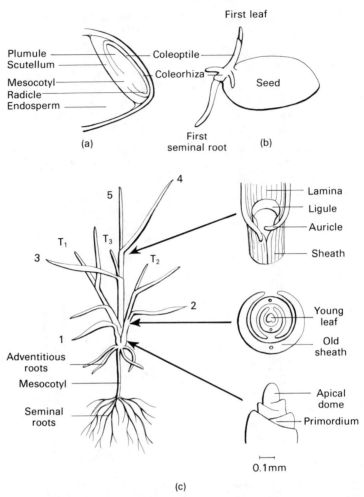

Figure 2.2 The grass seed (caryopsis) and young plant (a) Diagrammatic cross section of the embryo region of a seed; (b) germinating seed; (c) young plant with five leaves on the main stem, four of them fully expanded and three subtending daughter tillers (T_1, T_2 and T_3). Also (from the top): the junction of lamina and sheath, a cross section of the pseudo-stem (sometimes sheaths are folded rather than rolled) and the vegetative main stem apex. (After Armstrong, 1948; Langer, 1972.)

2.2.2 Germination

The germination of a viable seed is largely dependent on a favourable temperature and an adequate supply of water and oxygen. At temperatures below about 5 °C, germination rate and percentage are both impaired to an extent that varies with species (Chippindale, 1949). Above 5 °C only the rate of germination varies, increasing with temperature up to about 20 °C. Some weed grasses, such as *Poa annua*, are particularly responsive to large diurnal fluctuations in temperature and may fail to germinate under an insulating blanket of plant material, but do so under bare soil, thereby increasing their chances of emerging into an open patch within the community (Thompson *et al.*, 1977).

In some tropical species, such as *Bracharia ruziziensis*, phenolic compounds in retained structures (lemma, palea, rachilla and supernumerary sterile florets) inhibit germination by depriving the seed of oxygen, while in *Paspalum notatum* these same structures physically impede the passage of water. Neither of these problems occur in the UK forage grasses. Germination can be readily achieved by sowing viable seed, at a favourable time of year, between 0.5 and 3.0 cm deep in a well-prepared moist soil sufficiently compressed to ensure good contact between seed and soil without restricting aeration and drainage (Jones, 1972).

The first stage in germination is an entirely passive non-metabolic uptake of water occupying perhaps the first 12–24 h. Following this, cells in the aleurone layer surrounding the endosperm secrete enzymes, principally α-amylase, stimulated by gibberelins originating from the embryo (Paleg, 1960). The enzymes break down the endosperm starch into sugars, mainly maltose, which the scutellum absorbs and passes on through the mesocotyl to the embryo. There they supply the energy and carbon skeletons necessary for the growth of the plumule and radicle. When these break through the seed coat, germination can be considered complete (Fig. 2.2b).

2.2.3 Seedling emergence

The radicle grows downwards through the coleorhiza with one or more secondary roots – the number is characteristic of the species – arising at its base. These develop, through branching, into a highly efficient seminal root system on which the young seedling initially relies for its supply of water and inorganic nutrients. Generally, the seminal root system deteriorates as its functions are taken over by adventitious roots which arise from basal nodes on the main shoot and, later, on daughter tillers as they appear (Fig. 2.2c).

As the radicle grows downwards, the young shoot or plumule, and its ensheathing coleoptile, are carried towards the soil surface by the elongation of the mesocotyl (Arber, 1934; Brown, 1960). Mesocotyl elongation is a response to darkness. It ceases when the tip of the coleoptile emerges into light, thereby positioning the shoot apex just below the soil surface (Fig. 2.2c).

The growth of the unemerged seedling is totally dependent on the finite store of organic reserves in the endosperm. This varies with the size of the seed. Thus, potential mesocotyl elongation is least in small-seeded species such as timothy with an average seed weight of 0.4 mg (Armstrong, 1948), increasing through cocksfoot *Dactylis glomerata* (1.0 mg), to the large-seeded (1.7–2.1 mg) perennial ryegrass, Italian ryegrass (*Lolium multiflorum*), tall fescue (*Festuca arundinacea*) and meadow fescue (*Festuca pratensis*). Within a given species mesocotyl extension tends to be similar for seeds of different weight: lighter seeds produce a thinner structure (Jones, 1972).

With seed sown deeply, extension of the internode between the point of attachment of the coleoptile and that of the first true leaf can supplement mesocotyl elongation, but at the expense of an even greater depletion of endosperm reserves. Too deep sowing means the seedling may never emerge. From a depth of about 1 cm, emergence normally takes 6–8 days (Jones, 1971). During this period respiration associated with meeting the energy requirements of mobilizing reserves and transforming them into new plant tissue can account for more than one-third of the initial weight of the caryopsis (Anslow, 1962). Endosperm reserves are usually exhausted within about 5 days of the coleoptile breaking the soil surface. The young plant must by then have expanded sufficient photosynthetic surface to meet the demands of respiration and new growth.

2.3 THE VEGETATIVE SHOOT

2.3.1 Leaf growth on the main stem

At first the only site of leaf production is the meristematic apex of the main shoot or stem. Later, however, leaves also develop from the apices of daughter tillers produced from the axils of mature main stem leaves (Fig. 2.2c).

Considering only main stem leaves, the first produced is very small, especially in small-seeded grasses. Subsequent leaves increase progressively in length and width, and hence in area (and weight) until a plateau is reached. In S170 tall fescue, for example, lamina area increases about 20-fold between leaf 2 and leaf 7, but little thereafter (Robson, 1974).

The leaves have a limited lifespan which, depending on genotype and time of year, may be as little as 3 weeks. After a time, therefore, the oldest leaves begin to senesce and die. Soon the number remaining alive stabilizes, indicating that for each new leaf produced an old leaf has died. At first the old leaf is smaller than the one that replaces it, so the photosynthetic surface of the shoot continues to increase. Once leaf size has plateaued, however, the old leaf and the one that replaces it are of a similar size, with the total photosynthetic area of the shoot relatively constant. In a vegetative sward, most of the established tillers are in this state of dynamic equilibrium.

2.3.2 Leaf morphology and development

Each grass leaf consists of a blade or lamina connected at an angle to a generally shorter sheath, the junction being marked by a membranous ligule and, in some species, two claw-like auricles (Fig. 2.2c). What, to a casual observer, appears to be the stem is really a collection of sheaths rolled or folded one inside the other with the oldest sheath on the outside (Fig. 2.2c). Young leaves are extruded in succession through the centre of this pseudo-stem, disposing their laminae alternately left and right as they reach full size. The true stem is found at the base of the shoot, entirely concealed by encircling sheaths. The sheaths join the stem at nodes separated by internodes which normally remain short (<1 mm) while the shoot remains vegetative.

The leaves originate from a meristematic apical dome along the opposing flanks of which promordia are laid down in alternating order (Fig. 2.2c). As described by Langer (1972), 'Formation of the leaf primordium begins by rapid cell division in the dermatogen and hypodermis giving rise to a microscopically visible protuberance. Lateral spread of cell division changes this into a crescent and then a collar, and as this structure grows upwards it assumes the appearance of a cowl when eventually it overlaps the apical dome'.

The time interval between the production of successive leaf primordia, the plastochron, is generally shorter than that between their appearance as young growing leaves, with the result that leaf primordia accumulate – more so in those species that produce a spike when they flower than in those that produce a panicle (p. 35). The mechanism whereby a leaf primordium is triggered to begin rapid cell division and extension is not known, but it seems to be related to the stage of development reached by its immediate predecessors.

As the leaf primordium grows, cell division and expansion become restricted to a basal intercalary meristem divided in two by a band of parenchyma cells from which the ligule develops as an outgrowth of the epidermis. The upper portion of the meristem is associated with lamina growth, the lower with that of the sheath. Lamina extension largely precedes sheath extension, with the latter continuing until the ligule is exposed, by which time the tip of the next youngest leaf is already visible.

The restriction of growth to a basal meristem means that extension is already complete in that part of the leaf emerging from within the encircling sheaths of its predecessors. Moreover, the tip of a leaf is always older than the base – totally unlike the situation in most dicotyledonous plants, where the leaf is of a uniform age. Not surprisingly, when a grass leaf dies the most mature portion, the tip, dies first.

2.3.3 Environment and leaf growth

In the absence of water stress and nutrient deficiency, both of which restrict leaf expansion (see Chapters 5 and 6), and leaving aside the effects of grazing, light (irradiance) and especially temperature are the most important environmental variables affecting leaf growth.

(a) Irradiance

Although during a period of low irradiance less carbon is fixed than at high irradiance, a greater proportion is retained by the shoot at the expense of root (Ryle and Powell, 1976). Consequently, leaf growth is maintained with leaf area per unit leaf weight (i.e. specific leaf area) maximized (e.g. Templeton *et al.*, 1961; Friend *et al.*, 1965). Thus, leaves produced in low irradiance ('shade' leaves) tend to be longer and thinner, and generally more etiolated, than those produced in high irradiance ('sun' leaves). The rate of leaf appearance is much less sensitive to irradiance than is leaf morphology, unless very low levels are experienced over an extended period when leaf production may be severely curtailed (Silsbury, 1970). A number of studies suggest that at temperatures of 18–23 °C a four-fold reduction in PAR from 160 to 40 W m^{-2} adds only about one day to the leaf appearance interval.

Quite separately from its role during reproduction, photoperiod (i.e. the length of the light period, as distinct from its intensity or the total light energy received) also influences the potential size of leaf which can be developed (Ryle, 1966). Generally in the north European grasses, an increase in photoperiod over the midwinter minimum of 8 h of light per day increases leaf length and sometimes leaf width. The size of the effect varies markedly with species: in cocksfoot, doubling the photoperiod from 8 to 16 h day^{-1} may double final leaf area, whereas in ryegrass the leaf area increases only by 30–50%. These effects are primarily due to increased cell length and, to a lesser extent, cell number along the length of the leaf. Leaf width is affected less by photoperiod because it is primarily controlled by the number of cells around the circumference of the leaf primordium at the time of its inception (p. 33).

Few environmental factors act alone. Since, in the natural environment, changes in photoperiod and temperature are linked, the two factors act together and can only be separated experimentally. The effect of photoperiod is probably regulatory in its nature, ensuring that the maximum leaf size is only attained when other seasonal factors are optimal.

Under natural conditions short-term variation in irradiance both within and between days has little immediate effect on leaf growth; temporary storage of sugars in sheath bases, and other intermediate metabolites in the growing leaves themselves (Gordon *et al.*, 1977), provides a buffer against fluctuations in the supply of current assimilates. Thus, leaf growth reflects the underlying trend of the light environment to which a crop is exposed rather than its instantaneous state (Robson, 1981a).

(b) Temperature

Once carbon has been assimilated, its incorporation into new tissue, and its expression in terms of leaf expansion on the individual shoot, is more dependent on seasonal differences in temperature than on irradiance. This is despite the fact that in southern Britain, for example, the mean daily irradiance during the worst month of winter is almost exactly one-tenth that of the best summer month. Thus, Patel and Cooper (1961) found that plants grown throughout the year in a heated glasshouse, but otherwise experiencing natural fluctuations in daylength and irradiance, showed, at the very most, a two-fold difference in the rate of leaf appearance, whereas plants grown outside readily expand one leaf

The vegetative shoot

every 5–7 days in midsummer (Davies, 1977) but can slow to only one-tenth of this rate during the low temperatures of midwinter (Robson, 1967). Note that leaf senescence also slows so that the mean number of live leaves remains relatively constant.

Temperature affects not only the rate of leaf appearance and senescence, but also the final size and shape of a leaf, and its rate and duration of expansion. In general 'high temperature' leaves extend more rapidly, for a shorter period, to a greater final length; they tend to be longer in relation to their width, achieve a greater specific leaf area and have proportionately more lamina relative to sheath (Mitchell, 1953a, 1956; Mitchell and Lucanus, 1962; Cooper, 1964; Cooper and McWilliam, 1966; Robson, 1974). Their greater lamina length may be due more to a greater cell length than to an increase in number (Borrill, 1961; Cooper and Edwards, 1964) although the evidence is inconclusive (Ryle, 1966; Robson, 1974). The optimum temperature for most aspects of leaf growth tends to be in the region 20–25 °C for most north temperate grasses, with the night temperature equal to or slightly lower than that of the day (Evans et al., 1964; Cooper and Tainton, 1968; Robson, 1972, 1973c).

Although temperature has a major effect on the rate at which a grass shoot expands new leaf, it has little impact on the organization of the shoot apex or on the hierarchical relationship between a developing leaf and its successor and predecessor. For example, young plants of S170 tall fescue grown at 25 °C can produce leaf tissue on the main stem at four times the rate of those grown at 10 °C (Robson, 1969). This is achieved by a doubling of the frequency of leaf appearance (with a matching rise in primordia production) and by leaves extending at four times the rate but for only half the time, to twice the final length. Because both the time interval between the appearance of successive leaves and the duration of leaf extension are halved, the number of growing leaves remains constant (Fig. 2.3d). Alternative strategies which the plant could, in theory, adopt to bring about such an increase in leaf growth are shown in Fig. 2.3.

The rate of extension of the growing leaf is very sensitive to current temperature, responding to a change within minutes. The width of a leaf reflects the number of cell rows across it, and this in turn reflects the basal circumference of the shoot apex at the time when the leaf was still at a primordial stage (Abbe et al., 1941). Thus, temperature fluctuations from day to day, or diurnally, have no effect on the width of the growing leaf or, indeed, its successor (Robson, 1974); their only immediate influence on leaf area comes through their effect on leaf extension. Since the number of growing leaves per shoot is relatively constant (usually between one and two), the rate of extension of the youngest growing leaf has been used as an index of the rate of leaf area expansion of the whole shoot and even (if the tiller population also remains constant) of the crop itself (Roy and Peacock, 1972; Grant et al., 1983).

In using leaf extension as an indicator of the response of the crop to current temperature, it is important to realize that–depending on climate and crop

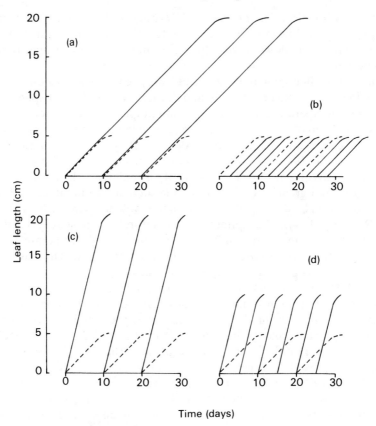

Figure 2.3 Shoots of *Festuca arundinacea* produce four times as much leaf at 25 °C (——) as at 10 °C (---) (Robson, 1969). Three ways in which this could, in theory, be achieved: (a) each leaf grows for 4× as long to 4× the final length, (b) successive leaves appear 4× as frequently, but grow at the same rate, for the same time, to the same final length, and (c) each leaf grows at 4× the rate to 4× the final length. In practice, (d) the plants tend to both a greater rate of leaf appearance (2×) and a greater rate of leaf extension (4×), but with the duration of extension halved to give leaves of 2× the final length, and with no change in the number of growing leaves.

structure, and varying with time of day – different plant parts experience temperatures that can differ by as much as 30 °C (Peacock, 1975a). As correlational analysis shows, and the experimental control of organ temperature confirms, the temperature controlling leaf growth is that of the vegetative shoot apex itself (Peacock, 1975b).

In spring many north temperate grasses are able to expand leaves faster, at a given temperature, than later in the year – or, to put it another way, they can achieve the same rate of leaf expansion as in the autumn, but at a

The reproductive shoot

lower temperature (Peacock, 1975c, 1976; Thomas and Norris, 1977) (see Chapter 4). The transition from an 'autumn' to a 'spring' mode of growth coincides with, although it may not be causally linked to, the earliest stages of reproductive development (Parsons and Robson, 1980), and with a major change in the carbohydrate strategy of the plant (see Fig. 2.12, below) as the emphasis shifts from the build-up of storage fructosans to their rapid remobilization (Pollok and Jones, 1979). It is important that breeders seeking enhanced leaf growth at low temperatures early in the year should be aware of this transition. Ranking orders established on plants showing the 'autumn' mode of growth may not apply in the spring.

2.4 THE REPRODUCTIVE SHOOT

2.4.1 Morphological events associated with reproduction

The change from a vegetative to a reproductive mode of growth has a major impact on the seasonal pattern of production of our north temperate grasses (see Chapters 3 and 4). The first morphological sign of the onset of reproduction occurs at the shoot apex, where the inception of new leaf primordia is accelerated and the length of the apex increases. Soon afterwards bud primordia develop in the axils of the older leaf primordia to give the characteristic 'double ridge' stage, the first unequivocal sign that the shoot is in transition to reproduction. Initially, leaf and bud primordia develop simultaneously, but the bud primordia quickly outstrip the leaf primordia, which become inhibited. These events along the flank of the shoot apex are accompanied by the development of more bud primordia at its tip, as the incipient inflorescence enters its most rapid period of growth (Cooper, 1951).

These developments at the shoot apex generally precede stem elongation which begins in young internodes in the sub-apical zone, moving upwards in a wave of cell division and expansion, progressively lifting the developing inflorescence above the surrounding foliage. The synchronous development of stem, inflorescence and the last-formed leaves culminates in the emergence of the mature inflorescence from the sheath of the final leaf, the flag leaf, and the first major phase of reproduction is complete.

The development of a varying number of alternate bud primordia along the shoot apex, with or without a terminal bud at its tip, is common to all grasses. Where the inflorescence develops into a spike, for example in the ryegrasses, each bud primordium develops into a spikelet – a single branched axis bearing 3–10 closely-packed florets. In grasses with panicles, e.g. tall fescue, the bud develops into a primary branch, which may further divide to give secondary or lower order branches. However, each branch finally terminates in one or more florets organized into a spikelet, the number of florets varying with the species (Fig. 2.1c).

2.4.2 Environmental control of reproduction

An initial period of vegetative growth is normal in the grasses, as the two to three-leaf primordia already formed in the embryo expand and develop following germination, or as parallel events occur in daughter tillers. Whether the shoot then becomes immediately reproductive depends on genotype and the environment to which it is exposed. Most north-temperate perennial grasses must experience winter conditions (low temperatures and/or short photoperiods) for a time if flower induction is to occur, although such requirements may be less-marked or entirely absent in biennials and annuals, respectively (Cooper, 1951). The perennial timothy is an exception, in that no winter requirement has been identified (Cooper, 1958). In general all grasses become 'induced to flower' if they are overwintered in the field or a cool glasshouse. It should be emphasized that this state can only be identified retrospectively, as no morphological change is induced in the shoot. Experimental temperatures between 0–10 °C (vernalization) and photoperiods of about 8 h day^{-1} are most effective in simulating a natural winter induction, although induction in artificial conditions is seldom as effective as that achieved in the natural environment.

Once flower induction is attained, the onset of reproductive growth at the shoot apex generally depends on a period of exposure to long photoperiods, but adverse conditions may delay or block the response. In artificial environments continuous light elicits the most rapid onset of reproduction; in the field most grasses develop inflorescences when a critical photoperiod is reached during early or late spring. This varies with species and cultivar: in ryegrasses it ranges from 8 to 13 h; in timothy, 15 or 16 h of light per day may be required (Cooper, 1951; Ryle and Langer, 1963). Even after the shoot apex has made the transition to reproductive growth, photoperiods of the appropriate length are still necessary for normal inflorescence development (Langer and Ryle, 1958). Warm or cool spring temperatures may accelerate or delay, by a few days, the development and emergence of the inflorescence.

Experimental exposure of the plant during this stage to high temperatures or short photoperiods may cause a reversion to vegetative growth, resulting in death of the inflorescence within the enveloping leaf sheaths or, if it has emerged, the development of leaf-like bracts and vegetative shoots on the inflorescence (Langer and Ryle, 1958). In extreme long-day species such as timothy these phenomena are occasionally seen in the field in late summer as the natural daylength declines.

Despite much elegant experimentation, little progress has been made in identifying the agent or agents which elicit the morphological changes which occur during reproduction (Evans, 1964a,b). The site of perception for low-temperature vernalization appears to be the shoot apex itself because the imbibed seeds of perennial ryegrass, for example, can be effectively treated in the dark. However, the stimulus persists and is apparently transmitted to subsidiary shoots which develop from the main shoot present in the embryo, for they too become induced to flower (Cooper, 1957). Short-day treatment, which can sometimes substitute for low temperatures (Cooper, 1960), is clearly perceived by the expanded leaf surface, again implying a transmissible stimulus. The induction of reproduction by long photoperiods is also perceived by the leaf surface, and presumably some stimulus moves to the shoot apex to elicit the appropriate morphological changes (Evans, 1960). In the grasses, the stimulus can also be transmitted to other organically-connected shoots, even though their own leaves have been held in short photoperiods.

2.5 TILLER PRODUCTION

So far we have concentrated on the growth and reproductive development of a single shoot, although frequent passing reference has been made to the production of daughter tillers which, more than anything else, enables the crop to establish itself rapidly from seed and ensures its subsequent perenniality.

Tillers arise from buds which develop in the axils of leaves (or leaf-like structures such as the coleoptile and prophyll, a coleoptile analogue preceding the first true leaf on a tiller) some two or three plastochrons after the subtending leaf is itself initiated. In tussock-forming grasses, such as the north temperate forages, tillers grow upward within the encircling sheaths of the subtending leaves to emerge at the ligule. In stoloniferous or rhizomatous grasses, such as rough-stalked meadow grass (*Poa trivialis*) or couch (*Agropyron repens*), the tiller may break through the base of the encircling sheath to give rise to a creeping stem with elongated internodes.

In perennial ryegrass the first tiller usually emerges from the axil of the first leaf on the main stem, once it and its successor are fully expanded and the third leaf is externally visible. Subsequent main stem tillers generally arise in acropetal succession keeping pace with leaf production on the main stem (Fig. 2.2c), and themselves producing leaves synchronously (Robson, 1974). In small-seeded grasses such as timothy, the early leaves of which are particularly small, the first two or three sites may remain unfilled with tillering beginning in the axil of the third or fourth leaf (Ryle, 1964). The coleoptile tiller is something of a 'wild card' in that it may emerge first, or later and hence out of sequence, or be absent altogether; prophyll tillers behave similarly. These apart, the pattern of tillering is remarkably regular. Tillers arising in the axils of main stem leaves are generally known as 'primaries'; they in turn produce secondary tillers, and so on. Thus, a hierarchy of tillers is formed connected by a complex system of anastomosing vascular bundles.

In favourable conditions, the early pattern of tillering in perennial ryegrass tends towards that shown in Fig. 2.4. The pattern in other north-temperate species is broadly similar, although it may differ a little in the precise interval between the appearance of a leaf and it subtending a tiller. The rate at which tiller numbers build up is very temperature dependent. High temperatures

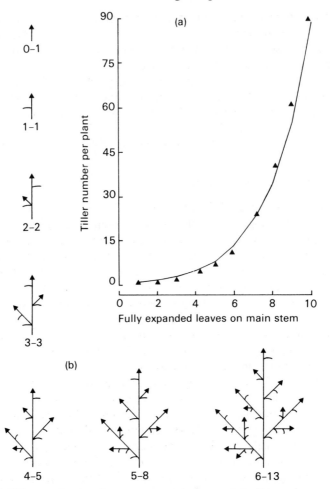

Figure 2.4 Tiller production in young spaced plants of *Lolium perenne* growing in favourable conditions (Robson, 1982b). (a) Increase in tiller number per leaf interval on the main stem (▲), approximates to the Fibonacci series (———): each number is the sum of its two predecessors (1, 1, 2, 3, 5, 8, 13, 21, etc). (b) Generally a tiller (▲) appears in a leaf axil once that leaf's successor is fully expanded. The numbers under each diagrammatic plant (e.g. 6–13) refer to fully expanded leaves on the main stem (6) and the total tiller number per plant (13). Note that at any stage the tiller complex in the axil of the first leaf on the main stem exactly mirrors that of the whole plant two leaf intervals earlier, while the rest of the plant mirrors the whole plant one leaf interval earlier (e.g. 4–5 added to 5–8 gives 6–13).

accelerate tiller production, but mainly through an increased rate of leaf, and hence axillary bud, production. If tiller number is plotted against leaf number on the main stem instead of against time, effects of temperature very largely disappear (Robson, 1974). High irradiance also enhances tillering, and shade

Biomass production

impairs it – but here the effect is not on the rate of site production, but rather on the extent to which sites are filled (Mitchell, 1953a). Under low irradiances a reduced supply of current assimilates is preferentially allocated to existing tillers at the expense of axillary buds. If, later, the light environment improves, sites which failed to produce a tiller may do so, although the older axillary buds may be less capable of responding (Mitchell, 1953b).

Low irradiances can be self-imposed, in the sense that as the plant grows larger and denser, self-shading increases. Inevitably, this leads to a reduction in tillering, the rate of which diverges from the exponential. Nevertheless, tiller production doesn't cease entirely. A single plant, growing outdoors, can easily produce 300 tillers in a single season, if competition from neighbours is prevented (Robson, 1968). When plants grow in communities, tillering is even more restricted and with a severity proportional to the plant population. Indeed, a common tiller density is usually achieved (Kays and Harper, 1974), independent of the plant density which in any case may fall in accord with the ubiquitous 3/2 self-thinning law (Harper, 1977) as inter-plant competition increases and some plants die. In long established grass communities, tiller populations fluctuate around a common baseline (different for different managements) as old tillers die and new ones are produced; each tiller produces *on average* only one daughter tiller (i.e. it replaces itself), but many leaves. Thus, in practice, the great majority of sites fail to subtend a tiller and the potential for tillering vastly exceeds its realization.

2.6 BIOMASS PRODUCTION

The rate of dry matter production of a young plant tends towards the exponential (Anslow, 1962), i.e. it tends to have an approximately constant relative growth rate (rate of dry weight increase per unit dry weight), R_W (symbols according to Warren Wilson, 1966). This holds as long as each increment in dry weight brings with it a proportionate increase in leaf area, so that the leaf area ratio (leaf area per unit plant dry weight), F_A, is unchanged, as long as the net assimilation rate or unit leaf rate (rate of dry weight increase per unit of leaf area), E_A, also remains constant. The latter depends strongly on the rate at which the leaf surfaces fixes carbon, and so is sensitive to any change in the illumination of the leaves or their photosynthetic efficiency. Both of these decline as within-plant shading and the mean age of the leaf surface increase, so that the period of exponential growth is inevitably brief.

2.6.1 Temperature

Temperature has a large effect on the growth of spaced grass plants. High temperatures promote the rapid expansion of relatively thin leaves at the expense of roots, thus raising F_A and so R_W, which can double between 10 and 20 °C (Robson, 1972). Although high temperature enhances the photosynthetic fixation of carbon by leaves (p. 44), it also raises the rate of its respiratory efflux from the whole plant (Robson, 1981a) so that, on balance, E_A is less affected than F_A. Again, the optimum temperature for R_W is about 25 °C, with the night temperature the same or slightly less (Robson, 1973c). It might be argued that a low night temperature would be advantageous in reducing dark respiration but, in spaced plants of tall fescue at least, this is more than offset by continued leaf expansion during the night leading to an increased photosynthetic surface the following day. A change in temperature has little immediate effect on R_W, since E_A is relatively insensitive and F_A can only change slowly.

2.6.2 Irradiance

Light dramatically affects both form and function. In an epic series of

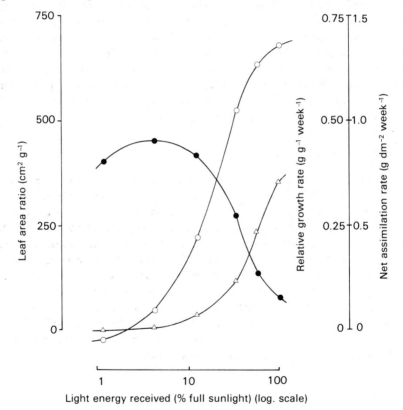

Figure 2.5 Mean relative growth rate (○) and net assimilation rate (△) during 4 weeks in August, and leaf area ratio (●) at the end of that period, of plants of *Festuca arundinacea* grown at irradiances ranging from 1.5 to 100% of full sunlight. (From Robson, 1965, 1969.)

Biomass production

experiments, Blackman and co-workers concluded that the F_A of spaced plants is positively and linearly related to the logarithm of the irradiance, reaching zero at the compensation point (Blackman and Wilson, 1951a). For F_A the correlation was negative but still linear, and could be extrapolated to a zero value, or 'extinction point', at some high irradiance. R_W, being the product of E_A and F_A – although strictly speaking only when instantaneous values are considered (Radford, 1967) – describes a parabola between the compensation and extinction points, with its apogee marking the irradiance at which R_W is maximal (Blackman and Wilson, 1951b). However, later work showed that the relationships of E_A and F_A with the logarithm of the irradiance are often more curvilinear than linear (Goodall, 1955; Njoku, 1960; Evans and Hughes, 1961; Jewiss and Sanderson, 1963), particularly if the range of irradiances actually investigated is widened (Blackman and Black, 1959).

This is true for most north temperate forage grasses. In tall fescue, for example, F_A reaches peak values close to the compensation point and shows signs of bottoming out at just above full daylight while E_A does the opposite (Fig. 2.5). Maximum values of R_W are achieved at close to or just above full daylight (Blackman and Black, 1959; Robson, 1965, 1969). Thus, even though shading to one-third of full daylight can increase the leaf area of spaced plants

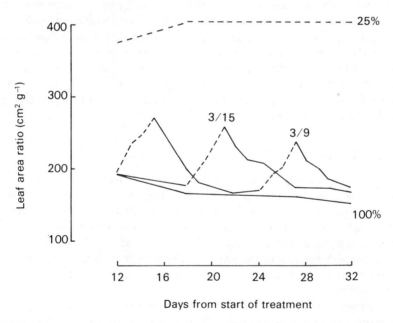

Figure 2.6 Changing leaf area ratio of plants of *Festuca arundinacea* grown continuously in 25% full sunlight, continuously in 100% full sunlight or switched between the two on either a 3/9 or a 3/15 day cycle: ——— 100% full sunlight; ---, 25% full sunlight. (From Robson, 1969.)

of tall fescue by trebling their F_A, they may still be only half the dry weight of plants grown in full daylight (Robson and Jewiss, 1968).

The input of solar radiation often varies irregularly from day to day as the degree of cloud cover varies. The photosynthetic rate of a plant, and hence its E_A, responds rapidly to a change in irradiance wherease F_A can only adjust more slowly. It follows that immediately after a change in the light regime R_W will depend on an E_A appropriate to the current conditions and on an F_A more typical of the previous conditions (Blackman and Wilson, 1954). Thus, the apparently surprising finding (Robson, 1969) that after a plant is transferred from low to high irradiances its increased E_A coupled with an existing high F_A give it a temporarily greater R_W than that of a plant grown continuously at a high irradiance. The reciprocal transfer leads to a plant with a lower R_W than that of one grown at a continuously low irradiance. Because of this, Ryle (1967) suggested, and Robson (1969) confirmed, that a more efficient use of light might be achieved by exposing plants to a high irradiance punctuated by short periods of low irradiance. Three 16 h days of 25% light, alternating with either 9 or 15 days of full light, led to plants with a greater average F_A than plants grown in high light throughout (Fig. 2.6), and with very similar relative growth rates.

Thus, it would seem likely that when a sward is open, as during establishment, and acts essentially as a collection of spaced plants, occasional cloudy days may not restrict growth too severely. Indeed, one would generally expect low irradiances to depress crop growth less during establishment than when the leaf canopy is closed; in the latter case an enhanced F_A brings no return in terms of increased light interception, while in the former it does. Conversely, low temperatures, by restricting leaf expansion more than net assimilation, should depress growth more during establishment than thereafter, and recent crop growth models predict precisely this (Johnson and Thornley, 1983).

2.7 LEAF PHOTOSYNTHESIS

When plant tissue is combusted the residual ash represents those mineral elements taken up from the soil via the roots. In grasses they generally make up between about 6 and 9% of the plant dry weight (Armstrong, 1948). The oxidized substances are carbohydrates and the carbon-based fractions of other organic compounds such as proteins. Thus, more than 90% of the plant dry weight stems directly from the photosynthetic assimilation of carbon. In the next few sections we will consider that assimilation and the effect on it of age and environment.

Determining the true rate of leaf photosynthesis from gas exchange measurements is difficult, because CO_2 production associated with respiration offsets some of the CO_2 uptake during photosynthesis. In north-temperate (C_3) grasses, CO_2 efflux in the light is due largely to photorespiration, which can account for as much as 40% of gross photosynthesis. It is likely that dark, mito-

Leaf photosynthesis

chondrial, respiration is suppressed, at least in part, in illuminated photosynthesizing tissues (Bidwell, 1983). In any case, the dark respiration of a grass leaf (although not of the whole plant or crop) is commonly less than 10% of net photosynthesis in high irradiances. Generally, the rates of photosynthesis presented here are net of both photorespiration and dark respiration (P_n). The Photosynthetic capacity (P_M) of a leaf we take to be its light-saturated rate at normal atmospheric CO_2 and O_2 concentrations and optimal temperature and humidity.

Leaf photosynthesis is influenced by the current environment, by the leaf age and by the environment the leaf experienced while its photosynthetic apparatus was being laid down.

2.7.1 Current environment and photosynthetic rate

The rate of net photosynthesis (P_n) of newly expanded leaves of such north temperate grasses as perennial ryegrass (Wilson and Cooper, 1969a; Prioul,

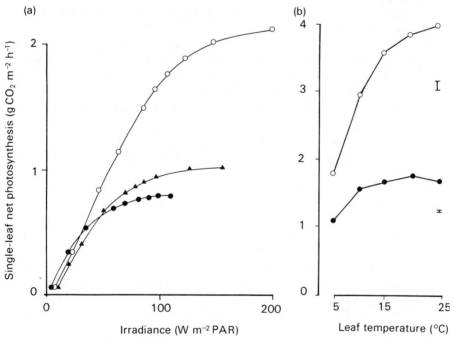

Figure 2.7 Net photosynthetic rate of single leaves of *Lolium perenne* at ambient CO_2 concentrations. (a) Effect of irradiance on leaves grown at $110\,W\,m^{-2}$ (PAR) (○), $45\,W\,m^{-2}$ (▲) and $16\,W\,m^{-2}$ (●). (Redrawn from Prioul, 1971.) (b) Effect of temperature on leaves measured at a water vapour saturation deficit of $3\,g\,m^{-3}$, and $250\,W\,m^{-2}$ (PAR) (○) or $50\,W\,m^{-2}$ (●); vertical bars represents s.e. (Redrawn from Woledge and Dennis, 1982.)

1971; Woledge, 1973), cocksfoot (Carlson et al., 1971) and tall fescue (Wilhelm and Nelson, 1978) follow similar light-response curves (Fig. 2.7a). In formally describing such data (converted to 'gross' photosynthesis (P_g) by adding dark respiration to net) the most widely used empirical curve is the now 'almost traditional' rectangular hyperbola (Johnson and Thornley, 1984), although it often gives a poor fit to experimental data (Marshall and Biscoe, 1980; Monteith, 1981). Much better and more versatile is the non-rectangular hyperbola first developed by Rabinovitch (1951), discussed most notably by Chartier (1970) and recently employed in a model of canopy photosynthesis by Johnson and Thornley (1984):

$$P_g = \frac{1}{2\theta} \{\alpha I + P_m - [(\alpha I + P_m)^2 - 4\theta\alpha I P_m]^{1/2}\} \tag{1}$$

Here I is the irradiance in W m^{-2} (PAR), P_g the rate of single leaf 'gross' photosynthesis (kg CO_2 m^{-2} s^{-1}), P_m the limiting value of P_g at saturating irradiance, α the initial slope, is the photochemical efficiency (kg CO_2 J^{-1}) and θ a dimensionless parameter ranging in value from zero, when Equation 1 reduces to the rectangular hyperbola, to unity, when a Blackman (1905) type limiting response is achieved (see Chapter 7).

The north-temperate grasses vary little in α (Charles-Edwards, 1978) with typical values for a number of C_3 species being close to 0.05 mol CO_2 mol^{-1} absorbed PAR at 30 °C and normal atmospheric concentrations of CO_2 and O_2 (Ehleringer and Bjorkman, 1977). Moreover, α appears to be relatively insensitive to variations in environmental factors such as temperature (Ku and Edwards, 1978).

On the other hand, P_m is genetically variable in C_3 grasses (Cooper and Wilson, 1970; Charles-Edwards, 1978). As we shall see, it is sensitive to age and environment, and it is on this aspect of the photosynthetic light-response curve that we concentrate. When measured at atmospheric CO_2 concentrations, near-optimal temperatures and high humidity the photosynthetic capacities of leaves of most north-temperate grasses grown in glasshouse or controlled environment are in the range 1.3–4.0 g CO_2 m^{-2} h^{-1} (Redman, 1974; Williams, 1974; Lloyd and Woolhouse, 1976; Williams and Kemp, 1978; Painter and Delting, 1978), generally lower than values quoted for C_4 grasses (Ludlow and Wilson, 1971; Bolton and Brown, 1980). Field-grown plants are capable of similar rates (Asay et al., 1974; Woledge and Leafe, 1976; Woledge, 1978), although leaves developed in a dense sward can have a reduced photosynthetic capacity (p. 46).

Apart from irradiance, current temperature is the most important environmental determinant of leaf photosynthesis. At light saturation the net photosynthetic rate of ryegrass leaves increases dramatically with temperature between about 5 and 25 °C (Fig. 2.7b). Similar results have been obtained for cocksfoot (Treharne and Eagles, 1970), tall fescue (Treharne and Nelson, 1975) and *Agropyron smithii* (Monson et al., 1982). At lower irradiances the

Leaf photosynthesis

effect of temperature is less, as light rather than enzymic reactions becomes limiting.

The response to temperature is often complicated by the fact that low air humidity (high saturation deficit) reduces photosynthesis through partial stomatal closure (Monson et al., 1982; Woledge and Parsons, unpublished) (see Chapter 6). In the natural environment, and in some experiments aimed at investigating temperature effects, high temperatures tend to be associated with high saturation deficits. Thus, an absence of an increase in net photosynthesis with increasing temperature may only imply a failure to prevent a rise in saturation deficit. In the natural environment low humidity may significantly restrict photosynthesis at higher temperatures.

2.7.2 Leaf age and photosynthesis

As a leaf ages its photosynthetic capacity declines, starting soon after full expansion and well before any visible sign of senescence (Jewiss and Woledge, 1967; Treharne and Eagles, 1970; Woledge, 1972). The tip, being older,

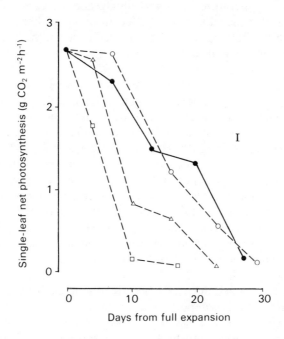

Figure 2.8 Effect of light environment on the decline with age of the photosynthetic capacity of leaves of *Lolium perenne*, measured at 20 °C, a water vapour saturation deficit of 3 g m^{-3} and an irradiance of 170 W m^{-2} (PAR). Leaves grown at 120 W m^{-2} (PAR) until fully expanded, and then either kept at this irradiance (●——●) or artificially shaded (---) to 57 W m^{-2} (○), 20 W m^{-2} (△) or 5 W m^{-2} (□): vertical bar indicates s.e. (Redrawn from Woledge, 1972.)

declines before the base (Prioul *et al.*, 1980a). A fall in both stomatal and residual conductance accompanies the photosynthetic decline (Woledge, 1972) as does a fall in RuBP carboxylase/oxygenase activity (Treharne and Eagles, 1970).

This age-related decline in photosynthetic capacity is slowed by low temperatures (Woledge and Jewiss, 1969) or shade (Woledge, 1971) and the leaves live for longer. However, it is accelerated if the leaf, or the whole plant, is kept in extremely low irradiances (20 W m^{-2} PAR or less) before measurement in high irradiances (Woledge, 1972) (Fig. 2.8). The rate of photosynthesis which a leaf exhibits in low irradiances declines more slowly with age than its light saturated rate (Woledge, 1971, 1972). Since many leaves in a grass community, and especially the older ones, experience only low irradiances, the loss to canopy photosynthesis due to leaf ageing is less than might otherwise be expected.

2.7.3 Environment and the development of the leaf's photosynthetic capacity

The irradiance, temperature, nitrogen supply and water stress that a leaf experiences during its development all affect its photosynthetic capacity when it reaches maturity. Of these, the effects of irradiance are the best understood and probably the most important.

(a) Irradiance

Leaves developed in low rather than high irradiances have a lower rate of photosynthesis per unit leaf area at light saturation, a similar or greater rate at low irradiance and a lower rate of dark respiration (Treharne and Eagles, 1969; Wilson and Cooper, 1969a,c; Prioul, 1971; Woledge, 1971, 1973) (Fig. 2.7a). A number of structural and biochemical features contribute to these differences in photosynthetic rate per unit leaf area. Both stomatal and residual conductances are greater in leaves grown in high irradiances (Prioul, 1971; Prioul *et al.*, 1975; Woledge, 1977; Woledge and Dennis, 1982). Such leaves may have more stomata (Wilson and Cooper, 1969c), are thicker (Prioul, 1971; Woledge, 1971), have more mesophyll cells per unit leaf area, more and larger chloroplasts and a greater activity of RuBP carboxylase/oxygenase (Prioul *et al.*, 1975, 1980a).

In the field the reduction in leaf photosynthetic capacity resulting from growth in low irradiances is important in vegetative swards. When a vegetative sward is allowed to grow for some weeks without interruption, as after establishment or when utilized by rotational grazing or infrequent cutting, developing leaves are increasingly shaded as the LAI of the sward increases and the light environment at its base worsens. As a result each tiller in the sward produces a succession of leaves with progressively lower photosynthetic capacities (Fig. 2.9a) so that when, eventually, they emerge into bright light they are unable to make full use of it (Woledge, 1973; Woledge and Leafe,

Leaf photosynthesis

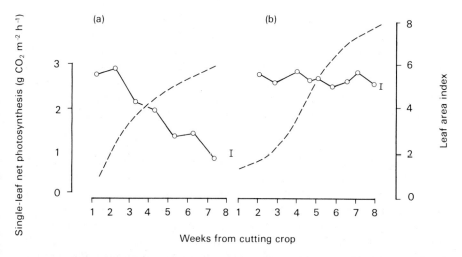

Figure 2.9 Photosynthetic capacity (○) of successive leaves of *Lolium perenne* taken, when newly expanded, from (a) a vegetative or (b) a reproductive field sward, and measured at 250 W m^{-2} (PAR), together with the LAI of the sward (---): vertical bars indicate s.e. (Redrawn from Woledge and Leafe, 1976.)

1976; Sheehy, 1977). This leads to a shortfall in carbon fixation by the crop canopy which, computer simulation (Monteith, 1965) suggests, could amount to 30% by the end of a 12-week growth period (Robson, 1973b).

In flowering swards, conversely, there is no reduction in the photosynthetic capacity of successive newly expanded leaves, even when high LAIs are achieved because flower stem extension keeps the developing leaves in high irradiances near the top of the canopy. Thus, each tiller produces a succession of leaves of equal photosynthetic capacity (Fig. 2.9b) (Woledge and Leafe, 1976), ensuring a high rate of canopy photosynthesis. This difference in the photosynthetic capacity of vegetative and reproductive swards is a major reason for the longer, steeper phase of dry matter production exhibited by the reproductive crop when compared with its vegetative counterpart (see Chapter 4).

An alternative explanation of the high photosynthetic rate of leaves on flowering tillers has been proposed, namely that it is induced by an increased demand for assimilates from the rapidly growing stem and ear (Silsbury, 1965; Behaeghe, 1976; Deinum, 1976). This presupposes that the photosynthesis of vegetative plants is normally restricted by a low demand for assimilates, but this does not appear to be the case. So long as their expanding leaves are protected from shading by neighbours, vegetative plants produce a succession of leaves whose photosynthetic capacity is as high as that of leaves on reproductive plants (Woledge, 1978, 1979). This is true even when protecting the growing leaf from shade also increases the assimilation of the rest of the plant, thus reducing the demand for assimilates from the leaf in question (Woledge, 1977). Conversely, the decline in the photosynthetic capacity of successive leaves in increasingly dense vegetative swards occurs even if the demand for assimilates from the sample leaf is raised by removing all the other leaves on the plant (Woledge, 1977). We can conclude, therefore, that it is a difference in illumination during leaf development that causes the difference in photosynthetic capacity between leaves of vegetative and reproductive swards, rather than any difference in assimilate demand.

It is usually assumed that it is the light conditions experienced by the developing leaf itself which determine its photosynthetic capacity. Certainly, chloroplast structure and RuBP carboxylase/oxygenase activity appear to be controlled by the amount of light reaching the leaf directly (Prioul *et al.*, 1980b). Conversely, the same authors suggest a role for light perceived by other parts of the plant – in that the number of cells in a grass leaf is determined early in development before it is directly exposed to light and, as we have seen, differences in cell numbers may contribute to the different photosynthetic capacities of leaves developed in contrasting light environments.

While growing leaves have a considerable ability to adapt their photosynthetic capacity as light conditions change (Woledge, 1971; Prioul *et al.*, 1980b), mature leaves can only readapt to a limited extent. Transfer from low to high irradiance increases leaf photosynthetic capacity by about one-quarter (Woledge, 1971). Grasses are less flexible in this respect than many other crop

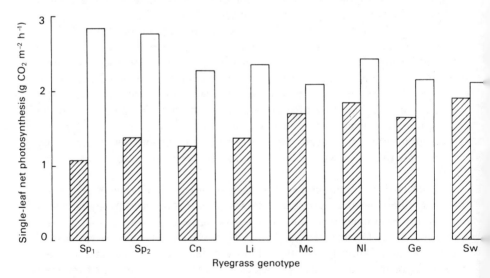

Figure 2.10 Genotypes of *Lolium perenne* differing in the sensitivity of their photosynthetic apparatus to development under shade. Net photosynthesis at 104 W m^{-2} (PAR) of leaves developed under 22 W m^{-2} (▨) or 50 W m^{-2} (□). (From Wilson, 1981; data of Wilson and Cooper, 1969c.)

Leaf photosynthesis

plants, but their rapid turnover of leaves results in a complete new canopy every month in summer. Thus, the plant as a whole, rather than the individual leaf, can adapt effectively to changed external light conditions.

There appears to be substantial genetic variation in the extent to which the photosynthetic apparatus of grass leaves is impaired by shade during development (Fig. 2.10). Some genotypes are relatively unaffected and, hence, might prove valuable under managements, such as infrequent cutting or rotational grazing, where a high LAI is sustained for an appreciable time. Others are very sensitive to shade but, when well illuminated, produce leaves of high photosynthetic capacity and so might be particularly useful under managements such as frequent cutting or continuous grazing, where a build-up of LAI is prevented. In ryegrass genotypes whose photosynthetic capacities were investigated by Wilson and Cooper (1969c) it is very noticeable that ranking orders drawn up for plants grown in shade differ dramatically from those for plants that had been fully illuminated during leaf development (Wilson, 1981). It has been argued (Robson, 1980, 1981b, 1983) that this phenomenon might in part explain the failure to increase grass yields by selecting for high photosynthetic rates *per se*.

(b) Temperature

The photosynthetic adaptation of grass leaves to the temperature at which the plants are grown (as distinct from an immediate response to the temperature of measurement) is by no means as clear-cut as their adaptation to light. It seems to take two forms (Fig. 2.11), one or both of which have been recorded for a number of pasture plants, (Charles-Edwards *et al.*, 1971; Treharne and Nelson, 1975; Woledge and Dennis, 1982). However, not all grasses respond to growth temperature in this way, and there are differences between species and even ecotypes within a species (Wilson and Cooper, 1969b; Treharne and Eagles, 1970). In some, growth at low temperature results in a greater photosynthetic capacity over a whole range of measurement temperatures, while in others, growth temperature has barely any effect (Treharne and Eagles, 1970).

These differences are not surprising. Net photosynthesis is the resultant of several processes, all of which can be influenced by both current and growth temperature, in different ways and to different extents. Thus, Treharne and Nelson (1975) found that a greater rate of photorespiration and dark respiration accounted for the lower temperature optimum for net photosynthesis of tall fescue plants grown at low temperature. Enhanced rates of dark respiration in leaves grown at low temperatures have also been reported by Woledge and Jewiss (1969) and Woledge and Dennis (1982). In some cases, factors such as mesophyll cell size (Wilson, 1970) or RuBP carboxylase/ oxygenase activity (Treharne and Eagles, 1970) appear to be involved in photosynthetic adaptation to temperature, although not in others (Wilson and Cooper, 1969b; Treharne and Nelson, 1975).

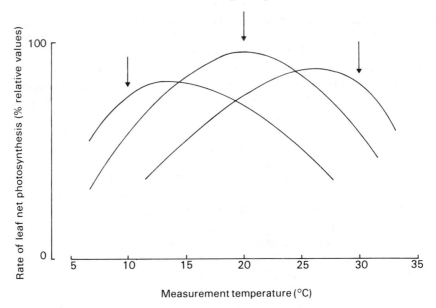

Figure 2.11 Generalized example of photosynthetic adaptation to growth temperature (10, 20 or 30 °C), assuming that maximum photosynthetic rates are achieved when leaves grown at 20 °C are measured at 20 °C. Arrows indicate temperature of growth. Note: (a) leaves grown at any one temperature have a greater photosynthetic rate at that temperature than leaves grown at any other, and (b) that the optimum temperature for photosynthesis tends away from 20 °C towards the growth temperature.

(c) Nitrogen

While nitrogen deficiency reduces grass yields, largely by restricting leaf expansion and, thus, light interception, it also impairs the photosynthetic capacity of the leaves (Woledge, 1975; Wilson, 1975; Powell and Ryle, 1978; Robson and Parsons, 1978) whenever the leaf nitrogen content falls below about 5% of the dry weight (Bolton and Brown, 1980). Thus, in the field, where nitrogen contents are usually very much lower than this, even in heavily fertilized crops (Whitehead, 1970), photosynthesis will often be limited. In infrequently harvested crops, nitrogen supply may have two conflicting effects. It may increase leaf photosynthesis early in the growth period but, by stimulating leaf expansion and mutual shading, it may indirectly depress it later (Woledge and Pearse, 1985).

2.8 MERISTEMATIC ACTIVITY AND THE UTILIZATION OF ASSIMILATES

The main function of mature leaves is to synthesize simple carbohydrates from carbon dioxide and water using the energy available from incident radiation.

Meristematic activity

Once formed, the carbohydrates are translocated to those centres of activity in the plant – the meristems – where new growth is occurring. The grasses, in common with all other flowering plants, possess four types of meristem, the locations of which are ordained by the way in which the different tissues are organized in the shoot. These meristems – leaf, branch, stem (intercalary) and root – may become active and expand to form a mature organ, or may remain suppressed, depending on genotype, ontogeny and environment.

We have already dealt with the growth of these organs, particularly the leaf, in some detail. Now we need to consider how their requirements for assimilates interact, and how the priorities accorded to them vary. Immediately following germination the main priorities for the young plant are the expansion of roots and of leaves: branch meristems are quiescent and internode elongation absent. Of the two active meristems, the roots gain the major share of assimilates. The first change in this pattern comes with the sequential activation of the branch meristems, leading to the production of daughter tillers. These very soon account for 30–35% of the current assimilates – largely at the expense of the roots whose share falls to 50%, while the leaf meristems account for 15–20%. As vegetative growth continues the root's share declines even further, while that of the leaf meristems rises to a maximum of 30%. If vegetative growth is particularly prolonged, then the intercalary meristems may become active – more so in annual than in perennial species and, again, at the expense of the roots (Ryle, 1970a,b). Nevertheless, the extent of internode elongation is normally very limited while the plant remains vegetative.

The onset of reproduction has major consequences for meristematic activity and assimilate utilization. First, the branch meristems, at the tip of the shoot, become converted into an inflorescence. Although it is sometimes assumed that this markedly increases the consumption of assimilate in that part of the shoot, the available evidence indicates that any increase is small (Ryle, 1972). The main significance of inflorescence development is that it terminates the shoot's ability to produce new stem segments, and its life history is thus destined to end once the ear has emerged (p. 26). Of more importance for patterns of assimilate utilization is the fact that reproduction triggers rapid stem elongation. The morphological changes associated with stem elongation have already been described (p. 35). At its peak, the wave of meristematic activity in the sub-apical internodes consumes 40–50% of the total available assimilate, while the proportion utilized in root growth exhibits a further decline to less than 10%, from which it never recovers (Ryle, 1970b). Once the peak of stem elongation is past, no more photosynthetic surface is expanded and root, stem and leaf meristem activity are all at a low ebb. Thus, *proportionately* more current assimilate becomes available for the tiller meristems – although very little more *absolutely* since, by now the shoot's assimilatory surface is ageing quickly and declining in both area and photosynthetic efficiency.

The general account of meristem activity outlined above is typical of a major shoot on a well-illuminated plant experiencing a favourable environment. The same general trends are evident in simulated swards grown in natural environments although the detail varies: in general, less carbon is exported to tillers and to roots (Parsons and Robson, 1981b). The same is true of field swards, although good data are very difficult to obtain. Nevertheless, enough is known from controlled environments to provide an account of the principles involved.

2.8.1 Environment and management

Grass swards are often dense. The presence of neighbours induces a struggle for light. The influence of this factor is illustrated in Table 2.1 for vegetative plants grown either in close-to-optimal irradiances or in a four-fold lower regime more akin to that found in a dense leaf canopy. The response is classic in the sense that when assimilate is in short supply, relatively more is devoted to leaf meristems (and to some extent, vegetative stem elongation), at the expense of branch (tiller) and root meristems. In other words, the shoot maximizes leaf growth at the expense of vegetative propagation (tillering) and particularly, root growth – a pattern characteristic of the perennial grass when grown in simulated swards and in the field, once the leaf canopy has closed (Parsons and Robson, 1981b; Colvill and Marshall, 1981; Parsons *et al.*, 1983).

TABLE 2.1 *Partitioning of* ^{14}*C-labelled assimilate between meristems after 24 h, in an annual ryegrass* (Lolium temulentum) *grown in high and low irradiances*

	Irradiance $W\,m^{-2}$ (PAR, 400–700 mm)			
	188		47	
Leaf exposed to $^{14}CO_2$	Youngest mature	Oldest	Youngest mature	Oldest
Photosynthetic rate ($g\,CO_2\,m^{-2}\,h^{-1}$)	3.20	2.14	1.50	1.13
Percentage assimilate exported to:				
leaf	17.6	3.2	23.8	12.3
stem	4.0	1.6	10.3	5.0
tillers	30.0	29.7	10.2	25.4
root	17.4	37.8	4.2	19.6

Some assimilate retained in donor leaf or (small proportion) located in other mature leaves.

Precisely the same principle operates if the grass shoot is defoliated. Assimilate from any residual leaf surface is exported to the leaf meristems to an extent dependent on the severity of the defoliation, even if the leaf surface concerned originally supplied root and tiller meristems (p. 53). If the defoliation is extreme, then export of assimilate to tillers and root ceases entirely

Meristematic activity

and root growth and respiration are severely curtailed, together with other processes such as ion uptake and cytokinin synthesis (Davidson and Milthorpe, 1966a,b). With the expansion of new leaf surface from the shoot apex the supply of assimilate to meristems gradually returns to the pattern characteristic of the undefoliated shoot (Ryle and Powell, 1975).

Generalizations about the influence of temperature on assimilate partitioning are difficult because this factor has a pervasive influence on the plant, acting on rates of cell division and expansion in all meristems. At low temperatures the rate of utilization of assimilate in meristematic growth may be decreased more than its rate of production by photosynthesizing leaves. Thus, competition between meristems for assimilate may be less than at higher temperatures, when the potential for meristematic growth may be closer to maximal. It is not entirely clear whether all plant meristems exhibit the same temperature optima. In the grasses low temperatures generally appear to favour root and tiller growth, but this may be more the result of the shoot's supply of assimilate more nearly satisfying, or even exceeding, the total potential requirements of all meristems, rather than of any real distinction between the temperature optima of different organs.

2.8.2 Source leaf and the dynamics of assimilate translocation

In the foregoing sections the assimilate which is used in meristematic growth has been discussed as though it were supplied from a single pool. In very young vegetative plants the leaves do appear to contribute to a common pool of assimilate but, with further growth, individual leaves develop specific roles in relation to the destination of their exported assimilate.

Most of the assimilate generated in the growing leaf is utilized in its own basal meristem, where new leaf or sheath tissue is being laid down, but a little passes through the zone of meristematic activity and is exported to other meristems. Mature leaves retain 10–25% of their assimilate for maintenance and other metabolic processes (Ryle, 1970a), with the destination of assimilate exported from them depending on the age of the leaf and its point of insertion on the shoot. In general, leaves export progressively more of their assimilate to leaf meristems the younger they are and the higher their insertion on the shoot (Ryle, 1970b). Conversely, older leaves at the base of the shoot export more to roots. The pattern of export to tillers (branches) depends on the relative positions of the leaf and the particular tiller. Generally, leaves export a large proportion of their assimilate to the tiller in their own, or the next lower, leaf axil. When intercalary meristems become active, they mostly utilize the assimilate from those leaves which would otherwise have exported predominantly to roots and tillers (Ryle, 1970b; Ryle and Powell, 1972).

When considering the role of leaves, it should be remembered that the leaves along a shoot differ in age, expanded surface and light interception. As a result the upper, younger, better-illuminated leaves photosynthesize more rapidly and generally contribute much more to the shoot's pool of assimilate than do older leaves at the base of the shoot.

Whereas leaves at particular stages of development may export preferentially to specific meristems, all leaves export to some extent to all active meristems in the same shoot. They may also export to adjacent shoots on the same plant to an extent dependent on the circumstances of the individual shoot and the nutrition of the whole plant. Generally, large-rooted shoots export assimilate to their daughter tillers until the latter develop their own leaf and root systems, although in young plants net export may continue beyond this stage (Ryle and Powell, 1972). A potentiality for bidirectional assimilate movement between organically connected shoots can be demonstrated by selective defoliation, in that the direction of assimilate translocation may be temporarily reversed (Sagar and Marshall, 1966).

Although little is known about the dynamics of assimilate translocation in the perennial grasses, radiocarbon studies demonstrate that, in good growing conditions, assimilate may be mobilized within 10–15 minutes of its formation, and that the velocity of its translocation along the lamina approaches 1 cm min^{-1}. The timing of assimilate arrival in the root indicates that translocation elsewhere in the plant is equally rapid (Ryle, unpublished).

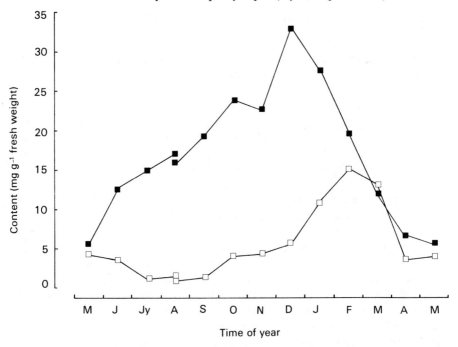

Figure 2.12 Seasonal pattern of accumulation and consumption of fructosans (■) and low molecular weight sugars, mainly sucrose (□) in the sheaths and stems of a high water-soluble carbohydrate selection line of *Lolium perenne*. (From Pollock and Jones, 1979.)

Respiration

Some of the assimilate formed in photosynthesis is temporarily retained in the leaf either as sucrose, starch or some other short-term storage product, only to be mobilized in the dark and exported, in the form of sucrose, to the meristems in the normal way. In most Gramineae the rate of export of assimilate from the leaf is more rapid in the light than in the dark. This does not imply a reduced availability of assimilate at the meristems in the dark, since not all of that arriving at meristems during a light period is immediately used in growth. Some is retained for use during the following night or the next day, so buffering the plant against day-to-day fluctuations in supply (Gordon et al., 1977). In much the same way – although over a longer timescale – perennial grasses accumulate and draw upon high-molecular weight, long-term storage compounds (Pollock and Jones, 1979). As we have seen earlier (p. 35), these compounds are accumulated during the autumn, reach maximal values in winter and decline as growth accelerates in the spring (Fig. 2.12).

2.9 RESPIRATION

Technical developments in the past two decades have made it possible to measure in detail, and routinely, the carbon fluxes into and out of individual plants and simulated canopies and, as a result, to construct accurate carbon balance sheets. Much of the pioneer work was done on the grass crop. As a result we know that of the carbon fixed in photosynthesis, one-quarter at most is ultimately harvested, the rest being lost, mostly through respiration. In simulated swards, under ideal conditions, the respiratory efflux of CO_2 accounts for half of the carbon fixed (Robson, 1973b). In the field, where conditions are often far from ideal, it can be much higher. Findings such as these have helped to create a renewed interest in all aspects of respiration, ranging from its biochemical basis to the practical possibilities of manipulating it with a view to increasing crop yields.

Respiration in plants has two functions, both essential to growth. First, it releases the energy required for the synthesis of new molecules and for other metabolic processes. Second, it provides the carbon skeletons, resulting from stepwise breakdown of substrates, from which new molecules and tissues can be synthesized. The 'classical' complete combustion of sugar molecules to CO_2 and H_2O during oxidative phosphorylation may be the exception rather than the rule.

In single plants and canopies it is not feasible to measure biochemical energy fluxes, so an understanding of respiration has been sought in measurements of CO_2 loss (or O_2 uptake), a difficult task in illuminated photosynthetic tissues (p. 42). Most early workers assumed implicitly or explicitly that respiration was linearly related to the weight of biomass. More detailed experimentation demonstrated that the respiration rates of plant canopies were not constant but varied with the environment, and led McCree and Troughton (1966) to postulate that respiration was linked with rate of photosynthesis, although a second component was, as before, related to plant biomass. McCree (1970) expressed these in a simple relationship:

$$R = kP_g + cW \qquad (2)$$

where R is the 24-h total of respiration, P_g is the daily gross photosynthesis, W is the dry weight of plant biomass, and k and c are constants. In McCree's experiments the total respiration was proportional to one-quarter of the gross photosynthesis ($k = 0.25$) plus an amount equivalent to 1.5% ($c = 0.015$) of the plant biomass present.

2.9.1 Synthetic and maintenance components

This conceptual view of respiration in terms of two components, one related to the consumption of assimilate in current growth (synthesis), and a second related to the respiration of mature tissues (maintenance), was provided with a quantitative biochemical framework by the contemporaneous studies of Penning de Vries and co-workers. They calculated the weight of plant products which could be generated from unit weight of assimilate by the most efficient known biochemical pathways, taking account of chemical, energetic and material requirements. Thus, if the substrate and the plant's final growth products were known, the numerical relationship between weight of substrate, weight of product, O_2 consumed, CO_2 respired and chemical energy required could all be calculated (Penning de Vries, 1972). This approach provided a firm basis from which the costs, in terms of assimilate respired, of the synthesis of new tissues and the maintenance of existing tissues could be calculated.

It should be emphasized that the distinction made here is basically that of respiration associated with the *de novo* synthesis of new organs, as opposed to that associated with the maintenance of existing organs. Although at the biochemical level the respiratory-chain energetics are probably identical, in the one case they are harnessed to the synthesis of new cells and tissues (biomass production), and in the other to sustaining metabolic integrity in existing cells (biomass maintenance) and hence have very different practical consequences.

The energy costs of producing a given weight of biomass are many times greater than those of maintaining that weight over a limited period, it is therefore not surprising that the meristematic shoot apex has a much higher rate of CO_2 efflux per unit dry weight than fully expanded leaf laminae (Robson and Parsons, 1981). Nevertheless, in neither tissue can respiratory activity be said to be of one form only. Indeed, the two activities are intermingled throughout the plant. Because of this, and because the CO_2 involved in both is chemically identical, it is not possible to distinguish between the CO_2 efflux from biosynthesis and that from maintenance using IRGA alone. This problem can be partially overcome by using radiocarbon to label assimilate formed in photosynthesis.

Respiration

As described earlier, the ^{14}C-labelled assimilate of mature leaves is quickly exported to meristems where some is respired to provide the energy for biosynthesis, while the remainder becomes incorporated as labelled carbon atoms in the tissues which are formed. As these new tissues mature and age, they too respire and $^{14}CO_2$ is evolved. Since the amount of ^{14}C-labelled assimilate can be measured, and that lost in the short term (synthetic) and long term (maintenance) can also be determined, this technique provides an alternative and, in some ways, more penetrating account of the magnitude and sequence of respiratory processes. The time course for $^{14}CO_2$ evolution derived from assimilate from a single leaf of ryegrass is depicted in Fig. 2.13. In principle the same technique can be used with whole plants or swards, albeit with some loss of detail.

Because of the basic nature of the biochemical processes involved, respiration in the grasses is similar to that of other plants, although the morphology and management of the crop impose some differences of detail. As far as the respiration linked to synthetic processes is concerned, the perennial grasses respire 20–35% of each unit weight of assimilate in coverting the remainder into new shoot and root, i.e. the conversion efficiency in biosynthesis is 65–80% (Robson, 1973b; Hansen and Jensen, 1977). On theoretical grounds the conversion efficiency may vary according to the chemical composition of the plant products and, for example, the chemical form of the source of nitrogen used by the plant (Penning de Vries, 1972), but these factors are unlikely to alter it substantially. Neither does temperature influence the conversion efficiency significantly; it affects the rate at which substrate is transformed into new tissues, i.e. it affects rate of growth, but the efficiency will not change unless the energy coupling in phosphorylation is affected. The rather wide band of efficiency quoted above primarily reflects the difficulty of measuring it accurately in whole plants and swards.

By contrast, maintenance respiration appears to be sensitive to a range of factors, and is more difficult to characterize. In the grasses, experimental values range between 6 and 90 mg $CO_2 g^{-1}$ (dry wt) day^{-1} (Robson, 1973b; Ryle *et al.*, 1976; Hansen and Jensen, 1977; Jones *et al.*, 1978; Parsons and Robson, 1982). In the short term, and no doubt over a limited range, the response to temperature exhibits a Q_{10} of about 2.0. In the long term, i.e. the life of individual organs, the relationship is not well understood, because little is known about the effect of temperature on molecular stability, turnover rates and energy coupling. Measured rates of maintenance respiration appear to vary with mineral nutrition (Robson and Parsons, 1978), stage of plant development (Jones *et al.*, 1978), and both the light (Woledge, 1971) and the temperature (Woledge and Jewiss, 1969) conditions experienced during growth. Grass tissue produced in early spring, for example, appears to have a higher rate of maintenance respiration, at a standard temperature, than tissue produced later in the year, when temperatures are higher but radiation receipts similar (Parsons and Robson, 1982).

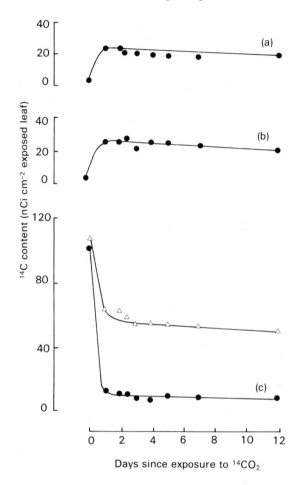

Figure 2.13 Decline, due to respiratory efflux, in the ^{14}C content of a uniculm barley plant (△) during 12 days from the exposure of the youngest fully expanded leaf to $^{14}CO_2$ in the light. Also, changes in the individual organs (●) reflecting, not only respiratory losses, but also import into (a) root and (b) shoot (excluding the fed leaf), and (c) export from the fed leaf. (From Ryle et al., 1976.)

As the foregoing discussion implies, the magnitude of maintenance respiration appears to be sensitive to the structure and composition of the crop. In *young plants* where accumulated biomass is small, maintenance respiration is also generally small relative to photosynthesis and growth. Total respiration at such a time amounts to 30–40% of gross photosynthesis, with less than half being assigned to maintenance processes. Total respiration, too, has a Q_{10} of about 2, at least over the temperature range 5–20 °C, and with an absolute value very responsive to previous environmental conditions. High irradiances and

Respiration

low temperatures, for example, increase substrate availability and so enhance the subsequent respiration rate, low irradiances and high temperatures do the opposite (Robson, 1981a).

In older plants or crops, as biomass builds up, the maintenance component also increases, but several factors limit the extent. Plant tissues have a finite life and they are shed, while those tissues which have a predominantly structural function, such as stem, respire only slowly after maturity. As biomass builds up in spring and the proportion of stem tissue rises, the respiration rate per unit dry weight declines – although much less so if expressed in terms of protein (Jones *et al.*, 1978). Thus, the size of the respiratory burden depends on the nature of the biomass as well as its weight. Factors such as developmental stage and crop management are important in this context.

2.9.2 Residual problems

So far respiration has been considered in terms of the whole plant and it has been implicitly assumed that growth, and respiration, occur only at the expense of carbohydrate produced in photosynthesis; that is, the links between substrate and growth, and between growth and respiration are tightly coupled. However, the light reactions of photosynthesis also generate ATP and reductants, and evidence is accumulating that they may participate in some synthetic functions, for example, in the metabolism of nitrogen and in metabolic accumulation and transport (Heber, 1974; Penning de Vries, 1975; Miflin and Lea, 1976). We have also implicitly assumed that the plant is homogeneous – that roots and shoots are metabolically very similar – but this may not be so. Roots accumulate and metabolize ions – processes which require energy, although it is not known how much. The products of these activities are utilized in shoot as well as in root growth, so that some root respiration 'supports' the growth of the shoot. Moreover, there is evidence of the cycling of metabolites from shoot to root. Unfortunately, not enough is known, at present, about respiratory efficiency and the distribution of metabolic functions between root and shoot to permit an accurate appraisal of the roles of their separate respiratory activities.

2.9.3 Respiration manipulation and yield improvement

Finally, it remains to ask whether, as respiratory losses are so large, it is possible to reduce them and thereby conserve carbon and raise yields. There seems little scope for improving the efficiency of synthetic metabolic processes since the theoretical estimates of Penning de Vries (1972, 1975), which accord with experimental data, already assume that the most efficient biochemical pathways are being followed. Thus, to reduce synthetic respiration would be to reduce growth.

Maintenance respiration is another matter. It is much less well defined. Beevers (1970) considers it to be associated with protein turnover in order cells and with the repair and maintenance of inherently unstable cell structures. He also identifies an 'idling' component associated with non-productive ATP

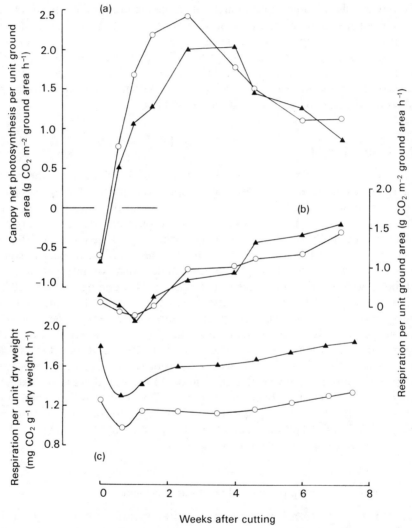

Figure 2.14 (a) Rates of canopy net photosynthesis, per unit ground area, at 20°C and 107 W m^{-2} (PAR) of selection lines of *Lolium perenne* with 'slow' (○) or 'fast' (▲) rates of 'mature tissue respiration' (developed by D. Wilson at the Welsh Plant Breeding Station) grown as established swards over an 8-week regrowth period. Also, rates of dark respiration at 15°C expressed per unit (b) ground area and (c) dry weight. (From Robson, 1982a.)

Respiration

hydrolysis but which de Wit *et al.* (1970) consider impractical to separate from 'maintenance'. Certainly, there is mounting evidence (e.g. Lambers, 1980; Lambers and Smakman, 1978; Lambers *et al.*, 1979; Lambers and Posthumus, 1980) for an, often active, alternative oxidative pathway of low phosphorylative capacity (only 1 ATP produced as against 3 via the normal cytochrome pathway). Not surprisingly, it has been suggested that if some of this respiratory efflux is wasteful, it might be amenable to reduction by breeding (McCree, 1974; Penning de Vries, 1974).

This has proved possible with S23 perennial ryegrass (Wilson, 1975). Lines selected to have a 'slow' rate of 'mature tissue' (maintenance) respiration outyield 'high' respiration lines (Robson, 1982a; Wilson and Jones, 1982) and, more importantly, their common parent by between 5 and 12%, depending on management (Wilson and Robson, 1981). The greater dry matter production results initially from a greater conservation of carbon (Robson, 1982a,b; Wilson, 1982) which, when reinvested, enables the 'slow' line to expand leaf more quickly, intercept more light and fix more carbon (Fig. 2.14a). The lower rate of respiration per unit dry weight (Fig. 2.14b) thus enables it to maintain a bigger 'factory' for the same respiratory costs per unit ground area (Fig. 2.14c).

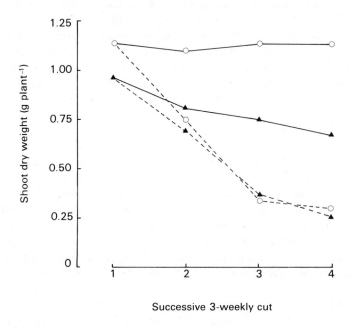

Figure 2.15 Harvestable yield of simulated swards of 'slow' (○) and 'fast' (▲) respiration lines of *Lolium perenne*, grown under either nitrogen-sufficient conditions (——) or increasing deficiency (– – –), during four successive 3-week regrowth periods. (Data from Robson *et al.*, 1983.)

The outlook is promising, although caution is needed. It may be that 'slow' respiration brings penalties with it, although none has yet been detected. This might be clearer if we understood better the biochemical basis of 'slow' respiration. At present all we know is that the differences between the lines are not due to different activities of the alternative oxidative pathway (Pilbeam *et al.*, 1986), nor does it seem likely that differences in the rate of protein turnover could be great enough to explain them, although differences in the costs of transporting assimilates from source to sink might be (Lambers and Day, pers. comm.).

The 'slow' lines appear to make more efficient use of fertilizer nitrogen, at high levels of input, whereas at deficient levels the yield advantage disappears (Fig. 2.15) (Robson *et al.*, 1983). The acid test will be whether the advantages of 'slow' respiration can survive the generations of seed multiplication necessary to get new varieties on to the farm and the subsequent management procedures they experience there.

2.10 THE PLANT AND THE COMMUNITY

So far we have dealt almost exclusively with the single plant or tiller. In bulk, these make up the dynamic community of the crop. Its responses to environment and management, including the presence of the grazing animal, determine its seasonal pattern of growth and development, and its total annual yield. These are covered in more detail in the next two chapters. Here we seek only to illustrate a few of the broad principles of how a crop is constructed and operates, in order to draw together some of the ideas that have gone before.

2.10.1 A model seedling sward

We begin by describing the progression, from a collection of spaced seedlings to a closed canopy at ceiling yield, of a model community – a uniform population of regularly spaced, vegetative plants, growing without interruption in constant and favourable environmental conditions, abundantly supplied with water and nutrients, and free from pests, diseases, grazing animals (which would harvest live leaves) and earthworms (which would remove dead leaves). The crop was S24 perennial ryegrass; the experimental conditions are described elsewhere (Robson, 1971, 1973a,b, 1982b; Woledge, 1973).

From seedling emergence, main stem leaves appeared one every 5 or 6 days, slowing to half that rate 12 weeks later (Fig. 2.16a), a decline common in communities (Robson and Deacon, 1978) but less evident in spaced plants (Robson, 1974). Concurrently, the time taken for successive leaves to reach full size doubled (Fig. 2.16b), so that the number growing at any one time remained constant at just under two. The rate of leaf extension peaked at about

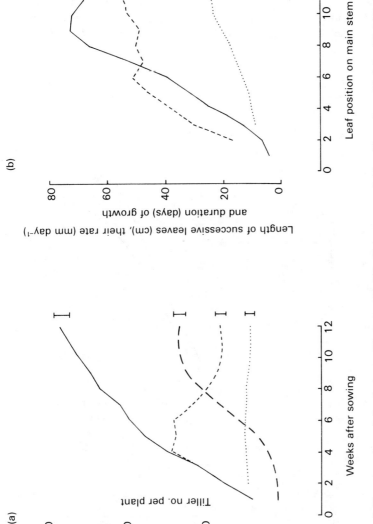

Figure 2.16 Leaf and tiller growth, and death, in seedling swards of *Lolium perenne* during 12 weeks at 22.5 °C day/12.5 °C night, 16-h days and 85 Wm^{-2} (PAR). (a) Tiller number per plant (———) together with the total number of main stem leaves (———), the number alive (– – –) and the number growing at any one time (. . . .). (b) Total length (lamina + sheath) of successive (1–12) main stem leaves (———), their rate of elongation from appearance to the appearance of their successor (– – –), and their duration of elongation (. . . .). Vertical bar indicates s.e. (From Robson, 1973a.)

50 mm day^{-1} for leaf 5 and the final leaf length (lamina + sheath) at about 70 cm for leaf 8 (Fig. 2.16b).

In this sward leaves accumulated on the main stem until there were five or six before the oldest leaf died. The number of live leaves then fell rapidly to level off at just under three. Meanwhile, tiller number rose, at first exponentially but then more slowly, until it too levelled off at about 26 per plant, or just over one per square centimetre of ground area (Fig. 2.16a). Taking leaf length as an

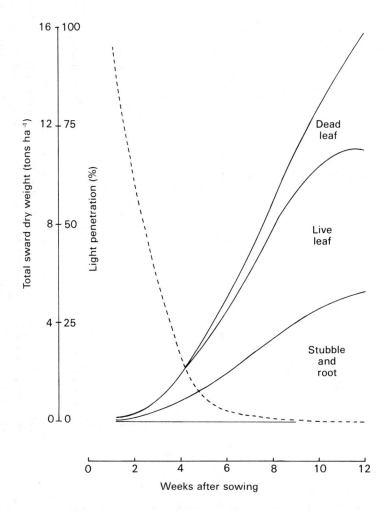

Figure 2.17 Ceiling yield in a seedling sward of *Lolium perenne*. Total dry matter production (———) partitioned between dead leaf, live leaf and stubble plus root during 12 weeks of growth; also percentage of light penetrating to the base of the sward (– – –). (From Robson, 1973a.)

index of leaf size, it is clear that, by the end of the experiment, the amount of live leaf on the main stem had reached a ceiling – as fast as new leaf was produced, an old one of the same size died. With tiller number by then constant, it is evident that the community as a whole also reached a ceiling in terms of live leaf tissue.

The cumulative dry weight data confirm this. The total crop biomass (which includes dead leaf) increased approximately exponentially (Fig. 2.17), until light interception was virtually complete (only 5% penetration to ground level; LAI 5), when it stabilized at a steady rate of about 270 kg ha^{-1} day^{-1}. As we have seen, leaf death began at about week 5 and increased rapidly until, by the 12th week, the rate of lamina death equalled the rate of production at about 200 kg ha^{-1} day^{-1}. Thus, a ceiling yield was achieved equivalent to about 10 t ha^{-1}, similar to values recorded for field swards grown under optimum conditions (Alberda and Sibma, 1968; Green et al., 1971) (Fig. 2.17).

Canopy net photosynthesis rose ten-fold between weeks 1 and 5 (Fig. 2.18a) as the sward increased its ability to intercept light, to a peak rate of about 2.4 g CO_2 m^{-2} ground area h^{-1}, similar to field values recorded by Leafe (1972) and

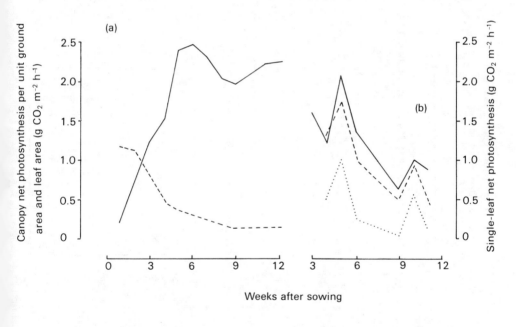

Figure 2.18 (a) Rate of canopy net photosynthesis at 22.5 °C and 85 W m^{-2} (PAR) of a seedling sward of *Lolium perenne*, per unit ground area (——) and leaf area (---). (Redrawn from Robson, 1973b.) (b) Rates of net photosynthesis of individual leaves from the same sward also measured at 22.5 °C and 85 W m^{-2}: youngest fully expanded leaves (——) and the next two older leaves (---) and (....). (Redrawn from Woledge, 1973.)

a little higher than those of Alexander and McCloud (1962) and Vickery *et al.* (1971). Thereafter, it declined somewhat, as the photosynthetic *capacity* (i.e. the rate at light saturation) of a successive youngest fully expanded leaves fell from a maximum of 2.0 to a minimum of 0.5 g $CO_2 m^{-2}$ leaf area h^{-1}, with the older leaves paralleling them (Fig. 2.18b). Note that the *actual* net photosynthetic rate per unit leaf area of the canopy declined from about 1.2 to <0.2 g $CO_2 m^{-2} h^{-1}$ as the LAI increased and the average amount of light received per unit of leaf area fell (Fig. 2.18a).

Canopy dark respiration rose 20-fold between weeks one and nine, levelling off when the weight of live tissue plateaued (Fig. 2.19a). However, the two did not increase proportionately. Thus, the rate of dark respiration, per unit dry weight, fell throughout the experiment. As we have seen, the respiratory efflux of CO_2 can be conveniently partitioned between a 'maintenance' component, broadly proportional to the existing mass of live tissue (Fig. 2.20a), and a 'synthetic' component broadly proportional to the rate of carbon assimilation and incorporation into new plant tissue (Fig. 2.20b). In the present example (Fig. 2.20c) the average rate of 'maintenance' respiration was $0.014 g g^{-1}$ day^{-1}, very close to values reported by McCree (1970) and Penning de Vries

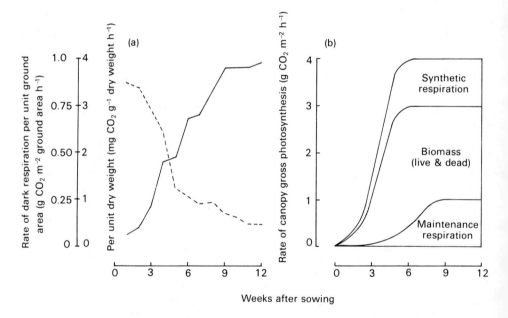

Figure 2.19 (a) Rate of dark respiration at 12.5 °C of a seedling sward of *Lolium perenne*, per unit ground area (——) and dry weight (– – –). (From Robson, 1973b.) (b) The partitioning by a mathematical model (Parsons, 1980), of canopy 'gross' photosynthesis, in the same sward, into synthetic respiration (assumed to be 25% of 'gross' photosynthesis), maintenance respiration (see text) and biomass. (From Robson, 1982b.)

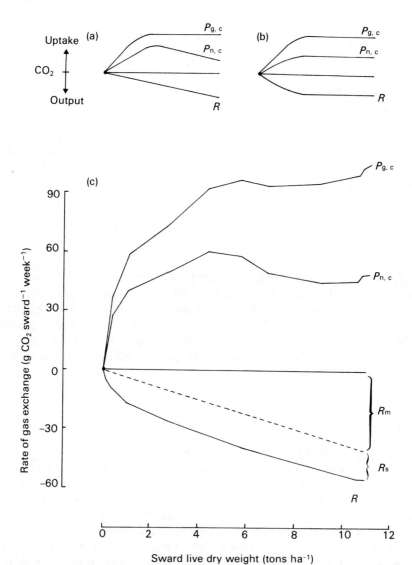

Figure 2.20 Canopy gross photosynthesis (P_g, c), net photosynthesis (P_n, c) and dark respiration (R), plotted against crop dry weight, assuming that respiration is proportional to (a) dry weight (Donald, 1961), (b) gross photosynthesis (McCree and Troughton, 1966) or (c) as observed in a seedling sward of *Lolium perenne* in which respiration could be related to both dry weight via a 'maintenance' component (R_m) and to gross photosynthesis via a 'synthetic' component (R_s); for values of R_m and R_s, see text. (From Robson, 1973b.)

(1972), while the average rate of 'synthetic' respiration indicated an overall conversion efficiency of 80%, at the upper end of the range of theoretical and experimental values referred to earlier (p. 57).

Together, and in roughly equal measure, 'synthetic' and 'maintenance' respiration accounted for 45% of the carbon fixed, integrated over the whole experimental period, rising from 25% very early on, when little mature tissue was present and maintenance costs were low, to a steady 50% later (Fig. 2.20c). These experimental data match well the output of a simple model (Parsons, 1980; Parsons and Robson, 1982) in which 25% of daily gross photosynthesis enters a labile fraction (and 50% a structural fraction) to be consumed at a set daily rate (e.g. 1% per day for 25 days) in order to maintain the integrity of the tissue of which it is a part. A key consequence of the model is that, on any one day, the CO_2 efflux consequent on 'maintenance' respiration is proportional to the total carbon fixed during the previous (in this example) 25 days. Thus, changes in the rate of gross photosynthesis are reflected in the rate of 'maintenance' respiration slowly, and in a buffered manner, as Fig. 2.19b shows.

A more obvious delay is that between the incorporation of recently fixed carbon in new plant tissue, and the death and detachment of that tissue some time later. Moreover, it is a delay which has important consequences for the way in which we harvest the crop and the readiness with which a ceiling yield is achieved (p. 73 and Chapter 4). However, before we consider this it is appropriate to examine those factors which determine the rate at which a canopy fixes carbon in the first place.

2.10.2 Determinants of canopy photosynthesis

If other environmental factors are non-limiting, and given a certain level of incoming radiation, three factors determine the rate of canopy photosynthesis: (a) the proportion of the incoming radiation that is intercepted, (b) how evenly it is distributed over the leaf surface and (c) how effectively the individual leaves make use of the light they receive. The last of these having been referred to earlier (p. 46), let us consider the other two, in reverse order.

The distribution of light over the leaf surface is a function of the architecture of the leaf canopy. When LAI is low and light interception incomplete, a prostrate plant habit is best in that it minimizes the amount of light passing between leaves to bare ground. When LAI is high and light interception complete, an erect habit is best in that it enables the leaves to be more uniformly illuminated. The form of the typical light-response curve of photosynthesis (Fig. 2.7a) necessarily means that uniform illumination leads to greater carbon fixation than patches of bright light and shade. From this it can be seen that a prostrate habit is best when a crop is frequently and severely cut or grazed, so that its LAI remains low, but an erect habit is best when it is not

The plant and the community

(Rhodes, 1969). Even so, the effects of growth habit on canopy photosynthesis appear to be smaller than those resulting from changes in the photosynthetic capacity of the individual leaves (Fig. 2.21). Moreover, a low photosynthetic capacity is *always* a disadvantage, whereas a particular growth habit may or may not be, depending on the density of the crop at the time.

The proportion of incoming radiation that a crop intercepts is a very important determinant of canopy photosynthesis, although the opportunity for improving light interception, and thereby raising yields, is not as great as in many annual crops such as sugar beet or potatoes (Watson, 1947). In an infrequently cut or rotationally grazed grass sward, for example, a fully light intercepting canopy may be maintained throughout most of the year; the progressive impairment of the photosynthetic capacity of successive leaves over each vegetative regrowth period is the main limitation to yield. The opposite is true, however, where continuous grazing maintains a high leaf photosynthetic

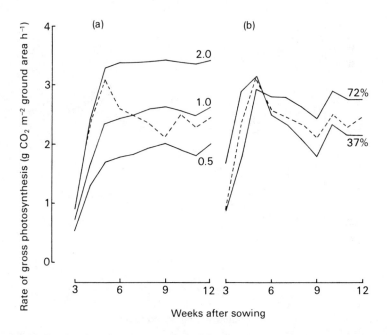

Figure 2.21 Rates of canopy gross photosynthesis of a seedling sward of *Lolium perenne* generated by the Monteith (1965) model (a) assuming that the net photosynthetic rate of the youngest fully expanded leaves remained constant at 2, 1 or $0.5 \text{ g CO}_2 \text{ m}^{-2}$ leaf area h^{-1} throughout the growth period (———), rather than declining from 2 to 0.5 mg $CO_2 \text{ m}^{-2}$ leaf area h^{-1} as was observed (---), or (b) assuming that the spatial arrangement of the leaf surface was such that each unit of LAI permitted either 37% (a prostrate canopy) or 72% (an erect canopy) of the incident radiation to pass through it during the entire growth period (———), rather than ranging between these values, as in fact occurred (---). (From Robson, 1973b.)

capacity, but a low LAI permits light to be wasted on bare ground and ineffective leaf sheaths (see Chapter 4).

Many recent studies relate light interception, in a linear fashion, to canopy net photosynthesis (Hesketh and Baker, 1967; Chu, 1970; Biscoe and Gallagher, 1977) or even to dry matter yield (Duncan *et al.*, 1973; Scott *et al.*, 1973). This is valid during the early stages of the establishment of a crop from seed, but it can be misleading if applied to an already established perennial grass sward undergoing regrowth. In one study of the carbon economy of a

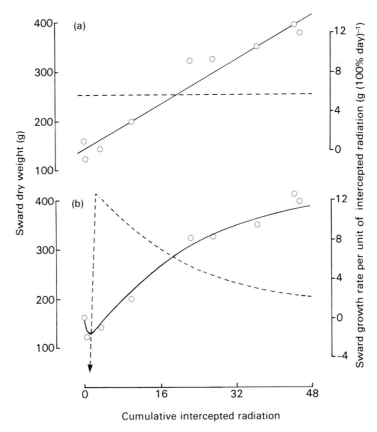

Figure 2.22 Changes in total dry weight during a 7-week regrowth period (○) of an established sward of *Lolium perenne* growing in constant environmental conditions, plotted against cumulative intercepted radiation (each unit on the x-axis equals 1 day at 100% interception or its equivalent, e.g. 2 days at 50%). (a) A linear regression (——) fitted to the data, gives a constant growth rate (---) of 5.7 g per unit intercepted radiation. (b) The actual trend of dry weight change (——), derived from detailed CO_2 exchange measurements, gives a growth rate (---) at first negative, then rising abruptly, finally declining in a curvilinear fashion from 12.0 g per unit intercepted radiation to close to 2.0 g. (Data from Robson, 1982a.)

The plant and the community

vegetative ryegrass crop (Robson, 1982a), the newly cut crop was initially below the compensation point, and hence lost weight until enough leaf area had been expanded to enable carbon fixation to offset respiration; it then achieved a positive carbon balance, regained the lost weight and moved ahead, rapidly at first but then more slowly as the respiratory burden built up and as the photosynthetic capacity of successive leaves declined. Superficially, the relationship between crop yield and cumulative light interception appeared linear, with a constant rate of dry matter production per unit of intercepted radiation (Fig. 2.22a) but, in reality, the picture was very different (Fig. 2.22b). Plotting two variables cumulatively, one against the other, and fitting a linear regression to them obscured important features of the relationship.

2.10.3 'Optimum leaf area index' and 'ceiling yield'

As a seedling grass crop growing in a constant environment increases in leaf area, it intercepts more light and fixes more carbon. Increasingly, however, new leaves shade older ones, so that each increment in leaf area brings a diminishing return until a point is reached when the extra carbon fixed only just balances the extra respiration generated by the new tissue. Kasanaga and Monsi (1954) termed this point the 'optimum leaf area index', at which the crop's gross rate of dry matter production is maximal. At a supra-optimal LAI the additional respiratory burden exceeds any increased carbon uptake, and the growth rate of the crop falls. Although there is broad acceptance of the above account, and although supra-optimal LAIs have been demonstrated in a number of studies (Watson, 1958; Davidson and Donald, 1958), it is often difficult in practice to establish a precise value for the 'optimum LAI' for a variety of reasons.

As the biomass of the young grass crop continues to increase, respiratory costs rise even further, reducing the carbon available for tissue production. Moreover, old tissues begin to die and become detached. The first tissues to die are the oldest but, because they were produced early in the life of the crop when its LAI, light interception and rate of tissue production were all low, the losses are slight. They soon rise, however, reflecting the earlier increase in the crop growth rate. Before long, if the crop is not harvested, the rate of tissue death equals the rate of production and a 'ceiling yield' of live tissue is reached (Fig. 2.17). There is no escaping this conclusion, although some workers (e.g. Alberda and Sibma, 1968) have had difficulty in reconciling it with the assumed capabilities of earthworms and other fauna to remove the quantities of dead leaf that must result.

Note that the 'ceiling yield' in grasses, and probably in many other crops, differs in an important respect from that visualized by Donald (1961) who first coined the term. He postulated that the crop biomass would continue to rise, through investment in non-photosynthetic organs, until the respiratory losses exactly equalled the photosynthetic gains, with no new tissue being produced.

Yield would be 'maximum and static'. It is difficult to envisage such a point ever being reached in any of our agricultural crops. Certainly, the grasses, having evolved to survive fire and grazing, are committed to a programme of continual leaf replacement which needs a supply of carbon to fuel it and which, in turn, ensures that live tissue never accumulates to the point where respiratory burdens imperil it. Whatever else a grass crop is, it is never 'static'.

In practice, even 'dynamic' ceiling yields are rarely allowed to develop in intensively managed agricultural grasslands – a crop held at a ceiling yield is wasting land and solar radiation. When, then, should a crop such as we have described be harvested? If environmental conditions are constant, and the

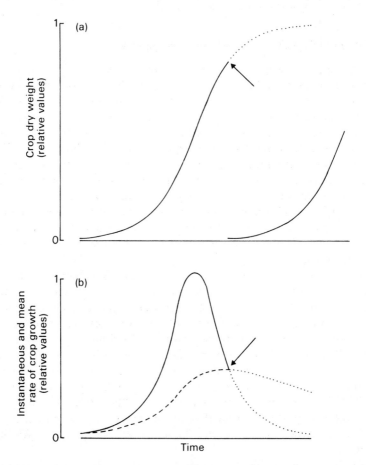

Figure 2.23 Timing of harvest for maximum yield. If the crop has a sigmoid pattern of dry weight increase and repeats that pattern from whenever it is harvested, as in (a), then the total yield, over a sequence of cuts, will be maximized by harvesting the crop when the *mean* growth rate, shown in (b), (- - -) is at a peak, i.e. when it equals the *instantaneous* growth rate (——). Arrows indicate the optimum time of harvest.

sigmoidal pattern of live dry matter production is repeated over successive regrowth periods, then the best time for harvesting is when the mean daily rate of dry matter accumulation is maximal (Watanabe and Takahashi, 1979), providing that the shape of the growth curve is relatively independent of the time of harvest – a dubious assumption. This must be correct mathematically (Fig. 2.23), although there are no obvious morphological or physiological markers that a farmer could use to tell when this point had been reached.

If environmental conditions, and hence rates of tissue production, are not constant but exhibit a steady improvement, or deterioration, interesting consequences ensue. As we have seen (p. 68), changes in the rate at which old tissue is lost lag behind changes in the rate at which new tissue is produced (Parsons and Robson, 1982). When incoming radiation rises steeply, as in the spring in the UK, so too does the rate at which the crop fixes carbon and expands new leaf tissue. Thus, at any one time tissue production exceed death (the latter reflecting an *earlier* and *lower* rate of production) and harvestable yield builds up (Fig. 4.7b). In the autumn, however, once a crop is fully light-intercepting, its rate of dry matter production slows as the level of incoming radiation falls. Tissue death (representing an *earlier* and *higher* rate of production) soon exceeds current production and harvestable yield declines (Fig. 4.7c), To prevent this, the farmer must cut the crop at an early stage, even though this means the loss of light to bare ground. Thus, the seasonal *pattern* of radiation (independent of any difference in the amount) makes the frequent harvesting of light vegetative crops in the autumn almost inevitable, while permitting a fully light-intercepting canopy to be retained in the spring until all the advantages of reproductive growth have been realised and a high yield has been achieved (Robson, 1981b).

REFERENCES

Abbe, E. C., Randolph, L. F. and Einset, J. (1941) The developmental relationship between shoot apex and growth pattern of leaf blade in diploid maize. *Am. J. Bot.*, **28**, 778–84.

Alberda, T. and Sibma, L. (1968) Dry matter production and light interception of crop surfaces III. Actual herbage production in different years as compared with potential values. *J. Br. Grassland Soc.*, **23**, 206–15.

Alexander, C. W. and McCloud, D. E. (1962) CO_2 uptake (net photosynthesis) as influenced by light intensity of isolated Bermuda-grass leaves contrasted to that of swards under various clipping regimes. *Crop Sci.*, **2**, 132–5.

Anslow, R. C. (1962) A quantitative analysis of germination and early seedling growth in perennial ryegrass. *J. Br. Grassland Soc.* **17**, 260–3.

Arber, A. (1934). *The Gramineae. A Study of Cereal, Bamboo and Grass*, Cambridge University Press, London.

Armstrong, S. F. (1948) *British Grasses and their Employment in Agriculture*, Cambridge University Press, London.

Asay, K. H., Nelson, S. J. and Horst, G. L. (1974) Genetic variability for net photosynthesis in tall fescue. *Crop Sci.*, **14**, 574–4.

Beevers, H. (1970) Respiration in plants and its regulation, in *Prediction and Measurement of Photosynthetic Productivity, Trebon*, (ed. I. Malek) Pudoc, Wageningen, pp. 209–14.

Behaeghe, T. J. (1976) Experiments on the seasonal variations of grass growth. in *Proc. XII Int. Grassland Congr., Moscow*, Vol. 1, Part 1. pp. 268–81.

Bidwell, R. G. S. (1983) Carbon nutrition of plants: photosynthesis and respiration, in *Plant Physiology – a Treatise*, Vol. VII (eds F. C. Steward and R. G.S. Bidwell) Academic Press, New York, pp. 287–457.

Biscoe, P. V. and Gallagher, J. N. (1977) Weather, dry matter production and yield, in *Environmental Effects on Crop Physiology* (eds J. J. Landsberg and C. V. Cutting), Academic Press, London, pp. 75–100.

Blackman, F. F. (1905) Optima and limiting factors. *Ann. Bot.*, **19**, 281–95.

Blackman, G. E. and Black, J. N. (1959) Physiological and ecological studies in the analysis of plant environment. XI. A further assessment of the influence of shading on the growth of different species in the vegative phase. *Ann. Bot.*, **23**, 51–63.

Blackman, G. E. and Wilson, G. L. (1951a) Physiological and ecological studies in the analysis of plant environment. VI. The constancy for different species of a logarithmic relationship between net assimilation rate and light intensity and its ecological significance. *Ann. Bot.*, **15**, 64–94.

Blackman, G. E. and Wilson, G. L. (1951b) Physiological and ecological studies in the analysis of plant environment. VII. An analysis of the differential effects of light intensity on the net assimilation rate, leaf area ratio, and relative growth rate of different species. *Ann. Bot.*, **15**, 374–408.

Blackman, G. E. and Wilson, G. L. (1954) Physiological and ecological studies in the analysis of plant environment. IX. Adaptive changes in the vegetative growth and development of *Helianthus annuus* induced by an alteration in light level. *Ann. Bot.*, **18**, 72–94.

Bolton, J. K. and Brown, R. H. (1980) Photosynthesis of grass species differing in carbon dioxide fixation pathways. *Pl. Physiol.*, **66**, 97–100.

Borrill, M. (1961) The developmental anatomy of leaves in *Lolium temulentum*. *Ann. Bot.*, **25**, 1–11.

Brown, W. V. (1960) The morphology of the grass embryo. *Phytomorphology*, **10**, 215–23.

Carlson, G. E., Hart, R. H., Hanson, C. H. and Pearce, R. B. (1971) Overcoming barriers to higher forage yields through breeding for physiological and morphological characteristics, in *Proc. XI Int. Grassland Congr., 1970*, pp. 248–51.

Charles-Edwards, D. A. (1978) An analysis of the photosynthetic productivity of vegetative crops in the United Kingdom. *Ann. Bot.*, **42**, 717–31.

Charles-Edwards, D. A., Charles-Edwards, J. and Sant, I. (1971) Leaf photosynthetic activity in six temperate grass varieties grown in contrasting light and temperature environments. *J. Exp. Bot.*, **25**, 715–24.

Chartier, P. L. (1970) A model of CO_2 assimilation in the leaf, in *Prediction and Management of Photosynthetic Productivity, Trebon*, (ed. I. Malek), Pudoc, Wageningen, pp. 307–16.

Chippindale, H. G., (1949) Environment and germination in grass seeds. *J. Br. Grassland Soc.*, **4**, 57–61.

References

Chu, Chung-Chi (1970) Solar radiation and photosynthesis of sugar cane in the field. *Taiwan Sugar,* **17,** 14–21.

Colville, K. E. and Marshall, C. (1981) The patterns of growth, assimilation of $^{14}CO_2$ and distribution of ^{14}C-assimilate within vegetative plants of *Lolium perenne* at low and high density. *Ann. Appl. Biol.,* **99,** 179–90.

Cooper, J. P. (1951) Studies on growth and development in *Lolium*. II. Pattern of bud development of the shoot apex and its ecological significance. *J. Ecol.,* **39,** 228–70.

Cooper, J. P. (1957) Developmental analysis of populations in the cereals and herbage grasses. II. Response to low temperature vernalization. *J. Agric. Sci.,* **49,** 361–83.

Cooper, J. P. (1958) The effect of temperature and photoperiod on inflorescence development in strains of timothy. (*Phleum* spp.) *J. Br. Grassland Soc.,* **13,** 81–91.

Cooper, J. P. (1960) The use of controlled life-cycles in the forage grasses and legumes. *Herbage Abstr.,* **30,** 71–9.

Cooper, J. P. (1964) Climatic variation in forage grasses. I. Leaf development in climatic races of *Lolium* and *Dactylis*. *J. Appl. Ecol.,* **1,** 45–61.

Cooper, J. P. and Edwards, K. J. R. (1964) Developmental genetics of leaf production. *Rep. Welsh Pl. Breeding Stn. 1963,* pp. 16–18.

Cooper, J. P. and McWilliam, J. R. (1966) Climatic variation in forage grasses. II. Germination, flowering and leaf development in Mediterranean populations of *Phalaris tuberosa*. *J. Appl. Ecol.,* **3,** 192–212.

Cooper, J. P. and Tainton, N. M. (1968) Light and temperature requirements for the growth of tropical and temperate grasses. *Herbage Abstr.,* **38,** 167–76.

Cooper, J. P. and Wilson, D. (1970) Variation in photosynthetic rate in *Lolium*. *Proc. XI Int. Grassland Congr., Surfers Paradise,* University of Queensland Press, St Lucia, pp. 522–7.

Davidson, J. L. and Donald, C. M. (1958) The growth of swards of subterranean clover with particular reference to leaf area. *Austral. J. Agric. Res.,* **9,** 53–72.

Davidson, J. L. and Milthorpe, F. L. (1966a) Leaf growth in *Dactylis glomerata* following defoliation. *Ann. Bot.,* **30,** 173–84.

Davidson, J. L. and Milthorpe, F. L. (1966b) The effect of defoliation on the carbon balance in *Dactylis glomerata*. *Ann. Bot.,* **30,** 185–98.

Davies, A. (1977) Structure of the grass sward. in *Proc. Int. Mts on Animal Production from Temperate Grassland* (ed. B. Gilsenan), Irish Grassland and Animal Production Association, Dublin, pp. 36–44.

Deinum, B. (1976) Photosynthesis and sink size: an explanation for the low productivity of grass swards in autumn. *Neth. J. Agric. Sci.,* **24,** 273–82.

Donald, C. M. (1961) Competition for light in crops and pastures. in *Mechanisms in Biological Competition* (ed. F. L. Milthorpe), Cambridge University Press, pp. 283–313.

Duncan, W. G., Shaver, D. N. and Williams, W. A. (1973) Insolation and temperature effects on maize growth and yields. *Crop Sci.,* **13,** 187–90.

Ehleringer, J. and Bjorkman, D. (1977) Quantum yields for CO_2 uptake in C_3 and C_4 plants, dependence on temperature, CO_2 and O_2 concentration. *Pl. Physiol.,* **59,** 86–90.

Evans, L. T. (1960) Inflorescence initiation in *Lolium temulentum* L. II. Evidence for inhibitory and promotive photoperiod processes involving transmissible products. *Austral. J. Biol. Sci.,* **13,** 429–40.

Evans, L. T. (1964a) Infloresence initiation in *Lolium temulentum* L. V. The role of auxins and gibberellins. *Austral. J. Biol. Sci.*, **17**, 10–23.

Evans, L. T. (1964b) Inflorescence initiation in *Lolium temulentum* L. VI. Effects of some inhibitors of nucleic acid, protein and steroid biosynthesis. *Austral. J. Biol. Sci.*, **17**, 24–35.

Evans, G. C. and Hughes, A. P. (1961) Plant growth and the aerial environment. I. Effect of artificial shading on *Impatiens paviflora*. *New Phytol.*, **60**, 150–80.

Evans, L. T., Wardlaw, I. F. and Williams, C. N. (1964) Environmental control of growth, in *Grasses and Grasslands* (ed. C. Barnard), MacMillan, London, pp. 102–25.

Friend, D. J. C., Helson, V.A. and Fisher, J.E. (1965) Changes in the leaf area ratio during growth of Marquis wheat, as affected by temperature and light intensity. *Can. J. Bot.*, **43**, 15–28.

Goodall, D. W. (1955) Growth of cacao seedlings as affected by illumination. *Rep. XIV Int. Hort. Congr., Wageningen*, pp. 1501–10.

Gordon, A. J., Ryle, G. J. A. and Powell, C. E. (1977) The strategy of carbon utilization in uniculm barley. I. The chemical fate of photosynthetically assimilated ^{14}C. *J. Exp. Bot.*, **28**, 1258–69.

Grant, S. A., Barthram, G. T., Torvill, L., King, J. and Smith, H. K. (1983) Sward management, lamina turnover and tiller population density in continuously stocked *Lolium perenne*-dominated swards. *Grass Forage Sci.*, **38**, 333–44.

Green, J. O., Corrall, A. J. and Terry, R. A. (1971) Grass species and varieties. Relationship between stage of growth, yield and forage quality. *Tech. Rep. No. 8, Grassland Res. Inst., Hurley*.

Hansen, G. K. and Jensen, C. R. (1977) Growth and maintenance respiration in whole plants, tops and roots of *Lolium multiflorum*, *Physiol. Pl.*, **39**, 155–64.

Harper, J. L. (1977) *Population Biology of Plants*, Academic Press, London.

Heber, V. (1974) Metabolite exchange between chloroplasts and cytoplasm. *A. Rev. Pl. Phsiol.*, **25**, 393–421.

Hesketh, J. and Baker, D. (1967) Light and carbon assimilation by plant communities. *Crop Sci.*, **7**, 285–93.

Jewiss, O. R. and Sanderson, J. F. (1963) The growth and flowering behaviour of S100 white clover. Effects of shading. *Experiments in Progress, 15: Annual Report for 1961–2*, Grassland Research Institute, Hurley, pp. 27–8.

Jewiss, O. R. and Sanderson, J. F. (1963) The growth and flowering behaviour of synthesis in tall fescue (*Festuca arundinacea* Schreb). *Ann. Bot.*, **31**, 661–71.

Johnson, I. R. and Thornley, J. H. M. (1983) Vegetative crop growth model incorporating leaf area expansion and senescence, and applied to grass. *Pl., Cell, Envir.*, **6**, 721–99.

Johnson, I. R. and Thornley, J. H. M. (1984) A model of Instantaneous and Daily Canopy Photosynthesis. *J. Theor. Biol.*, **107**, 531–45.

Jones, L. (1971) The development and morphology of seedling grasses. Pt 2. *A. Rep. Grassland Res. Inst., Hurley, 1970*, pp. 151–7.

Jones, L. (1972) Principles of establishment of grass, in *Grasses and Legumes in British Agriculture* (eds C. R. W. Spedding and E. C. Diekmahns), Commonwealth Agricultural Bureaux, Farnham Royal, pp. 51–4.

Jones, M. B., Leafe, E. L., Stiles, W. and Collett, B. (1978) Patterns of respiration of a perennial ryegrass crop in the field. *Ann. Bot.*, **42**, 693–703.

References

Kasanga, H. and Monsi, M. (1954) On the light transmission of leaves and its meaning for production of dry matter in plant communities. *Jap. J. Bot.*, **14**, 304–24.

Kays, S. and Harper, J. L. (1974) The regulation of plant and tiller density in a grass sward. *J. Appl. Ecol.*, **62**, 97–105.

Ku, S.-B. and Edwards, G. E. (1978) Oxygen inhibition of photosynthesis. III. Temperature dependence of quantum yield and its relation to O_2/CO_2 solubility ratio. *Planta*, **140**, 1–6.

Lambers, H. (1980) The physiological significance of cyanide-resistant respiration. *Pl., Cell, Envir.*, **3**, 293–302.

Lambers, H. and Posthumus, F. (1980) The effect of light intensity and relative humidity on growth rate and root respiration of *Plantago lanceolata* and *Zea mays*. *J. Exp. Bot.*, **31**, 1621–30.

Lambers, H. and Smakman, G. (1978) Respiration of the roots of flood-tolerant and flood-intolerant *Senecio* species: affinity for oxygen and resistance to cyanide. *Physiol. Pl.*, **42**, 163–6.

Lambers, H., Noord, R. and Posthumus, F. (1979) Respiration of *Senecio* shoots; inhibition during photosynthesis, resistance to cyanide and relation to growth and maintenance. *Physiol. Pl.*, **45**, 351–6.

Langer, R. H. M. (1956) Growth and nutrition of timothy (*Phleum pratense*). I. The life history of individual tillers. *Ann. Appl. Biol.*, **44**, 166–87.

Langer, R. H. M. (1972) How grasses grow. *Institute of Biology's Studies in Biology No. 34*, Edward Arnold, London, p. 7.

Langer, R. H. M. and Ryle, G. J. A. (1958) Vegetative proliferations in herbage grasses. *J. Br. Grassland Soc.*, **13**, 29–33.

Leafe, E. L. (1972) Micro-environment, carbon dioxide exchange and growth in grass swards, in *Crop Processes in Controlled Environments* (eds A. R. Rees, K. R. Cockshull, D. W. Hand and R. G. Hurd), Academic Press, London, pp. 157–74.

Lloyd, N. D. H. and Woolhouse, H. W. (1976) The effect of temperature on photosynthesis and transpiration in populations of *Sesleria caerulea* (L.) Ard. *New Phytol.*, **77**, 553–9.

Ludlow, M. M. and Wilson, G. L. (1971) Photosynthesis of tropical pasture plants 1. Illuminance, carbon dioxide concentration, leaf temperature and leaf-air vapour pressure difference. *Austral. J. Biol. Sci.*, **24**, 449–70.

Marshall, B. and Biscoe, P. V. (1980) A model for C_3 leaves describing the dependance of net photosynthesis on irradiance. II. Application to the analysis of flag leaf photosynthesis. *J. Exp. Bot.*, **31**, 41–8.

McCree, K. J. (1970) An equation for the rate of respiration of white clover plants grown under controlled conditions, in *Prediction and Measurement of Photosynthetic Productivity, Trebon* (ed. I. Malek), Pudoc, Wageningen, pp. 221–9.

McCree, K. J. (1974) Equations for the rate of respiration of white clover and grain sorghum, as functions of dry weight, photosynthetic rate and temperature. *Crop Sci.*, **14**, 509–14.

McCree, K. J. and Troughton, J. H. (1966) Prediction of growth rate at different light levels from measured photosynthesis and respiration rates. *Pl. Physiol., Lancaster*, **41**, 559–66.

Miflin, B. J. and Lea, P. J. (1976) The pathway of nitrogen assimilation in plants. *Phytochemistry*, **15**, 873–5.

Mitchell, K. J. (1953a) Influence of light and temperature on the growth of ryegrass (*Lolium* spp.). I. Pattern of vegetative development. *Physiol. Pl.,* **6,** 21–46.

Mitchell, K. J. (1953b) Influence of light and temperature on the growth of ryegrass (*Lolium* spp.). II. The control of lateral bud development. *Physiol. Pl.,* **6,** 425–43.

Mitchell, K. J. (1956) Growth of pasture species under controlled environments. I. Growth at various levels of constant temperature. *N.Z. J. Sci. Technol.,* **38A,** 203–16.

Mitchell, K. J. and Lucanus, R. (1962) Growth of pasture species under controlled environments. III. Growth at various levels of constant temperature with 8 and 16 hours of uniform light per day. *N.Z. J. Agric. Res.,* **5,** 135–44.

Monson, R. K., Stidham, M. A., Williams, G. J. III, Edwards, G. E. and Uribe, E. G. (1982) Temperature dependence of photosynthesis in *Agropyron smithii* Rydb. *Pl. Physiol.,* **69,** 921–8.

Monteith, J. L. (1965) Light distribution and photosynthesis in field crops. *Ann. Bot.,* **29,** 17–37.

Monteith, J. L. (1981) Does light limit crop production? in *Physiological Processes Limiting Plant Productivity* (ed. C. B. Johnson), Butterworths, London, pp. 23–8.

Njoku, E. (1960) An analysis of plant growth in some West African species. II. The effect of shading. *J. W. Afr. Sci. Assoc.,* **4,** 1–17.

Painter, E. L. and Delting, J. K. (1978) Effects of defoliation on net photosynthesis and regrowth of western wheatgrass. *J. Range Mgmt,* **34,** 68–71.

Paleg, L. G. (1960) Physiological effects of gibberellic acid. II. On starch hydrolizing enzymes of barley endosperm. *Pl. Physiol., Lancaster,* **35,** 902–6.

Parsons, A. J. (1980) The physiological basis of seasonal differences in the growth of perennial ryegrass. Ph.D. Thesis, University of Reading.

Parsons, A. J. and Robson, M. J. (1980) Seasonal changes in the physiology of S24 perennial ryegrass (*Lolium perenne* L.) I. Response of leaf extension to temperature during the transition from vegetative to reproductive growth. *Ann. Bot.,* **46,** 435–44.

Parsons, A. J. and Robson, M. J. (1981a) Seasonal changes in the physiology of S24 perennial ryegrass (*Lolium perenne* L.) II. Potential leaf and canopy photosynthesis during the transition from vegetative to reproductive growth. *Ann. Bot.,* **47,** 249–58.

Parsons, A. J. and Robson, M. J. (1981b) Seasonal changes in the physiology of S24 perennial ryegrass (*Lolium perenne* L.) III. Partition of assimilates between root and shoot during the transition from vegetative to reproductive growth. *Ann. Bot.,* **48,** 733–44.

Parsons, A. J. and Robson, M. J. (1982) Seasonal changes in the physiology of S24 perennial ryegrass (*Lolium perenne* L.) IV. Comparison of the carbon balance of the reproductive crop in spring and the vegetative crop in Autumn. *Ann. Bot.,* **50,** 167–77.

Parsons, A. J., Leafe, E. L., Collett, B., Penning, P. D. and Lewis, J. (1983) The physiology of grass production under grazing II. Photosynthesis, crop growth and animal intake of continuously grazed swards. *J. Appl. Ecol.,* **20,** 127–39.

Patel, A. S. and Cooper, J. P. (1961) The influence of seasonal changes in light energy on leaf and tiller development in ryegrass, timothy and meadow fescue. *J. Br. Grassland Soc.,* **16,** 299–308.

Peacock, J. M. (1975a) Temperature and leaf growth in *Lolium perenne* I. The thermal

References

microclimate, its measurement and relation to crop growth. *J. Appl. Ecol.*, **12**, 99–114.

Peacock, J. M. (1975b) Temperature and leaf growth in *Lolium perenne* II. The site of temperature perception. *J. Appl. Ecol.*, **12**, 115–23.

Peacock, J. M. (1975c) Temperature and leaf growth in *Lolium perenne* III. Factors affecting seasonal differences. *J. Appl. Ecol.*, **12**, 685–97.

Peacock, J. M. (1976) Temperature and leaf growth in four grass species. *J. Appl. Ecol.*, **13**, 225–32.

Penning de Vries, F. W. T. (1972) Respiration and Growth, in *Crop Processes in Controlled Environments* (eds A. R. Rees, K. E. Cockshull, D. W. Hand and R. G. Hurd), Academic Press, London, pp. 327–47.

Penning de Vries, F. W. T. (1974) Substrate utilization and respiration in relation to growth and maintenance in higher plants. *Neth. J. Agric. Sci.*, **22**, 40–4.

Penning de Vries, F. W. T. (1975) The cost of maintenance processes in plant cells. *Ann. Bot.*, **39**, 77–92.

Pilbeam, C. J., Robson, M. J. and Lambers, H. (1986) Respiration in mature leaves of *Lolium perenne* as affected by nutrient supply and cutting. *Physiol. Pl.*, **66**, 53–57.

Pollock, C. J. and Jones, T. (1979) Seasonal patterns of fructosan metabolism in forage grasses. *New Phytol.*, **83**, 9–15.

Powell, C. E. and Ryle, G. J. A. (1978) Effect of nitrogen deficiency on photosynthesis and the partitioning of ^{14}C-labelled assimilate in unshaded and partially shaded plants of *Lolium temulentum*. *Ann. Biol.*, **90**, 241–8.

Prioul, J. L. (1971) Réaction des feuilles de *Lolium multiflorum* à l'éclairement pendant la croissance et variation des résistances aux éxchange gazeux photosynthétiques. *Photosynthetica*, **5**, 364–75.

Prioul, J. L., Reyss, A. and Chartier, P. (1975) Relationship between carbon dioxide transfer resistances and some physiological and anatomical features, in *Environmental and Biological Control of Photosynthesis* (ed. R. Marcelle), Junk, The Hague, pp. 17–28.

Prioul, J. L., Brangeon, J. and Reyss, A. (1980a) Interaction between external and internal conditions in the development of photosynthetic features in a grass leaf. I. Regional responses along a leaf during and after low-light or high-light acclimation. *Pl. Physiol.*, **66**, 762–9.

Prioul, J. L., Brangeon, J. and Reyss, A. (1980b) Interaction between external and internal conditions in the development of features in a grass leaf. II. Reversibility of light-induced responses as a function of development. *Pl. Physiol.*, **66**, 770–4.

Rabinovitch, E. I. (1951) External and internal limiting factors in photosynthesis, in *Photosynthesis and Related Processes 2(1)*, Interscience, New York, pp. 838–55.

Radford, P. J. (1967) Growth analysis formulae – their use and abuse. *Crop Sci.*, **7**, 171–5.

Redman, R. E. (1974) Photosynthesis, respiration and water relations of *Agropyron dasystachyum* measured in the laboratory. *Canadian Committee for the IBP-Matador Project, Tech. Rep. No. 47.*

Rhodes, I. (1969) The yield, canopy structure and light interception of two ryegrass varieties in mixed culture and mono-culture. *J. Br. Grassland Soc.*, **24**, 123–7.

Robson, M. J. (1965) An investigation into the physiology of certain indigenous and Mediterranean ecotypes of tall fescue (*Festecua arundinacea* Schreb.) and other grasses. Ph.D. Thesis, University of Reading.

Robson, M. J. (1967) A comparison of British and North African varieties of tall fescue (*Festuca arundinacea*). I. Leaf growth during winter and the effect on it of temperature and day length. *J. Appl. Ecol.*, **4**, 475–84.
Robson, M. J. (1968) The changing tiller population of spaced plants of S.170 tall fescue (*Festuca arundinacea*). *J. Appl. Ecol.*, **5**, 575–90.
Robson, M. J. (1969) Light, temperature and the growth of grasses. *A. Rep. Grassland Res. Inst., Hurley, 1968*, pp. 111–23.
Robson, M. J. (1971) The use of simulated swards in growth rooms. *A. Rep. Grassland Res. Inst., Hurley, 1970*, pp. 158–68.
Robson, M. J. (1972) the effect of temperature on the growth of S170 tall fescue (*Festuca arundinacea*). I. Constant temperature. *J. Appl. Ecol.*, **9**, 647–57.
Robson, M. J. (1973a) The growth and development of simulated swards of perennial ryegrass. I. Leaf growth and dry weight change as related to the ceiling yield of a seedling sward. *Ann. Bot.*, **37**, 487–500.
Robson, M. J. (1973b) The growth and development of simulated swards of perennial ryegrass. II. Carbon assimilation and respiration in a seedling sward. *Ann. Bot.*, **37**, 501–18.
Robson, M. J. (1973c) The effect of temperature on the growth of S170 tall fescue (*Festuca arundinacea*). II. Independant variation of day and night temperature. *J. Appl. Ecol.*, **10**, 93–105.
Robson, M. J. (1974) The effect of temperature on the growth of S170 tall fescue (*Festuca arundinacea*). III. Leaf growth and tiller production as affected by transfer between contrasting regimes. *J. Appl. Ecol.*, **11**, 265–79.
Robson, M. J. (1980) A physiologists' approach to raising the potential yield of the grass crop through breeding, in *Opportunities for Increasing Crop Yields* (eds R. G. Hurd, P. V. Biscoe and C. Dennis), Pitman Advanced Publishing Programme, Boston, pp. 33–49.
Robson, M. J. (1981a) Respiratory efflux in relation to temperature of simulated swards of perennial ryegrass with contrasting soluble carbohydrate contents. *Ann. Bot.*, **48**, 269–73.
Robson, M. J. (1981b) Potential production – what is it and can we increase it? in *Plant Physiology and Herbage Production* (ed. C. E. Wright), British Grassland Society, Hurley, pp. 5–18.
Robson, M. J. (1982a) The growth and carbon economy of selection lines of *Lolium perenne* cv. S23 with 'fast' and 'slow' rates of dark respiration. I. Grown as simulated swards during a regrowth period. *Ann. Bot.*, **49**, 321–9.
Robson, M. J. (1982b) The growth and carbon economy of selection lines of *Lolium perenne* cv. S23 with 'fast' and 'slow' rates of dark respiration. II. Grown as young plants from seed. *Ann. Bot.*, **49**, 331–9.
Robson, M. J. (1983) All flesh is grass – bigger yields from Britain's most important crop. *A. Rep. Br. Grassland Res. Inst., Hurley, 1982*, pp. 132–49.
Robson, M. J. and Deacon, M. J. (1978) Nitrogen deficiency in small closed communities of S24 ryegrass. II. Changes in the weight and chemical composition of single leaves during their growth and death. *Ann. Bot.*, **42**, 1199–213.
Robson, M. J. and Jewiss, O. R. (1968) A comparison of British and North African varieties of tall fescue (*Festuca arundinacea*). III. Effects of light, temperature and day length on relative growth rate and its components. *J. Appl. Ecol.*, **5**, 191–204.
Robson, M. J. and Parsons, A. J. (1978) Nitrogen deficiency in small closed

References

communities of S24 ryegrass. I. Photosynthesis, repiration, dry matter production and partition. *Ann. Bot.*, **42**, 1185–97.

Robson, M. J. and Parsons, A. J. (1981) Respiratory efflux of CO_2 from mature and meristematic tissue of uniculum barley during eighty hours of continuous darkness. *Ann. Bot.*, **48**, 727–31.

Robson, M. J., Stern, W. R. and Davidson, I. A. (1983) Yielding ability in pure swards and mixtures of lines of perennial ryegrass with contrasting rates of 'mature tissue' respiration. in *Efficient Grassland Farming* (ed. A. J. Corrall), British Grassland Society, Hurley, pp. 291–2.

Roy, M. G. and Peacock, J. M. (1972) Seasonal forecasting of the spring growth and flowering of grass crops in the British Isles, in *Weather Forecasting for Agriculture and Industry* (ed. J. A. Taylor), David and Charles, Newton Abbott, pp. 99–114.

Ryle, G. J. A. (1964) A comparison of leaf and tiller growth in seven perennial grasses as influenced by nitrogen and temperature. *J. Br. Grassland Soc.*, **19**, 281–90.

Ryle, G. J. A. (1966) Effects of photoperiods in growth cabinets on the growth of leaves and tillers in three perennial grasses. *Ann. Appl. Biol.*, **57**, 269–79.

Ryle, G. J. A. (1967) Growth rates in *Lolium temulentum* as influenced by previous regimes of light energy. *Nature*, **213**, 309–11.

Ryle, G. J. A. (1970a) Partition of assimilates in an annual and a perennial grass. *J. Appl. Ecol.*, **7**, 217–27.

Ryle, G. J. A. (1970b) Distribution patterns of assimilated ^{14}C in vegetative and reproductive shoots of *Lolium perenne* and *L. temulentum*. *Ann. Appl. Biol.*, **66**, 155–67.

Ryle, G. J. A. (1972) A quantitative analysis of the uptake of carbon and of the supply of ^{14}C-labelled assimilates to areas of meristematic growth in *Lolium temulentum*. *Ann. Bot.*, **36**, 497–512.

Ryle, G. J. A. and Langer, R. H. M. (1963) Studies on the physiology of flowering of timothy. I. Influence of day length and temperature on initiation and differentiation of the inflorescence. *Ann. Bot.*, **27**, 213–31.

Ryle, G. J. A. and Powell, C. E. (1972) The export and distribution of ^{14}C-labelled assimilate from each leaf on the shoot of *Lolium temultentum* during reproductive and vegetative growth. *Ann. Bot.*, **36**, 363–75.

Ryle, G. J. A. and Powell, C. E. (1975) Defoliation and regrowth in the graminaceous plant: The role of current assimilate. *Ann. Bot.* **39**, 297–310.

Ryle, G. J. A. and Powell, C. E. (1976) Effect of rate of photosynthesis on the pattern of assimilate distribution in the graminaceous plant. *J. Exp. Bot.*, **27**, 189–99.

Ryle, G. J. A., Cobby, J. M. and Powell, C. E. (1976) Synthetic and maintenance respiratory losses of $^{14}CO_2$ in uniculum barley and maize. *Ann. Bot.*, **40**, 571–86.

Sagar, G. R. and Marshall, C. (1966) The grass plant as an integrated unit – some studies of assimilate distribution in *Lolium multiflorum* Lam. *Proc. IX Int. Grassland Congr.*, Vol. I, pp. 493–7.

Scott, R. K., English, S. D., Wood, D. W. and Unsworth, M. H. (1973) The yield of sugar beet in relation to weather and length of growing season. *J. Agric. Sci., Camb.*, **81**, 339–47.

Sheehy, J. E. (1977) Microclimate, canopy structure and photosynthesis in canopies of three contrasting temperate forage grasses. III. Canopy photosynthesis, individual leaf photosynthesis and the distribution of current assimilate. *Ann. Bot.*, **41**, 593–604.

Silsbury, J. H. (1965) Interrelations in the growth and development of *Lolium*. I. Some effects of vernalisation on growth and development. *Austral. J. Agric. Res.*, **16**, 903–13.

Silsbury, J. H. (1970) Leaf growth in pasture grasses. *Trop. Grasslands*, **4**, 17–36.

Templeton, W. C., Mott, G. O. and Bula, R. J. (1961) Some effects of temperature and light on growth and flowering of tall fescue, *Festuca arundinacea* Schreb. I. Vegetative development. *Crop Sci.*, **1**, 216–9.

Thomas, H. and Norris, I. B. (1977) the growth responses of *Lolium perenne* to the weather during winter and spring at various altitudes in mid-Wales. *J. Appl. Ecol.*, **14**, 949–64.

Thompson, K., Grime, J. P. and Mason, G. (1977) Seed germination in response to diurnal fluctuations of temperature. *Nature*, **267**, 147–9.

Treharne, K. J. and Eagles, C. F. (1969) Effect of growth at different light intensities on photosynthetic activity of two contrasting populations of *Dactylis glomerata* L, in *Progress in Photosynthesis Research*, Vol. 1 (ed. H. Metzner), International Union of Biological Science, Tubingen, pp. 377–82.

Treharne, K. J. and Eagles, C. F. (1970) Effect of temperature on photosynthetic activity of climatic races of *Dactylis glomerata* L. *Photosynthetica*, **4**, 107–17.

Treharne, K. J. and Nelson, C. J. (1975) Effect of growth temperature on photosynthetic and photorespiratory activity in tall fescue. in *Environmental and Biological Control of Photosynthesis* (ed. R. Marcello), Junk, The Hague, pp. 61–9.

Vickery, P. J., Brink, V. C. and Ormrod, D. P. (1971) Net photosynthesis and leaf area index relationships in swards of *Dactylis glomerata* under contrasting defoliation regimes. *J. Br. Grassland Soc.*, **26**, 85–90.

Warren Wilson, J. (1966) Effect of temperature on net assimilation rate. *Ann. Bot.*, **30**, 753–61.

Watanabe, K. and Takahashi, Y. (1979) Effects of fertilisation level on the regrowth of orchard grass. I. Changes of yield and growth with time. *J. Jap. Soc. Grassland Sci.*, **25**, 195–202.

Watson, D. J. (1947) Comparative physiological studies on the growth of field crops. I. Variation in net assimilation rate and leaf area between species and varieties, and within and between years. *Ann. Bot.*, **11**, 41–76.

Watson, D. J. (1958) The dependance of net assimilation rate on leaf area index. *Ann. Bot.*, **22**, 37–54.

Whitehead, D. C. (1970) *The Role of Nitrogen in Grassland Productivity, Bulletin 48*, Commonwealth Bureau of Pastures and Field Crops, Hurley, Berkshire.

Wilhelm, W. W. and Nelson, C. J. (1978) Irradiation response of tall fescue genotypes with contrasting levels of photosynthesis and yield. *Crop Sci.*, **18**, 405–8.

Williams, G. J. III (1974) Photosynthetic adaption to temperature in C_3 and C_4 grasses. A possible ecological role in shortgrass prairie. *Pl. Physiol.*, **54**, 709–11.

Williams, G. J. III and Kemp, P. R. (1978) Simultaneous measurement of leaf and root gas exchange of shortgrass prairie species. *Bot. Gaz.*, **139**, 150–7.

Wilson, D. (1970) Starch and apparent photosynthesis of leaves of *Lolium perenne* grown at different temperatures. *Planta*, **91**, 274–8.

Wilson, D. (1975) Variation in leaf respiration in relation to growth and photosynthesis of *Lolium*. *Ann. Appl. Biol.*, **80**, 323–38.

References

Wilson, D. (1981) The role of phsiology in breeding herbage cultivars adapted to their environment, in *Plant Physiology and Herbage Production* (ed. C. E. Wright), British Grassland Society, Hurley, pp. 95–108.

Wilson, D. (1982) Response to selection for dark respiration rate of mature leaves of *Lolium perenne* L. and its effect on growth of young plants. *Ann. Bot.*, **49**, 303–12.

Wilson, D. and Cooper, J. P. (1969a) Effect of light intensity and CO_2 on apparent photosynthesis and its relationship with leaf anatomy in genotypes of *Lolium perenne* L. *New Phytol.*, **68**, 627–44.

Wilson, D. and Cooper, J. P. (1969b) Effect of temperature during growth on leaf anatomy and subsequent light saturated photosynthesis among contrasting *Lolium* genotypes. *New Phytol.*, **68**, 1115–23.

Wilson, D. and Cooper, J. P. (1969c) Effect of light intensity during growth on leaf anatomy and subsequent light-saturated photosynthesis among contrasting *Lolium* genotypes. *New Phytol.*, **68**, 1125–35.

Wilson, D. and Jones, J. G. (1982) Effects of selection for dark respiration of mature leaves on crop yields of *Lolium perenne* cv. S.23. *Ann. Bot.*, **49**, 313–20.

Wilson, D. and Robson, M. J. (1981) Varietal improvement by selection for dark respiration rate in perennial ryegrass, in *Plant Physiology and Herbage Production* (ed. C. E. Wright), British Grassland Society, Hurley, pp. 209–11.

Wit, C. T. de, Brouwer, R. and Penning de Vries, F. W. T. (1970) The simulation of photosynthetic systems, in *Prediction and Measurement of Photosynthetic Productivity, Trebon*, (ed. I. Malek), Pudoc, Wageningen, pp. 47–70.

Woledge, J. (1971) The effect of light intensity during growth on the subsequent rate of photosynthesis in leaves of tall fescue (*Festuca arundinacea* Schreb). *Ann. Bot.*, **35**, *311–22.*

Woledge, J. (1972) The effect of shading on the photosynthetic rate and longevity of grass leaves. *Ann. Bot.*, **36**, 551–61.

Woledge, J. (1973) The photosynthesis of ryegrass leaves grown in a simulated sward. *Ann. Appl. Biol.*, **73**, 229–37.

Woledge, J. (1975) Photosynthesis and respiration of single leaves. *A. Rep. Grassland Res. Inst., Hurley, 1974*, p. 39.

Woledge, J. (1977) Effects of shading and cutting treatments on the photosynthetic rate of ryegrass leaves. *Ann. Bot.*, **41**, 1279–86.

Woledge, J. (1978) The effect of shading during vegetative and reproductive growth on the photosynthetic capacity of leaves in a grass sward. *Ann. Bot.*, **42**, 1085–9.

Woledge, J. (1979) Effect of flowering on the photosynthetic capacity of ryegrass leaves grown with or without natural shading. *Ann. Bot.*, **44**, 197–207.

Woledge, J. and Dennis, W. D. (1982) The effect of temperature on photosynthesis of ryegrass and white clover leaves. *Ann. Bot.*, **50**, 25–35.

Woledge, J. and Leafe, E. L. (1976) Single leaf and canopy photosynthesis in a ryegrass sward. *Ann. Bot.*, **40**, 773–83.

Woledge, J. and Jewiss, O. R. (1969) The effect of temperature during growth on the subsequent rate of photosynthesis in leaves of tall fescue (*Festuca arundinacea* Schreb.). *Ann. Bot.*, **33**, 897–913.

Woledge, J. and Pearse, P. J. (1985) The effect of nitrogenous fertiliser on the photosynthesis of leaves in a ryegrass sward. *Grass Forage Sci.*, **40**, 305–309.

CHAPTER 3
The regrowth of grass swards

Alison Davies

3.1 INTRODUCTION

The reaction of the grass sward to defoliation is principally determined by the position of its component organs in relation to the height of defoliation. Indeed, the success of the grasses on a global basis is attributable largely to their growing points and meristematic regions normally being below defoliation height, except during reproductive development. Thus, the grass plant maintains two options: to reproduce vegetatively (by the production of new tillers) or to reproduce from seed. Perennial grasses from cool temperate regions (e.g. perennial ryegrass) are characterized by a short flowering period followed by active tiller production. Grasses from regions of summer drought (e.g. Westerwold ryegrass *Lolium multiflorum*) behave as annuals (Cooper and Saeed, 1949) and depend almost entirely on seed set for their survival.

As discussed in Chapter 2, in the vegetative tiller new leaves and sheaths develop from buds on the shoot apex, which is generally situated at or close to ground level (Fig. 3.1a). Moderate or lax defoliation removes portions of the older, fully-emerged laminae and of any developing leaves emerging from within the sheaths of the older leaves. Severe defoliation may remove all of the emerged laminae and even parts of the pseudostem (de Lucia Silva, 1974). Regrowth depends on the growth of young leaves which are still in the process of emerging from the enclosing sheath or sheaths. Growth of these leaves takes place in the basal meristematic regions (Davidson and Milthorpe, 1966a), so that the upper, fully grown portions already present inside the sheath are exserted. Little or no further growth occurs in fully emerged leaf laminae or their sheaths.

During the annual cycle of growth the meristematic apices of some of the tillers become reproductive and the intercalary meristem, which is initially very short, begins to extend. The transformed apex is carried upwards to form a terminal inflorescence, and thus becomes liable to removal when the plant is defoliated (Fig. 3.1b). Once the apex has been removed the tiller dies, so regrowth depends on the presence of other tillers with intact apices or on the production of new tillers. Thus, the pattern of regrowth of grass following

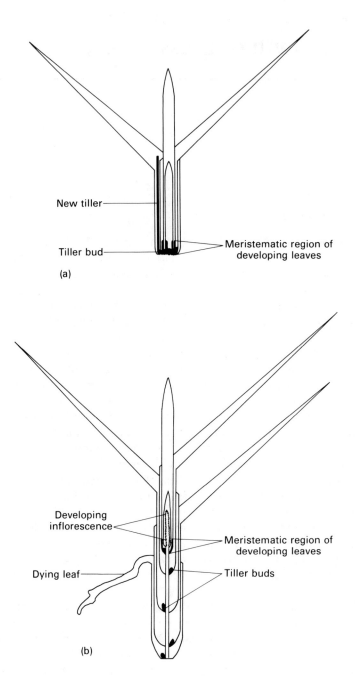

Figure 3.1. A diagrammatic representation of (a) a vegetative tiller and (b) a reproductive tiller of perennial ryegrass. (After Davies, 1973b.)

Regrowth in vegetative plants

defoliation is strongly dependent on whether the tillers are in a vegative or reproductive state when the plant is defoliated.

3.3 REGROWTH IN VEGETATIVE PLANTS

The response of the plant to defoliation at any time depends on the plant material which remains to initiate regrowth. The stubble factors which influence regrowth fall 'into two classes – those that affect photosynthesis directly and those, such as the availability of labile reserves . . . that affect the ability of the sward to regenerate new photosynthetic tissue' (King et al., 1979).

If sufficient photosynthetic tissue is left to provide for the total respiratory needs (both growth and maintenance, see Chapter 2) of the remaining plant tissue, then the plant will begin to accumulate weight immediately after it is cut. In these circumstances the relationship between the logarithm of dry weight $\ln W$ and time t will approximate to a straight line with slope

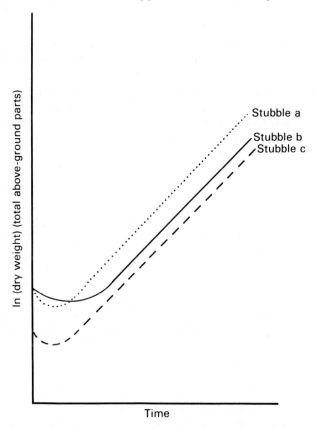

Figure 3.2 Weight of total above-ground parts during regrowth from stubbles of different weight and efficiency (see text for details).

$(\ln W_1 - \ln W_0)/(t_1 - t_0)$. This is exactly the same as the mean relative growth rate (R_w) for the period $t_1 - t_0$, and may vary with variety or genotype. If defoliation results in insufficient photosynthetic tissue to meet respiratory requirements, then the plant will experience a period in negative carbon balance during which it will draw on existing carbon resources to form new leaf, and will thus lose dry weight. This interruption in the course of dry matter increase may be of shorter or longer duration, as shown in examples a and b or c and b in Fig. 3.2, which illustrates possible regrowth patterns in identical genetic material subjected to different prior managements. In these examples the effects of different initial stubble weight (examples a and c) and efficiency in promoting regrowth (examples b and c, or a and b) can be distinguished not only from the different periods of weight loss, but also by the relationship between initial and final weights, i.e. the mean R_w over the whole period. This model is helpful in understanding defoliation responses in the context of short-term experiments and in devising practically feasible ways of evaluating the regrowth potential of different stubbles (Davies, 1974).

Invesitigations have been carried out to determine the possible effects on regrowth of a number of changes in stubble composition. These include the proportion and activity of the remaining photosynthetic tissue (laminae and sheaths), the percentage of readily available carbohydrates in the stubble and the proportion of associated root material (the root/shoot ratio). No critical study has been made of the relative efficiencies of stubble composed of many small tillers or few large tillers, as the problem of obtaining variation in this aspect of stubble morphology without concomitant variations in other characters remains unsolved.

3.2.1 Leaf area

In the sward the importance of maintaining a leaf area sufficient to intercept most of the incident light for photosynthesis and the rapid recovery of this leaf area after defoliation was illustrated by the classic studies of Brougham (1956). In his experiment a ryegrass sward containing some red and white clover was defoliated in spring to heights of 2.5, 7.6 and 12.7 cm (1, 3 and 5 inches). Regrowth increased as the residual LAI increased. Figure 3.3 shows the general relationship between the crop growth rate (CGR) and the current LAI. The increase in CGR was observed to continue up to the point at which virtually all of the incoming light was being intercepted and the sward formed a closed canopy.

While Brougham's experiment establishes the importance of LAI in regrowth of swards, it does not indicate the relative importance of leaf area as a component of stubble weight for individual plants. More-detailed experiments conducted on small groups of tillers growing in water culture show that the R_w is little affected by defoliation as long as one intact lamina remains (Table 3.1). It also suggests that the removal of older laminae may have less effect than the

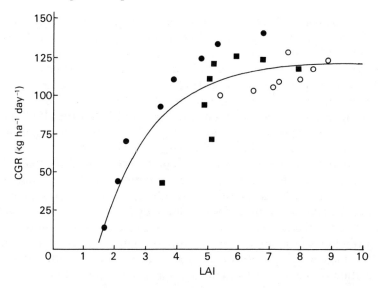

Figure 3.3 The relationship between crop growth rate (CGR) and leaf area index (LAI) in spring in swards dominated by short-rotation ryegrass defoliated to (●) 2.5 cm, (■) 7.6 cm, (○) 12.7 cm. Mean standard error of estimate = 18 kg ha^{-1} day^{-1}. (After Brougham, 1956.)

TABLE 3.1 *Effects of removal of leaf laminae on R_w (relative growth rate) (shoots + roots) in perennial ryegrass (after Davies, 1974)*

Laminae removed	None	L1	L2	L3	L1 & 2	L2 & 3	All	s.e. ±
R_w	0.092	0.085	0.085	0.090	0.084	0.091	0.064	0.002

L1 is the emerging leaf lamina, L2 and L3 the youngest and next-youngest fully-emerged leaf laminae. L4, if present, is included with L3.

removal of younger laminae. However, removal of all laminae leads to a marked reduction in R_w and to reduced development of tiller buds.

The comparative absence of an effect of partial defoliation on R_w has also been reported by Ryle and Powell (1975) for uniculm barley. They ascribed the compensation effect to an increased export of carbon from remaining leaf tissue to the developing leaves rather than to the root. Defoliation may also increase photosynthesis per unit area of lamina relative to that in control plants. Examples include the work of Jewiss *et al.* (1969) on perennial ryegrass, and Gifford and Marshall (1973) on *Lolium multiflorum*. Generally the effects of removing older leaves are less marked than the effects of removing younger leaves in both perennial ryegrass (Davies, 1974; Vine, 1977) and in *Phalaris arundinacea* (Begg and Wright, 1964; Brown *et al.*, 1966; Mittra and Wright, 1966), but this may depend on nutrient status. Removal of the emerging leaf in

Dactylis glomerata reduced extension (of leaves with no visible ligules on the main tiller) more than the removal of the fully emerged leaves at high nutrient level, but the reverse was the case at low nutrient level (Davidson and Milthorpe, 1966a). The suggestion was made that in the latter instance the older leaves supplied the developing leaves with essential minerals.

The contribution of exposed green leaf sheaths to the net carbon balance after cutting is not well documented. Davidson and Milthorpe (1966b) calculated that the rate of photosynthesis per unit area of leaf sheaths was about one-third of that of leaf laminae in cocksfoot (*Dactylis glomerata*); Thorne's (1959) data on barley at about mean ear emergence suggested that sheath photosynthesis per unit area of exposed photosynthetic surface was about 85% of that of the associated leaf lamina. Regrowth of perennial ryegrass plants with most of the leaf laminae removed was reduced by 20–30% when the sheath and pseudostem was shaded (Davies *et al.*, 1983). In a continuously grazed sward the contribution of sheath to canopy net photosynthesis was found to be less than 5% (Parsons *et al.*, 1983a). However, this result was attributed to an accumulation of older sheaths (many of which were senescent or even dead) around the younger sheaths.

3.2.2 Carbon reserves

A plant which has insufficient assimilatory surface to supply its current respiratory needs for maintenance and growth must make use of available carbon resources, the most important of which (in temperate grasses) are the non-structural water-soluble carbohydrates. These are glucose, fructose, sucrose and the fructose polymers (fructosans), which constitute a major carbohydrate source and/or sink. The work of Sullivan and Sprague (1943) and Sprague and Sullivan (1950) established that fructosan levels in stubble (and in roots) of perennial ryegrass and cocksfoot plants fell rapidly after defoliation. This fall is presumably arrested when sufficient current photosynthate is produced to meet demands. The return to the original level was found to be slower, but was normally complete within 4–5 weeks, though it may be slower at low temperatures (Davies, 1965). Changes in other soluble carbohydrates were less marked, and no real changes at all were observed in non-carbohydrate components during regrowth in the light. The conclusion drawn from this work was that fructosans acted as reserve substances which were utilized during regrowth.

This conclusion was reinforced by observations that in various grass species frequent defoliation can lead to low stubble carbohydrates, poor regrowth and even death (Graber *et al.*, 1928; McCarty and Price, 1942; McIlvanie, 1942). It was, however, challenged by May (1960), who observed that none of the available evidence demonstrated unequivocally that the level of carbohydrate substances in the stubble materially affected regrowth. There is, in fact, ample evidence to show that other substances may be utilized in regrowth (Davidson

Regrowth in vegetative plants

and Milthorpe, 1966b; Alberda, 1966; Jewiss *et al.*, 1969, and unpublished). The latter found that in perennial ryegrass plants grown in the growth room only one-quarter of the weight loss from the stubble could be accounted for in terms of water-soluble carbohydrate, the rest being hemicellulose, proteins, organic acids and cellulose. However, it should be pointed out that in this investigation both the light level during regrowth and the carbohydrate percentage were low.

Evidence that carbohydrate levels may directly influence regrowth was provided by Alberda (1966), who held perennial ryegrass plants in the dark for 3.5 days to reduce the level of carbohydrate in the stubble before defoliation. Regrowth in these plants was substantially less than in control plants receiving continuous light. The weight of other substances, the tiller number and the residual leaf area were all unaffected. In the high-carbohydrate plants the new leaf appeared to be formed almost entirely at the expense of stubble carbohydrates. The suggestion (Davies, 1965) that carbohydrate levels will only influence regrowth if they fall below a critical level is supported by the results of a similar type of experiment reported by Davies *et al.* (1972) in which plants of perennial ryegrass were subjected to exposure to two different temperatures for 2 days before defoliation. Here there was no difference in regrowth at 20 °C, although the percentages of water-soluble carbohydrates in the stubble differed (20.4% at 10 °C and 15.2% at 30 °C). However, when the experiment was repeated with the exception that pretreatments were preceded by an initial period of 2 days of exposure to darkness, regrowth from the 10 °C pretreatment was 60% greater than from the 30 °C pretreatment. Carbohydrate levels in the stubble were 12.5% (10 °C) and 6.5% (30 °C), and the weight of other substances in stubble and roots was scarcely affected. This experiment shows that regrowth is reduced when levels of stubble carbohydrate are low.

The effect of carbohydrate resources in the stubble on R_w over a period of regrowth will depend on its level. If the level is high, then some of the carbohydrate will simply act as ballast and R_w will be lower than in plants with less carbohydrate (Alberda, 1970). If the level is low, R_w may well be less than in a plant with the optimum percentage of stubble carbohydrate for regrowth in the circumstances of the experiment. Thus R_ws for an S.24 perennial ryegrass clone with stubble carbohydrate levels of 25, 20, 10 and 5% growing for 11 days at 17 °C were, respectively, 0.082, 0.095, 0.111 and 0.082 g g^{-1} day^{-1}. Further experimentation is needed to establish the levels of carbohydrate needed for maximum regrowth in different species or varieties and in different circumstances.

3.2.3 Relative importance of leaf area and carbon reserves

The relative effects on regrowth of the amount of leaf tissue in the stubble and the carbohydrate reserves have received limited attention. Ward and Blaser (1961) used dark and light pretreatments extending over a period of 3.5 days to give levels of fructose-based compounds of 8% and 2% in cocksfoot plants, which were then defoliated either to leave 55 mm or to completely remove each fully-emerged leaf blade. A full interpretation of their results is difficult because no starting weights are given, but it appears that dark or light pretreatment had a stronger effect than leaf area during the first 25 days of regrowth. The low R_w (0.03 g g^{-1} day^{-1}) between days 5 and 15 in plants with low carbohydrate and low leaf area suggests that this stubble was inefficient in promoting regrowth (R_w in the other three treatments was 0.11–0.12 g g^{-1} day^{-1}).

The interpretation of the data of Booysen and Nelson (1975) on tall fescue presents fewer problems. Their pretreatments involved exposure to high or low light levels for 3 days, giving water-soluble carbohydrate contents of 11.8% and 6.1%, all or half of each fully-emerged leaf being removed at the time of cutting. The removal of all of the fully emerged leaves reduced both regrowth and R_w. A direct comparison of low WSC/high leaf area plants with high WSC/low leaf area plants (starting weights being very similar in both cases) showed that leaf tissue was more effective than an equal weight of WSC in promoting regrowth.

In some circumstances the relationship between the weight of stubble and the amount of regrowth may be very close. Thus, in an experiment in which the amount and constitution of the stubble in field plots of vegetative perennial ryegrass was varied by a range of pretreatments (Davies, 1966) the correlation coefficients on stubble weight over the first 6 weeks of regrowth ranged between 0.998 and 0.976 ($P < 0.001$). Leaf weight in the stubble ranged from 1 to 15%, while carbohydrate ranged from 8 to 25%. Finally, as 76% of the variation in stubble weight was accounted for by differences in tiller numbers, it is evident that tiller density may strongly influence regrowth.

3.2.4 Roots and regrowth

Although mention is made in the literature of reserves in stubble and roots (Weinmann, 1948), there is no evidence that roots can supply carbon to shoots after defoliation in non-rhizomatous grasses. No movement of this kind was detected in the ^{14}C radiocarbon tracer studies of Marshall and Sagar (1965, 1968), and published data suggest that root respiration may well constitute one of the major burdens on the carbon economy of the plant after defoliation (Davidson and Milthorpe, 1966b). The effect of defoliation seems to be to stop root growth (Crider, 1955) and depress mineral uptake (Oswalt *et al.*, 1959), effects associated with a decline in root respiration (Davidson and Milthorpe, 1966b). Roots often continue to lose weight after shoots have begun to gain weight. Tracer studies using ^{14}C (Clifford and Langer, 1975; Ryle and Powell, 1975) show a reduction in photosynthetic carbon transported to roots of defoliated plants of *Lolium multiflorum* and uniculm barley immediately after defoliation. The latter authors concluded, 'It seems unlikely that such severely

Regrowth in vegetative plants

stressed tissues (viz. the roots) could provide a source of carbon compounds for regrowth of the aerial tissues, whatever the level of defoliation'.

Root pruning has variable effects on regrowth. Jacques and Edmond (1952) reported that weekly removal of new roots (leaving six functioning roots only) tended to increase regrowth of soil-grown perennial ryegrass and cocksfoot over a period of several weeks of repeated defoliation, but found that continued removal was detrimental. Davies and Troughton (1971, and unpublished) found that removal of one-third of the root system had very little effect on the regrowth of perennial ryegrass grown in nutrient solution, but that removal of two-thirds reduced regrowth after a fortnight by 26% compared with the controls. Any immediate advantage which may result from the removal of part of the respiratory burden is probably negated by the need to grow more roots to restore the lost absorbing surface. Root-pruned plants normally produce more roots than controls (Jacques and Edmond, 1952; Ješko and Troughton, 1978). The conclusion appears to be that equal weights of stubble supported by different weights of root material are likely to grow equally well unless damage to the roots (e.g. by pests) is fairly severe.

3.2.5 Species and variety differences

Different species and varieties (Davies *et al.*, 1972; Jones, 1983) and genotypes (Harris, 1973) show different reactions to variations in cutting height and frequency. Initially these will result from differences in stubble composition and activity; in the longer term the speed and nature of changes in sward structure will also be important. Selection for low mature tissue respiration in perennial ryegrass has resulted in improved regrowth and in tiller populations which gradually increased relative to those of high respiration selections (Wilson and Robson, 1981; see also Chapter 2). Other factors, for example the time taken to resume tillering after cutting (Davies, 1965), may also affect longer-term reactions. At present, however, relatively little analytical information exists to suggest why some species or varieties tolerate more frequent or closer cutting than others. Cocksfoot, in particular, seems to be capable of withstanding very frequent cutting in some instances (Davidson and Milthorpe, 1966a).

The results in Table 3.2 show the behaviour of five grass species in two experiments. In the first experiment plants grown in nutrient solution were cut either twice in two days (i.e. on days 0 and 2) or once on day 2. Regrowth was compared 3 weeks later. In the second experiment soil-grown vegetative plants were cut every 5 days and the regrowth after the fifth cut on 30 July was compared with regrowth after the first cut on 9 July. Deaths were scored after 11 cuts. In both experiments regrowth was better sustained in cocksfoot than in the other species and there were fewer deaths. Deaths were high in the ryegrasses (especially in Italian ryegrass), and regrowth was sustained less well

TABLE 3.2 *Response to short cutting intervals in different grass species (after Davies et al., 1972; Davies, unpublished).*

Species	Experiment 1 Regrowth on day 23 after cuts on day 0 and day 2 as a percentage of regrowth after a single cut on day 2	Experiment 2	
		Regrowth after 5 cuts at 5 day intervals as a percentage of regrowth after the first cut	Deaths after 11 cuts at 5-day intervals (%)
Italian ryegrass	63	12	70
Perennial ryegrass	70	22	45
Tall fescue	65	36	16
Timothy	86	42	31
Cocksfoot	91	43	4
Defoliation heights (cm)	4	2	2

than in either cocksfoot or timothy. With longer cutting intervals the ryegrasses, and especially Italian ryegrass, outyielded the other species (Davies *et al.*, 1972).

3.3 CHANGES IN SWARD STRUCTURE DURING REGROWTH

The principal morphological changes occurring as a sward accumulates herbage mass after defoliation are the effects on extension growth (lengths of leaves and sheaths), on the extent of tissue losses and on tiller numbers. These changes affect the amount and composition of the stubble left when regrowth is terminated by further defoliation.

3.3.1 Leaf and sheath length

The length of new leaves (and their rate of appearance) is influenced to a large extent by the length of the sheath tube through which the leaves emerge (Grant *et al.*, 1981a). Begg and Wright (1962) observed that extension growth is strongly retarded as the leaves emerge into the light, and suggested that this response may be phytochrome mediated. If the sheath tube is left intact (Youngner, 1972; Davies, 1974) the emerging leaves will be relatively long and may appear more slowly than in comparable undefoliated material. If, on the other hand, the cut removes the upper part of the sheaths the new leaves will be shorter and may appear as quickly as in undefoliated plants.

Changes in sward structure

In the spring and summer the sheath length of leaves increases with increasing herbage mass and new leaves appear at successively increasing heights above ground, so that the green canopy moves upwards as the older leaves die. These changes are well illustrated in an experiment reported by Jackson (1976). He found that the heights of insertion of the lowest green leaf blades on tillers of S.23 perennial ryegrass at constant cutting heights of 3, 6, 9 and 12 cm were, on average, 2.6, 4.3, 6.6 and 9.2 cm, respectively. Weights of green leaf blade were very similar for all cutting heights. Interrupting the sequence of cuts by a single 3-cm cut resulted in an immediate lowering of the height of the lowest green leaf blade to just under 3 cm, viz. the ligule of the first leaf to emerge after cutting was, on average, just below cutting height. On reversion to the original cutting height the height at which successive new leaves appeared increased and the green leaf canopy moved upwards. During the period of readjustment (Fig. 3.4) green leaf developed in the intermediate horizon. As these leaves senesced and were replaced by younger leaves carried at a greater height, the weight of green leaf blade tended once again to equalize at the different cutting heights. Adjustment was complete even in the 12/3/12 cm treatment after the fourth harvest. Lax cutting also induced extension of basal internodes, carrying tiller apices above ground level and placing considerable numbers at risk from decapitation when a 3 cm cut was taken.

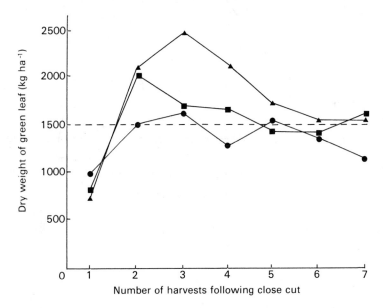

Figure 3.4 Effect of interrupting a sequence of cuts at (●) 6 cm, (■) 9 cm and (▲) 12 cm with a single close (3 cm) cut on total green leaf present before subsequent cuts (overall mean for constant cutting heights given as broken line for comparison). (After Jackson, 1976.)

Stubble from swards maintained at greater heights is thus characterized by leaflessness, reduced tiller numbers and an accumulation of dead leaf sheaths.

3.3.2 Leaves per tiller

As herbage accumulates during regrowth the number of intact new leaves on each tiller increases. In vegetative swards of perennial ryegrass the mean number of living leaves per tiller rarely exceeds three (Hunt, 1965; Alberda and Sibma, 1968) and the production of a fourth leaf tends to be counterbalanced by the loss of the first one (Davies and Calder, 1969; Davies, 1971). Unless the most recently-formed leaf is appreciably larger and heavier than the one that it replaces (or stem development occurs) the net gain in weight of living leaf will be small (see Chapters 2 and 4). In many instances, and especially in the second half of the year or in winter, any net gain of this kind is counterbalanced by the

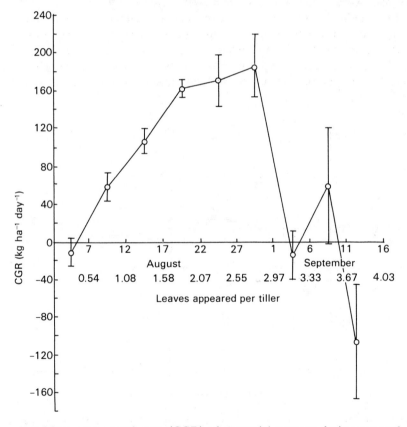

Figure 3.5 Mean crop growth rate (CGR) of perennial ryegrass during regrowth in August/September and number of leaves appeared since cutting: vertical bars indicate s.e. (After Davies, 1971.)

Changes in sward structure

loss of whole tillers, so net crop growth rate becomes zero and a ceiling yield is reached. Assuming that no intact leaves remain after defoliation, the ceiling yield can be expected to be attained after three leaf-appearance intervals (tables of mean rates of leaf appearance in perennial ryegrass are presented in Davies, 1977). The situation is illustrated in Fig. 3.5 for a well-fertilized sward of perennial ryegrass growing in August and September. Ceiling yields in a vegetative sward thus depend on seasonal variations in leaf length (see Davies, 1977) and tiller numbers (Section 3.3.3). Estimates based on this information and on weight per unit length of leaf (approx. $0.8\,\mathrm{mg\,cm^{-1}}$ for the same experiments) correspond well with observed values (Davies, 1960; Alberda, 1965; Davies and Simons, 1979). It is probable that a similar situation exists in other grasses, though it does not follow that the maximum number of living leaves per tiller is the same in all cases.

3.3.3 Tiller density and turnover

Recent observations by Hodgson *et al.* (1981) have shown that tiller numbers and tiller weights in swards maintained at different herbage masses by continuous grazing are related to each other through the so-called minus 3/2 law established by Yoda *et al.* (1963). This law, initially applied to plant numbers and plant weights during a period of self-thinning, has been shown by Kays and Harper (1974) to apply equally well to tiller numbers and tiller weights in perennial ryegrass. The relationship is perhaps best understood in the present context by noting that it predicts that a sward with twice (or n times) the biomass will have tillers which are spaced twice (or n times) as far apart. Its operation in mature swards is confirmed in Fig. 3.6, which shows the results of Grant *et al.* (1981b) obtained from perennial ryegrass swards grazed to maintain herbage mass between 700 and $2800\,\mathrm{kg\,ha^{-1}}$, together with values from hand-planted perennial ryegrass swards under intermittent defoliation at Aberystwyth. The latter values are restricted to those from experiments in which swards had attained a closed canopy and tiller numbers had either reached a maximum or were declining, i.e. in which the conditions laid down by White and Harper (1970) had been satisfied. The self-thinning law does not cover circumstances in which tiller numbers are increasing or where (as in very short, hard-grazed swards) there is a loss of plant cover.

Tiller numbers also vary with time of year, and when maximum numbers from intermittently defoliated vegetative (or predominantly vegetative) swards are set out against the month in which they were recorded a further general relationship seems to emerge (Fig. 3.7). The shape of this curve indicates that tiller numbers may also depend on solar radiation levels, a suggestion supported by the results of shading experiments (Spiertz and Ellen, 1972; Thomas and Davies, 1978). The figure shows that, although herbage mass is less in winter, there are fewer rather than more tillers (about half the number present in summer). Mean tiller numbers obtained from perennial ryegrass

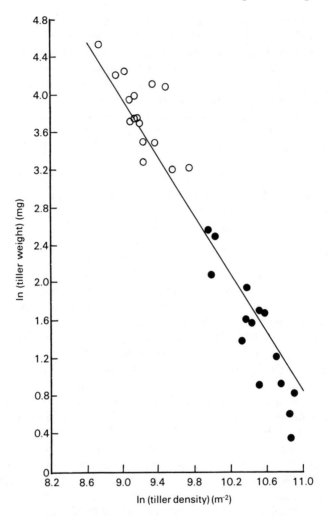

Figure 3.6 Relationship between ln (tiller weight) and ln (tiller density) in perennial ryegrass/perennial ryegrass dominant swards. ○, Cut swards, 30 June–17 November. (After Davies, unpublished). ●, Grazed swards, 15 July–4 September. (After Grant et al. 1986). Slope of line = $-\frac{3}{2}$.

swards subjected to monthly cuts (Garwood, 1969) are given for comparison. Current work suggests that numbers in rotationally grazed swards may well be very similar (Collett et al., 1981; Jones et al., 1982). Seasonal variation seems to be less marked in continuously grazed swards (Arosteguy, 1982), probably because it is masked in the context of a general upward trend in tiller numbers (Jones, 1981; Parsons et al., 1983a). This may well be linked with other gradual adaptive changes in sward structure.

Changes in sward structure

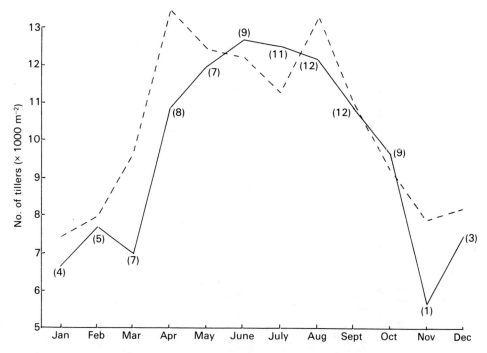

Figure 3.7 Mean maximum tiller numbers in perennial ryegrass swards over the course of the year. (Data from Garwood, 1969, from monthly cuts are given by broken line for comparison.) (*n*) = Number of experiments.

Swards may respond to defoliation by net gains, net losses or no overall changes in tiller populations, depending on initial tiller numbers at the time of cutting and the time of year. During the spring the tendency will be to net gains, but during autumn the opposite will be the case. In the latter part of the summer a sward defoliated intermittently (say every 4 weeks) may develop a stand of tillers sufficiently dense to permit rapid canopy development and suppress further tillering (Davies, 1977).

The operation of the various factors which control tiller production and death can be described most usefully by a specific example, namely an S.23 perennial ryegrass sward sown in spring and defoliated on 26 July after two different pretreatments (Fig. 3.8). Immediately after cutting, and especially if the crop harvested is a heavy one, some tillers may die. Deaths are also increased if some sheath tissue is removed (de Lucia Silva, 1974). At this time, too, few new tillers may be produced (Davies, 1974) since relatively little carbon seems to be diverted to tiller production (Clifford and Langer, 1975). This phase is followed by a longer or shorter period of tiller increase during which the rate of tillering may be rapid, with tiller buds developing into visible tillers as quickly as they are laid down (Davies and Thomas, 1983). With

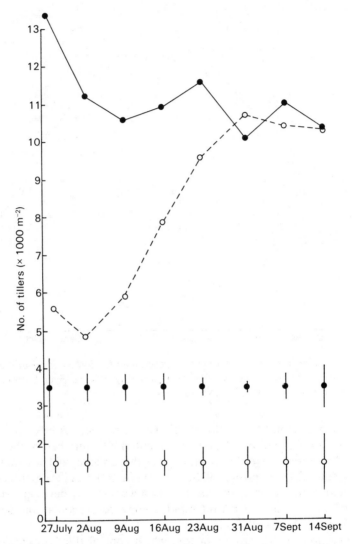

Figure 3.8 Changes in tiller numbers in dense and open swards of S.23 perennial ryegrass (resulting from different N× frequency of cutting managements) during regrowth. ○, Open swards, cut 10 and 26 July (no N applied). ●, Dense swards, cut 2, 16, 25, 30 June, 7, 26 July (26 kg N applied 3, 21 June; 6, 20 July). All swards fertilized fortnightly at 50 kg nitrogen ha^{-1} after final cut).

increasing herbage accumulation, however, base-shading increases and tillering slows down (Mitchell and Coles, 1955). In the dense swards in Fig. 3.8 this stage was reached far earlier than in the open swards, but open and dense swards both stabilized at very similar tiller densities and levels of herbage mass. Recent work (Davies et al., 1983) has served to confirm the findings of Mitchell and Coles (1955) and to demonstrate that more dry matter is partitioned to the growth of existing tillers and less to the development of new tillers in base-shaded plants. A reduction in the proportion of radiocarbon translocated to tiller buds in shaded plants has been reported in one experiment conducted by Ryle and Powell (1976). Ultimately tillers die (Ong, 1978), most probably as a result of being overtopped and shaded by other tillers (Alberda and Sibma, 1982). The accumulation of dead material may also have this effect (Simons et al., 1974; Tainton, 1974). Tillers may be suppressed to the point at which they will die when the sward is cut (Alberda, 1966). Wade (1979) observed that tiller losses in grazed sward were high in the first week after defoliation and that subsequent tiller production was inversely proportional to LAI. When tillers cease to develop leaf sheaths accumulate, and these may die and persist after the sward has been defoliated. With continued lax defoliation tubes of dead sheaths may accumulate (Jackson, 1974) and suppress further tiller development (Davies et al., 1983). The controlling factor seems to be the attenuation of the incident light through the canopy, influencing tillering via changes in the proportions of red/far-red light and involving the phytochrome system (Deregibus et al., 1983).

In continuously grazed swards herbage weight is normally maintained well below the level which would be attained if the swards were allowed to grow without interruption. Accordingly, tiller numbers are much higher than in swards cut intermittently (Jones et al., 1982) or grazed rotationally (Tallowin, 1981). Numbers recorded in sheep-grazed pastures dominated by perennial ryegrass have risen to 50 000 tillers m^{-2} (Grant et al., 1981b) or even 70 000 tillers m^{-2} in mid-season and after a number of years of hard grazing (Parsons et al., 1983a). Tiller populations in cattle-grazed swards at equivalent sward height/herbage mass appear to be lower than in swards grazed by sheep (Boswell and Crawford, 1979; Arosteguy et al., 1983). High tiller densities have also been recorded in densely sown and frequently defoliated box swards (Grant et al., 1981a), but except under strict green-keeping managements (Canaway, pers. comm.) such densities are rarely, if ever, maintained in cut swards in the field.

3.4 REGROWTH IN DIFFERENT DEFOLIATION SYSTEMS

The changes which take place in a sward as herbage mass accumulates largely determine the effects of different systems of defoliation on regrowth in both the shorter and the longer term. In the short term the most important factor is the relationship between rates of dry matter production and loss, the greater rate of increase at high LAI being increasingly counterbalanced by losses of leaves and even of whole tillers. This relationship determines the optimum length of the regrowth period and strongly influences response to changes in cutting height and frequency (see Chapter 4). In the longer term defoliation practices also affect stubble quantity and stubble conditions, and therefore influence rates of regrowth.

3.4.1 Factors influencing the optimum height of defoliation

Although the relationship between crop growth rate and LAI (see Fig. 3.3) suggests that greater dry weights of harvested herbage might be expected from increased heights of defoliation many authors (for example, Reid, 1959, 1962; Reid and MacLusky, 1960; Tayler and Rudman, 1966; Plancquaert, 1967) have observed greatly increased production from more closely defoliated swards. The reasons for this lie in the changed patterns of leaf and tiller formation and loss.

The effects of two different cutting heights on net herbage production on a single tiller are illustrated diagrammatically in Fig. 3.9. Although the increase in total weight of the more leniently defoliated tiller after the first cut was greater than in the more severely defoliated tiller, not all of the extra material would be harvestable if a second cut were taken at the same height (the extra leaf present below the cutting level would be lost to death and decay in the normal manner). At still greater cutting heights losses would be potentially greater and might include some virtually intact leaves. The potential advantages of a higher LAI in terms of net photosynthesis are therefore not automatically realized in greater weights of harvested herbage (King *et al.*, 1979) (see Chapter 4).

The other important consequence of lax defoliation (and of the greater

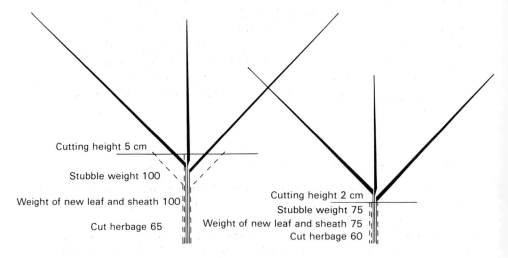

Figure 3.9 Influence of cutting height on harvestable herbage. —, new leaves and sheaths; --- old leaves and sheaths. (Based on data of Simons *et al.*, 1972.)

Different defoliation systems

accumulation of herbage mass) is a reduced tiller population (for example, Appadurai and Holmes, 1964; Harris, 1973) (see Section 3.3.3). Production of aerial tillers, which are vulnerable both to detachment and removal when grazed, is also a characteristic of lax defoliation regimes (Jackson, 1974; Simons et al., 1974). In the longer term reductions in tiller number and increases in the height of the green leaf canopy may combine to produce a situation in which weights of green leaf are virtually the same over a wide range of cutting heights (see Section 3.3.1).

Detailed sward observations thus serve to clarify the general observation that maximum production of harvestable herbage dry matter is obtained when swards are allowed to grow for some weeks between cuts, but are defoliated quite closely (Brougham, 1959; Huokuna, 1960; Bryant and Blaser, 1961; Frame and Hunt, 1971). Exceptions occur – possible reasons include damage (particularly when cattle are allowed to graze long herbage) and overgrazing of some areas when mean grazing height falls to 2.5 cm. This can be particularly damaging in areas where summers are relatively dry (Brougham, 1960; Weeda, 1965; Campbell, 1969; Tainton, 1974).

The optimum height of defoliation may also vary with species (Tayler and Rudman, 1966; Davidson, 1969; Lawrence and Ashford, 1969) and even with genotype. Table 3.3. shows the effects of a 4-month period (May–August) of 3-weekly cuts at two defoliation heights in two perennial ryegrass genotypes selected for differences in leaf length, and also illustrates something of the range of sward structural adaptations which must, in their entirety, determine reactions to heights of defoliation.

Cutting at 10 cm increased the height of the leaf canopy relative to its height when swards were cut at 2.5 cm, and decreased the number of tillers, particularly when aerial tillers (of which there were a considerable number in

TABLE 3.3 *After-effects of cutting two perennial ryegrass genotypes at different heights*

	Long-leaved genotype		Short-leaved genotype		±s.e.
Cutting height (cm)	2.5	10	2.5	10	
Point of insertion of uppermost leaf (cm)	3.7	8.0	2.3	4.5	0.27
Tillers m^{-2} ($\times 10^2$)	18.2	16.2	11.6	5.7	0.89
Tillers m^{-2} with growing points at ground level	12.4	7.8	8.7	3.7	0.57
Percentage leaf blade in stubble	3.7	11.1	23.9	30.2	1.04
Weight of stubble (kg ha^{-1})	1922	3799	1097	692	120.4
Regrowth after 3 weeks (kg ha^{-1})	960	1010	327	48	50.9

the long-leaved genotype) were excluded. Although the weight of stubble was greater at 10 cm than at 2.5 cm in the genotype with long leaves, there was no corresponding difference in regrowth above cutting height. In the genotype with short leaves the low tiller number at 10 cm was associated with a low stubble weight, and there was very little harvestable regrowth.

3.4.2 Defoliation in swards continuously grazed to different fixed heights

The effects of defoliation height on regrowth are particularly well illustrated by recent grazing studies (Arosteguy *et al.*, 1983; Bircham and Hodgson, 1983; Grant *et al.*, 1983a) in which stocking rates were varied to maintain different fixed sward heights. In these studies rates of herbage growth and senescence and (by difference) the rate at which herbage was being consumed by the animals were estimated using tissue turnover methods (for details see Davies, 1981). Essentially similar results, which are discussed in greater detail in Chapter 4, have been obtained from carbon balance studies by Parsons *et al.* (1983a,b).

Results from these studies show that shorter swards have more, smaller tillers (see Section 3.4.3) and a slower rate of growth per tiller than longer swards, but that provided the chosen sward height allows the maintenance of a fairly complete canopy the rate of growth (i.e. the product of tiller number and

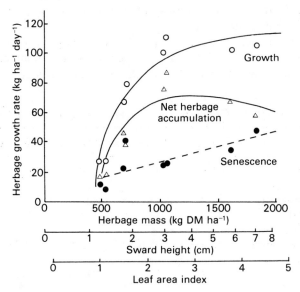

Figure 3.10 Relationships between biomass (kg ha^{-1} organic matter) sward height (cm) and LAI and rates of herbage growth, senescence and net herbage accumulation (all kg ha^{-1} DM) in swards continuously stocked by sheep. (After Bircham and Hodgson, 1983.)

Different defoliation systems

tiller growth) remains much the same. Increasingly inefficient harvesting of this growth by the animals at greater mean sward height leads, however, to the loss of a greater percentage of the herbage grown to senescence. This results in a reduction in the rate of net herbage accumulation (Fig. 3.10), which in this instance equals herbage consumption by the animal since there is no change in the height or weight of standing green herbage. Between the limits set, on the one hand, by lack of sufficient ground cover to intercept light and, on the other hand, by inefficient harvesting differences in maintained sward height or weight do not greatly influence what the animal consumes. The range of sward heights or weights of standing green herbage over which this holds true appears to include levels from 700 to 1700 kg ha^{-1} (Arosteguy *et al.*, 1983; Bircham and Hodgson, 1983) and sward heights from 2 to 6 cm (Grant *et al.*, 1983a). Further adaptation to very close grazing may be possible where tiller populations are built up gradually over a period of some years (Parsons *et al.*, 1983a).

Efficiency of harvesting by the animal seems to vary also from experiment to experiment. The mean proportion of herbage grown which is lost to senescence while harvesting efficiency remains high has varied from 20 to 25% (Grant *et al.*, 1983a; mature wethers) to 50% (Bircham and Hodgson, 1983; Arosteguy *et al.*, 1983; ewes and lambs), 50–60% (Parsons *et al.*, 1983; yearling wethers) and 60% (Arosteguy *et al.*, 1983; yearling cattle). The subject is one which needs further investigation.

3.4.3 Transfer experiments

The effect of changes in herbage mass on subsequent rates of net herbage accumulation is of particular interest, since steady-state swards are rarely met in practice. Bircham (1981) and Bircham and Hodgson (1984) investigated rates of growth and senescence in high- and low-mass swards which were respectively either grazed down to low mass (HL) or allowed to grow until a higher mass had accumulated (LH). The period of adjustment allowed before measurements commenced was 3.5 weeks. Net herbage accumulation was similar from both steady-state treatments and the LH treatment, but was greatly reduced in treatment HL in which there were fewer tillers and growth per tiller was at the reduced level characteristic of a low-mass sward. Later evidence (Grant *et al.*, 1983b) shows that the short, densely tillered LH swards may, in the shorter term, combine high rates of growth per tiller with high tiller populations before tillers die in response to the increase in herbage mass. Successful exploitation of this temporary advantage may underlie reports of increased production from variable cutting height systems (Smith, 1968; Ollerenshaw and Hodgson, 1977).

3.4.4 Long-term effects of different defoliation systems on stubble composition

The effects of different defoliation systems on the stubble can now be summarized. Where previous defoliation has been frequent and/or the sward

TABLE 3.4 *Effects of pretreatment (10 August–24 September) on stubble in a vegetative sward of S.24 perennial ryegrass*

	Cut once		Cut every 14 days		
	7.6 cm (3 inches)	1.3 cm (½ inches)	7.6 cm (3 inches)	1.3 cm (½ inches)	s.e.
High N					
Percentage leaf	12.2	12.3	22.6	35.4	3.70
Tillers m^{-2}	7298	6588	8051	9838	360.6
Percentage WSC	14.7	12.6	12.2	10.8	—
Dry wt g m^{-2}	109.1	98.2	116.3	135.7	6.88
Low N					
Percentage leaf	27.0	33.4	35.6	42.0	3.70
Tillers m^{-2}	6039	6620	6028	6179	360.6
Percentage WSC	21.2	20.8	13.2	10.4	—
Dry wt g m^{-2}	114.5	127.1	120.1	93.3	6.88

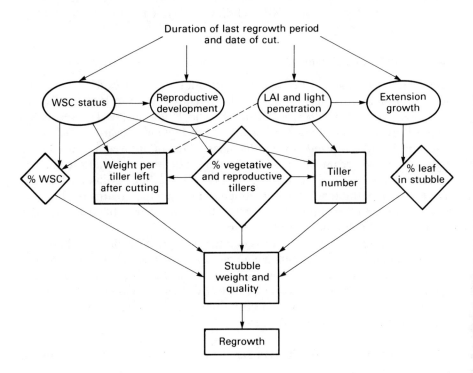

Figure 3.11 A conceptual model of factors affecting regrowth in relation to previous management.

Reproductive swards

was cut short, the stubble will tend to be relatively leafy and tiller numbers will be high. The growth habit of the plants may also be more prostrate. The percentage carbohydrate levels in stubble will be lower than in more leniently defoliated plants. If previous defoliations have been infrequent and/or at an increased cutting height, a comparable stubble cut back to the same height as in the first instance will contain less leaf and fewer tillers and may well be of a lower total dry weight. An example of these interactive effects is shown in Table 3.4, and similar trends can also be seen in data of King et al., (1979) and Grant et al. (1981a). A conceptual model is presented in Fig. 3.11. The ellipses in the top line show the aspects of growth in the last regrowth period which are influenced by its timing and duration: these in turn determine the weight and character of the stubble and its regrowth potential. Factors influencing stubble weight and its qualitative aspects are shown, respectively, in rectangular and diamond-shaped boxes in the second line.

3.5 THE REGROWTH OF REPRODUCTIVE SWARDS

The importance of the location of the apex in relation to defoliation has been recognized by various authors (Branson, 1953; Davies, 1956; Neiland and Curtis, 1956; Langer, 1959). During the annual cycle of growth the meristematic apices of some tillers become reproductive, and basal internodes begin to elongate, carrying the developing inflorescence upwards and exposing it to decapitation (Fig. 3.1). Once the apex has been removed the tiller dies, so regrowth in reproductive swards depends partly on the contribution made by the tiller bases of these reproductive tillers and partly on the number and developmental status of any accompanying undecapitated tillers. The timing of cuts during the reproductive phase is thus critical in relation to the quantity and character of the regrowth.

3.5.1 Stubble composition and regrowth

As decapitated tillers cannot continue growth, immediate regrowth depends on the tillers which have escaped decapitation and can continue to form new leaves and/or stem material. New tillers are, however, developed on the cut stubs of the decapitated tillers, which may accumulate large amounts of reserve carbohydrate (Jewiss and Powell, 1966; Awopetu, 1979). Ample evidence exists to show that when these stubs are cut short fewer new tillers are formed (Davies, 1977; Awopetu, 1979; Davies et al., 1981) and regrowth is reduced. In a mixed stubble of decapitated and undecapitated tillers regrowth can therefore be shown to be related to the carbon reserves in the stubs and to the weight of the remaining undecapitated tillers (Jewiss and Powell, 1966). The delay normally experienced before new tillers are formed from reproductive stubs means that mixed stubble is likely to regrow more slowly than a stubble composed solely of vegetative tillers. The R_w has been found to be lower in perennial ryegrass when the percentage of flowering tillers is greater (Davies *et al.*, 1981), but the greater weight of reproductive stubs may compensate for their reduced efficiency on a per tiller basis.

Observations have shown that in dry conditions many of the vegetative tillers in a reproductive sward of perennial ryegrass die before or as a consequence of defoliation (Davies, unpublished). This seems to occur to a lesser extent in vegetative swards, so that their regrowth is less drought-sensitive. New tillers in reproductive swards of Italian ryegrasses appear more slowly in dry than in moist conditions (Nyirenda, 1982), and tillering may cease altogether if water is withheld (Thomas, pers. comm) (see Chapter 6), so that this component of regrowth is temporarily suppressed, although not necessarily more so than the regrowth of existing tillers. The tiller buds at the base of the decapitated stubs may develop at a later date (Davies *et al.*, unpublished; Brereton, pers. comm.), but whether they can then make equally efficient use of the carbon reserves in the stubs is not certain.

The tillers produced from reproductive stubs frequently emerge from old buds exposed by the decay of their subtending sheaths or break through the old sheaths (extravaginal tillers). They are considerably smaller than the old vegetative tillers (Davies *et al.*, 1981). Regrowth may then consist of a dense mat of small tillers with exposed green sheaths which can, in turn, provide a highly efficient basis for the next regrowth cycle (Davies and Evans, 1982).

3.5.2 Regrowth in relation to time of defoliation

The importance of the timing of spring cuts in relation to stage of development was clearly shown by Davies (1956) in his comparison of regrowth in an early- and a late-flowering timothy. On 8 May, when the swards were first cut, the tiller apices in the early variety were above cutting level, so the stubble consisted almost exclusively of cut stem bases. Regrowth, which was first observed some 10 days later, consisted of small tillers developing from buds on the stem bases. The tiller apices of the late variety, which were at an earlier stage of development, were substantially undamaged by the cut and regrew immediately. Later studies (Davies, 1969) confirmed that cuts which remove all or most of the tiller apices result in an almost complete absence of photosynthetic surface for at least a week in timothy varieties, and illustrated an inverse relationship between yield at the first and second cuts (Fig. 3.12). After very late first cuts in June or July, however, Jewiss and Powell (1966) showed that regrowth in timothy increased relative to earlier cuts, in response to an increase in the carbohydrate content of the stem bases and to an increase in the number and weight of subsidiary tillers capable of further growth after defoliation.

Although only the actual reproductive tillers elevate their apices in perennial ryegrass (unlike timothy, in which the accompanying vegetative apices are also elevated at the time of flowering) the situation is very similar. Regrowth after a

Reproductive swards

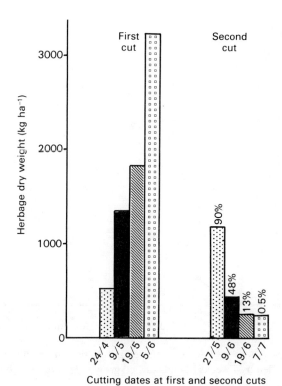

Figure 3.12 First and second cut herbage dry matter production in S.352 timothy. Percentages above second cut histograms show the proportion of tillers present on 10 April left undecapitated at the first cut and which developed in the second cut regrowth. (Based on data of Davies, 1969.)

cut on 19 May which removed 4% of the tiller apices was 4240 kg ha^{-1} dry wt. whereas a cut on 8 June, which removed 32% of the tiller apices (and probably destroyed virtually all of the season's reproductive tillers) produced only 1930 kg ha^{-1} dry wt. of regrowth (Davies, 1973a). Similar trends have been demonstrated by Binnie *et al.*, (1980). Behaeghe (1974) has linked changes in regrowth of reproductive swards of perennial ryegrass and cocksfoot with changes in mean apex height and percentage tiller mortality. The regrowth curves obtained by Behaeghe (Fig. 3.13) show clear evidence of a lag phase after mature reproductive swards were cut in early June. Comparable swards subjected to several earlier cuts were almost completely vegetative in character and regrew immediately. There was little difference in regrowth between these two swards at a later cut, and it seems possible that once swards have recovered from a cut during the main reproductive phase their further total growth may be fairly independent of their earlier cutting regime (Davies, 1969; Binnie *et al.*, 1980).

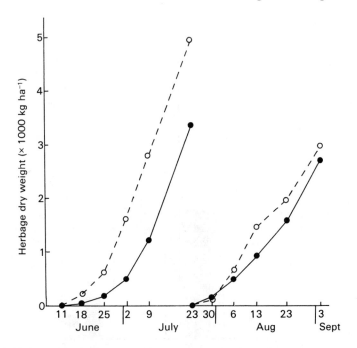

Figure 3.13 Increases in herbage dry matter production after cuts on 11 June and 23 July. ●, preceded by 7 weeks of undisturbed growth; ○, preceded by frequent cuts. (After Behaeghe, 1974.)

If reproductive tillers are not decapitated they may respond in various ways, depending on the number of undeveloped leaves remaining at the apex (Davies, 1973b). In extreme cases defoliation may remove virtually all of the leaf tissue, leaving the tiller to continue growth as a leafless reproductive stem. Sometimes, when the reproductive apex is very close to the defoliation height the tiller may die, but this propensity may vary in different varieties or species (Davies, 1973b).

3.5.3 Tiller production in reproductive swards

In spring, as reproductive development proceeds, tiller numbers cease to increase (Laude et al., 1968) and may even fall (Davies and Calder, 1969; Davies and Simons, 1979). Fewer new tillers are produced (Langer, 1958, 1959; Garwood, 1969), though production may be resumed after ear emergence (Jewiss and Powell, 1966; Clifford and Langer, 1975). The suppression of tiller production is related to the presence of the reproductive apex on the tiller (Laidlaw and Berrie, 1974) and the tendency of the tiller to retain carbon rather than to export it to subsidiary tillers (Ryle, 1970; Ong et al., 1978). The latter

Reproductive swards

authors also showed that increased tiller mortality in the reproductive sward was accounted for almost exclusively in terms of the death of smaller and younger tillers. The results of these influences is that, when reproductive development is allowed to proceed to ear emergence, tiller numbers are lower than in comparable material in either the pre- or post-reproductive stage (Davies *et al.*, 1981). Numbers at mean ear emergence in S.24 and S.23 perennial ryegrass average some 7500 tillers m^{-2} (nine experiments) and 10 500 tillers m^{-2} (seven experiments), both appreciably lower than the maxima shown in Fig. 3.7. In swards well supplied with water and nitrogen the loss of tillers by decapitation is quickly made up for by development of new tillers, though heavy fertilization of the reproductive crop can both increase deaths of non-flowering tillers after defoliation and reduce formation of new tillers (Dawson *et al.*, 1983). Each reproductive stub in perennial ryegrass seems to be capable of producing two to three new tillers (Davies *et al.*, 1981) and under favourable conditions the number of living tillers may double in 2 weeks. In this way the sward rapidly increases its tiller population towards the maximum characteristic for the time of year.

3.5.4 Head numbers

Responses to defoliation in the reproductive phase are related both to the duration of the phase and to the percentage of tillers which become reproductive. In general, there seems to be an upper limit to the number of seed heads which can develop in a well-managed sward. Although the chances of flowering are decreased by shading or in a dense stand (Ryle, 1961, 1965; Spiertz and Ellen, 1972; Darwinkel, 1978) and are increased by the practice of mechanically opening up the sward or 'gapping' (Lambert, 1963, 1964; Lewis, 1968) considerable variation in management may be possible before numbers are affected. Thus, nitrogen applications appear to be effective at or before apical differentiation, but not subsequently (Lewis, 1969; Nordestgaard, 1980) and high fertilizer applications may bring about no further increases in head numbers (Hebblethwaite *et al.*, 1980) and can even reduce numbers in timothy (Evans, 1954; Nordestgaard, 1980). Autumn and early spring defoliations appear to have little influence on the number of heads (Roberts, 1965), though they may influence the number of vegetative tillers and therefore the percentage of tillers which flower (Davies and Simons, 1979).

Comparisons of data from different experiments, harvest years and cutting treatments for S.24 perennial ryegrass confirm that the total number of tillers which become reproductive remains similar over a wide range of circumstances. The maximum of 3650 heads m^{-2} recorded by Hebblethwaite *et al.*, (1980) in a developing seed crop tallies well with mean ear emergence totals from three box experiments (Davies *et al.*, 1981; Davies, unpublished) and two field experiments (Davies, unpublished). In the field experiments the total numbers of reproductive decapitated apices were of a very similar order over the course of three harvest years and under cutting treatments ranging in frequency from a cut at mean ear emergence plus aftermath cuts to cuts every 3 weeks either at 2.5 or 10 cm. The only exception to this was the most severe treatment in the dry spring of 1975, and this may have been a consequence of increased tiller deaths in the main category of head-producing tillers. Very frequent close defoliation can, however, substantially reduce the total number of reproductive tillers in a way that cannot be accounted for by tiller deaths. Box-grown swards of S.23 perennial ryegrass cut every 10 days on average at heights of 3 and 2 cm contained only 13% and 8% of the total number of decapitated tillers observed when the swards were cut at mean ear emergence. Similarly low totals have been recorded in continuously grazed swards (Grant, pers. comm.).

The number of vegetative tillers associated with the reproductive tillers during the phase of reproductive development varies considerably with management. Although in perennial ryegrass the percentage of vegetative tillers at mean ear emergence normally lies between 67 and 50%, extremes of 73 and 40% have been recorded. Cutting in spring delays reproductive development (Davies, 1969), and later-formed tillers tend to be both lighter and less leafy (Davies, 1975).

3.5.5 Species and variety differences

Species and varieties may differ in relation to a number of factors which can influence regrowth in the reproductive phase. The most obvious of these factors are the proportions of tillers with vulnerable apices and whether non-reproductive tillers also elevate apices. Species such as timothy, which elevate all or nearly all tiller apices, would seem to include crested and beardless wheatgrasses (*Agropyron desertorum* and *A. inerme*) and Russian wild rye (*Elymus junceus*) (Hyder, 1972). Close grazing of crested wheatgrass in Oregon terminates stem dominance and initiates growth of a second crop of tillers, which are mostly vegetative (Hyder and Sneva, 1963). Rhodes grass (*Chloris gayana*) develops successive cycles of tillers with elevated apices and regrowth is delayed increasingly as the cutting interval is extended because fewer tillers capable of regrowth remain after cutting (Dovrat *et al.*, 1980). Ratios of reproductive and non-reproductive tillers vary widely in prairie grasses (Neiland and Curtis, 1956) and even in varieties within a single species (cocksfoot) when assessed 10 days after mean ear emergence (Davies, 1976).

Differences in the speed of reproductive development lead to differences in the time during which apices are vulnerable to decapitation (Branson, 1953). Differences in the length of this period (and in its timing) have been demonstrated for *Hyparrhenia hirta*, *Themeda triandra* and *Tristachyia hispida* by Booysen *et al.* (1963), and for timothy, cocksfoot and perennial ryegrass by Davies (1969). The leafiness of the regrowth after removing a reproductive crop is influenced by variations in the number of undeveloped leaves en-

The influence of the animal

sheathing the apex just before it is elevated to cutting height, and the fate of tillers which only just escape decapitation, some of which may die (Davies, 1973b).

Heading in later regrowth is related to flowering requirements (Cooper and Saeed, 1949), which show considerable species and variety variability (Cooper, 1960) (see Chapter 2). The flowering period may extend well into the later parts of the summer if there is little or no winter requirement, and a very high proportion of tillers may ultimately flower. Tillers of Italian ryegrass formed in early August may flower in September, but the likelihood that a tiller will flower falls off steeply with tillers formed from June onwards. For tillers formed before this the chances are in excess of 94% (Davies, unpublished).

3.6 THE INFLUENCE OF THE ANIMAL ON SWARD REGROWTH

The regrowth of grass swards is influenced by the grazing animal and by grazing management in a number of ways. In a grazing system herbage may be removed over several days followed by a rest period (rotational grazing) or grazing may be continuous; but individual tillers in all systems are in effect grazed rotationally with greater or lesser frequency (see Section 3.6.3 and Chapter 4). There are three main areas in which the presence of the grazing animal further modifies sward growth: these are the effects of excreta, the effects of treading and the effects of selective removal of different sward components. The effects of dung and urine on sward growth have been considered both separately and in combination, and a good deal of work has been done on problems of rejection caused by cattle dung. Rejection problems caused by sheep dung, which is usually deposited as pellets scattered over a relatively wide area or in night camps, have received less attention, but current work (Forbes and Hodgson, 1985) suggests that sheep are less sensitive to the presence of their own dung than cattle are. Treading has mainly been studied in rather artificial circumstances, and information on the relative influence of sheep and cattle is scanty. Cattle exert about 1½ times as much pressure per unit area of hoof as sheep (Frame, 1976) and, although there is no evidence that they cause more damage on drier soils (Edmond, 1970), observation suggests that this may not be equally true for wet conditions.

Selection by the animal may operate in three ways: they may reject certain areas within the sward in favour of others, alter the depth of the grazing horizon or select within that horizon (Hodgson, 1982). Cattle appear to be less selective than sheep, probably because the tearing action which they employ in grazing does not allow them the same freedom of choice as sheep, which have mobile lips and a biting action. Correspondingly, sheep may be less liable to uproot whole tillers when grazing pressure is heavy (see Section 3.4.3). The combination of uprooting and greater hoof pressure may account for the higher losses of tillers from cattle-grazed swards observed by Arosteguy et al. (1983).

Information on these influences has been reviewed by Watkin and Clements (1978). Marsh and Campling (1970) and Hodgson (1982) have reviewed the effects of dung and selective grazing, respectively, and Brown and Evans (1973) have reported on the treading experiments of the late D. B. Edmond.

3.6.1 Excreta

Excreta return nutrients to the sward, those of importance being P, Ca, Mg, N and K in dung and N and K in urine (Watkin and Clements, 1978). Dung is, however, relatively unimportant as a source of N (Marsh and Campling, 1970). Because the breakdown of dung is relatively slow, yield responses from dung are observed over a longer period than those from urine (Norman and Green, 1958), though they may be observable in favourable conditions after as little as 15 days (MacDiarmid and Watkin, 1971). The natural return of dung and urine has a favourable effect on herbage production in swards dominated by grasses (Sears et al., 1948; Herriott and Wells, 1963; Curll and Wilkins, 1983), but this may be negated in swards with a high clover content because this content is reduced (Sears and Thurston, 1953; Watkin, 1957; Wheeler, 1958). In the longer term the beneficial effects of dung alone on production are of the same order as those of urine alone in a grass/clover sward, but the combination is more effective than either of the two separately (Sears et al., 1948).

The size of a urine patch has been reported to average $0.03 \, m^2$ for wether sheep and $0.42 \, m^2$ for cows, the influence extending outwards to about 4–5 cm beyond the wetted edge (Doak, 1952). A cattle dung pat is about 25 cm in diameter (Bastiman and van Dijk, 1975) and its influence extends gradually outwards, being most evident in the first 15 cm from the edge of the pat (MacDiarmid and Watkin, 1971).

Unless the grass is scorched by the urine or rank growth is allowed to accumulate on the affected area, it will be rejected only during the current grazing period in the rotation. After as little as 4 weeks it has no adverse effect on intake and may even be preferred by the animal (Norman and Green, 1958). Dung has more-lasting effects. Experiments show that 75% of grass tillers underneath a cattle dung pat die within 15 days (MacDiarmid and Watkin, 1971), leaving bare patches which may persist 6–12 months before recolonization (Weeda, 1967). The time needed for a dung pat to disappear varies with the conditions, and is normally faster in warm, wet weather. In the Pennines, Bastiman and van Dijk (1975) observed that dung pats disappeared in 70 days in spring and 36 days in summer, but in the drier climate of New Zealand Weeda (1967) found greater variability. Sloppy dung could disappear in as little as 1 month of wet weather, but dung which had dried out and formed a hard crust might take as long as 16 months.

Dung is a major cause of herbage rejection by cattle. For example, Marten and Donker (1966) reported that 93% of rejected areas had dung present on them. Rejection seems to be a function of smell (Watkin and Clements, 1978)

The influence of the animal

and the area rejected is about five to six times the area actually covered by the dung (Norman and Green, 1958; Marsh and Campling, 1970; Bastiman and van Dijk, 1975).

The period of rejection for a cattle dung pat and its surrounds (0.29 m² on average) may be as long as a year (Norman and Green, 1958), but at increased stocking rates both the period during which herbage is rejected and the area affected by rejection are reduced (Greenhalgh *et al.*, 1966; Bastiman and van Dijk, 1975). Because of the influence of weather and stocking rate, the total area of sward which is rejected varies considerably (Table 3.5), but it can be as high as 35–40% (Tayler and Rudman, 1966; Bastiman and van Dijk, 1975). At the other extreme rejection by dairy cows is scarcely observable under continuous grazing if the sward height is not allowed to exceed 5 cm (Baker, pers. comm.). Sheep will graze areas rejected by cattle (Arosteguy, 1982; Nolan, 1983) and vice versa (Forbes and Hodgson, 1985). This is one reason why mixed grazing can lead to appreciable increases in animal production (Nolan, 1983).

TABLE 3.5 *Total area of herbage rejected because of dung fouling during the season (from Marsh and Campling, 1970)*

Reference	Area refused (%)	Type of grazing animal
Ivins (1954)	35 and 19	Dairy cows
Tayler and Large (1955)	38 and 45	Beef cattle
van der Kley and van der Ploeg (1955)	25	Dairy cows
Arnold and Holmes (1958)	15 and 26	Dairy cows
MacLusky (1960)	15	Dry and milking cows
Tayler and Rudman (1966)	10 and 12	Beef cattle
Greenhalgh and Reid (1969)	23 and 24	Dairy cows
MAFF (1969)	25	Beef cattle

3.6.2 Treading

Treading by sheep or cattle can greatly reduce herbage production in some circumstances, leaves being crushed, bruised, covered in mud or buried. Treading damage is maximized at high fertility levels or when the sward is wet, particularly if it has been puddled (Brown and Evans, 1973). Yield reductions are not, however, always observed and swards may recover after considerable damage has been inflicted (Campbell, 1966). Decreases in both dry matter production and tiller numbers may be quite small at realistic stocking rates (Curll and Wilkins, 1983), though at the very high stocking rate of 50 ewes ha^{-1} they were reduced, respectively, by 10 and 14% compared with untrodden controls. Species vary in their reaction to treading: perennial ryegrass is very resistant, while timothy and cocksfoot are less tolerant (Edmond, 1964).

Obviously, the disadvantages of treading may be outweighed in practice by the beneficial effects of animal return. Treading may also have effects which are beneficial. In cut swards new leaves and new tillers tend to emerge through the old sheaths so that green material is, in general, borne at or above the level of the cutting horizon (Section 3.4.1). Hoof action in grazed swards tends to split old leaf sheaths (Grant, pers. comm.) and allows new tillers to appear at the sward base. It is doubtful whether swards could adapt as successfully as they do to constant low-level grazing in the absence of such a mechanism, which operates both to permit the development of green leaf material below grazing height and to stimulate tiller production.

3.6.3 Selective grazing, grazing frequency and severity

Animals graze selectively, removing leaf rather than stem and younger, more accessible leaves rather than older leaves (Hodgson, 1966; Watkin and Clements, 1978; de Lucia Silva, 1974). Taller tillers are preferred to shorter ones (Hodgson and Ollerenshaw, 1969; Morris, 1969) and dead matter is rejected (Thomson, 1978), so the fractions consumed tend to be the most digestible ones (Hamilton et al., 1973). To some extent at least, this simply reflects selectivity in respect of the depth of the grazing horizon; Barthram (1981), for example, having shown that sheep will reduce herbage intake rather than graze into horizons of the sward containing leaf sheath and dead leaf. Selection for green leaf within a grazing horizon has, however, been demonstrated by Chacon and Stobbs (1976) in a *Setaria sphacelata* sward grazed by cattle.

The frequency of defoliation of individual tillers seems, in general, to be well related to the metabolic live weight of stock per unit area (Wade and Baker, 1979). In swards carrying animal liveweights of approximately 2000 and 3000 kg ha^{-1} tillers were, on average, defoliated at frequencies of 11 and 6 days (Hodgson, 1966; Greenwood and Arnold, 1968; Hodgson and Ollerenshaw, 1969; McIvor and Watkin, 1973; Wade, unpublished). A similar relationship was established by Curll and Wilkins (1983), who observed defoliation frequencies of 4.8, 5.6, 7.6 and 9.3 days at stocking rates of 55, 45, 35 and 25 sheep ha^{-1}. Less information exists on effects of grazing severity (percentage of green leaf length removed by grazing between observations). Data from two continuous-grazing trials gave values ranging from 67% at 91 sheep ha^{-1} to 13% at 29 sheep ha^{-1} (Hodgson and Ollerenshaw, 1969) and from 62% at 55 sheep ha^{-1} to 48% at 25 sheep ha^{-1} (Curll and Wilkins, 1983). The corresponding frequencies were, respectively, 5.5 and 9.5 days and 4.8 and 9.3 days. Higher percentages can be removed in a rotational system (de Lucia Silva, 1974).

Swards grazed at low stocking rates exhibit higher variability than swards grazed at high stocking rates (Hodgson and Ollerenshaw, 1969; Arosteguy, 1982).

3.7 CONCLUSIONS

In this chapter factors which affect and control regrowth of grasses after cutting or grazing have been considered and, in particular, the influence of changes in sward structure on regrowth. Some indication has been given of how regrowth depends on the initial resources of leaf area and, when demands cannot be met by rates of current photosynthesis, on carbon compounds present in the stubble. An account has been given of how and when the accumulation of dry matter in new leaves is offset by losses of older leaves present either in the stubble or accumulating during regrowth. Evidence has been presented showing how tiller populations are related to the level of herbage mass which is allowed to accumulate, and particular attention has been drawn to the effect of defoliation at different stages of reproductive development on tiller survival and production (and hence on regrowth).

Much of the evidence which has been presented relates to perennial ryegrass, information on other temperate grasses being much less complete. Less still is known about tropical grasses, though it is clearly evident that their response in the reproductive phase of development can vary widely. No reference has been made to mixed swards. Many of the experimental data discussed are derived from cutting experiments, though the necessary collateral information about grazed swards is rapidly being accumulated. It is, however, abundantly clear that the changing morphology of the plant strongly influences its response to defoliation in both the shorter and the longer term. It is also evident that the grass sward is capable of a high degree of adaptation to management (see also Chapter 4).

The framework which has been assembled does not, yet, permit accurate quantification of defoliation responses or prediction of all likely interactions between seasonal growth pattern and defoliation strategies. Many questions remain to be answered. Even so, sufficient progress has been made to enable agronomists to understand more readily why management systems fail or succeed and plant breeders to suggest how selection may be applied most effectively.

REFERENCES

Alberda, Th. (1965) The problems of relating greenhouse and controlled environmental work to sward conditions. *J. Br. Grassland Soc.*, **20**, 41–8.

Alberda, Th. (1966) The influence of reserve substances on dry-matter production after defoliation. *Proc. 10th Int. Grassland Congr.*, Helsinki, pp. 140–7.

Alberda, Th. (1970) The influence of carbohydrate reserves on respiration, photosynthesis and dry matter production of intact plants. *Proc. 11th Int. Grassland Congr.*, Queensland, pp. 517–22.

Alberda, T. and Sibma, L. (1968) Dry matter production and light interception of crop surfaces. 3. Actual herbage production in different years as compared with potential values. *J. Br. Grassland Soc.*, **23**, 206–15.

Alberda, Th. and Sibma, L. (1982) The influence of length of growing period, nitrogen fertilization and shading on tillering of perennial ryegrass (*Lolium perenne* L.). *Neth. J. Agric. Sci.*, **30**, 127–35.

Appadurai, R. R. and Holmes, W. (1964) The influence of stage of growth, closeness of defoliation and moisture on the growth and productivity of a ryegrass white clover sward. *J. Agric. Sci., Camb.*, **62**, 327–332.

Arnold G. W. and Holmes, W. (1958) Studies in grazing management. 7. The influence of strip grazing v. controlled free grazing on milk yields, milk composition and pasture utilisation. *J. Agric. Sci., Camb.*, **51**, 248–56.

Arosteguy, J. C. (1982) The dynamics of herbage production and utilisation in swards grazed by cattle and sheep. Ph.D. Thesis, University of Edinburgh.

Arosteguy, J. C., Hodgson, J., Souter, W. G. and Barthram, G. T. (1983) Herbage growth and utilisation on swards grazed by cattle and sheep, in *Efficient Grassland Farming* (ed. A. J. Corrall), Occasional Symposium, No. 14, British Grassland Society, Hurley, pp. 155–8.

Awopetu, J. A. (1979) Patterns of tiller formation, regrowth and changes in percentage water-soluble carbohydrate during the reproductive phase of different grass types. M.Sc. Thesis, University College of Wales, Aberystwyth.

Barthram, G. T. (1981) Sward structure and the depth of the grazed horizon. *Grass, Forage Sci.*, **36**, 130–1.

Bastiman, B. and van Dijk, J. P. F. (1975) Muck breakdown and pasture rejection in an intensive paddock system for dairy cows. *Exp. Husb.*, **28**, 7–17.

Begg, J. E. and Wright, M. J. (1962) Growth and development of leaves from intercalary meristems in *Phalaris arundinacea* L. *Nature, Lond.*, **194**, 1097–8.

Begg, J. E. and Wright, M. J. (1964) Relative effectiveness of top and basal leaves for the growth of vegetative shoots of reed canarygrass (*Phalaris arundinacea* L.). *Crop Sci.*, **4**, 607–9.

Behaeghe, T. J. (1974) Experiments on the seasonal variations of grass growth. *Proc. XIIth Int. Grassland Cong.*, Moscow, Vol. 1, pp. 268–81.

Binnie, R. C., Chestnutt, D. M. B. and Murdoch, J. C. (1980) The effect of time of defoliation and height of defoliation on the productivity of perennial ryegrass swards. *Grass, Forage Sci.*, **35**, 267–73.

Bircham, J. S. (1981) The effects of a change in herbage mass on herbage growth, senescence and net production rates in a continuously stocked mixed species sward. in *Plant Physiology and Herbage Production* (ed. C. E. Wright), Occasional Symposium, No. 13, British Grassland Society, Hurley, pp. 85–7.

Bircham, J. S. and Hodgson, J. (1983) The influence of sward condition on rates of herbage growth and senescence in mixed swards under continuous stocking management. *Grass, Forage Sci.*, **38**, 323–31.

Bircham, J. S. and Hodgson, J. (1984) The effects of change in herbage mass on rates of herbage growth and senescence in mixed swards. *Grass, Forage Sci.*, **39**, 111–5.

Booysen, P. de V. and Nelson C. J. (1975) Leaf area and carbohydrate reserves in regrowth of tall fescue. *Crop Sci.*, **15**, 262–6.

Booysen, P. de V., Tainton, N. M. and Scott, J. D. (1963) Shoot-apex development in grasses and its importance in grassland management. *Herbage Abstr.*, **33**, 209–13.

Boswell, C. C. and Crawford, A. J. M. (1979) Changes in the perennial ryegrass component of grazed pastures. *Proc. N.Z. Grassland Assoc., 1978*, **40**, 125–35.

References

Branson, F. A. (1953) Two new factors affecting resistance of grasses to grazing. *J. Range Mgmt*, **6**, 165–71.
Brougham, R. W. (1956) Effects of intensity of defoliation on regrowth of pasture. *Austral. J. Agric. Res.*, **7**, 377–87.
Brougham, R. W. (1959) The effects of frequency and intensity of grazing on the productivity of a pasture of short-rotation ryegrass and red and white clover. *N.Z. J. Agric. Res.*, **2**, 1232–48.
Brougham, R. W. (1960) The effects of frequent hard grazings at different times of the year on the productivity and species yields of a grass-clover pasture. *N.Z. J. Agric. Res.*, **3**, 125–36.
Brown, K. R. and Evans, P. S. (1973) Animal treading. A review of the work of the late D. B. Edmond. *N.Z. J. Exp. Agric.*, **1**, 217–26.
Brown, R. H., Cooper, R. B. and Blaser, R. E. (1966) Effects of leaf age on efficiency. *Crop Sci.*, **6**, 206–9.
Bryant, H. T. and Blaser, R. E. (1961) Yields and stands of orchardgrass compared under clipping and grazing intensities. *Agron. J.*, **53**, 9–11.
Campbell, A. G. (1966) Effects of treading by dairy cows on pasture production and botanical structure, on a Te Kowhai soil. *N.Z. J. Agric. Res.*, **9**, 1009–24.
Campbell, A. G. (1969) Grazing interval, stocking rate and pasture production. *N.Z. J. Agric. Res.*, **12**, 67–74.
Chacon, E. and Stobbs, T. H. (1976) Influence of progressive defoliation of a grass sward on the eating behaviour of cattle. *Austral. J. Agric. Res.*, **27**, 709–27.
Clifford, P. E. and Langer, R. H. M. (1975) Pattern and control of distribution of ^{14}C assimilates in reproductive plants of *Lolium multiflorum* Lam. var. Westerwoldicum. *Ann. Bot.*, **39**, 403–11.
Collett, B., Lewis, J. and Parsons, A. J. (1981) The regrowth of a grazed sward, in *Plant Physiology and Herbage Production*, (ed. C. E. Wright), Occasional Symposium, No. 13, British Grassland Society, Hurley, pp. 207–8.
Cooper, J. P. (1960) The use of controlled life-cycles in the forage grasses and legumes. *Herbage Abstr.*, **30**, 71–9.
Cooper, J. P. and Saeed, S. W. (1949) Studies on growth and development in *Lolium*. 1. Relation of the annual habit to head production under various systems of cutting. *J. Ecol.*, **37**, 233–59.
Crider, F. J. (1955) Root-growth stoppage resulting from defoliation of grass. *USDA Tech. Bull. No. 1102*, 3–23.
Curll, M. L. and Wilkins, R. J. (1983) The comparative effects of defoliation, treading and excreta on a *Lolium perenne–Trifolium repens* pasture grazed by sheep. *J. Agric. Sci., Camb.*, **100**, 451–60.
Darwinkel, A. (1978) Patterns of tillering and grain production of winter wheat at a wide range of plant densities. *Neth. J. Agric. Sci.*, **26**, 383–98.
Davidson, J. L. (1969) Growth of grazed plants. *Proc. II Austral. Grassland Congr.*, pp. 125–37.
Davidson, J. L. and Milthorpe, F. L. (1966a) Leaf growth in *Dactylis glomerata* following defoliation. *Ann. Bot.*, **30**, 173–84.
Davidson, J. L. and Milthorpe, F. L. (1966b) The effect of defoliation on the carbon balance in *Dactylis glomerata*. *Ann. Bot.*, **30**, 185–98.
Davies, A. G. (1960) Conditions influencing primary growth and regrowth of perennial ryegrass. *Rep. Welsh Pl. Breeding Stn for 1959*, pp. 110–6.

Davies, A. (1965) Carbohydrate levels and regrowth in perennial ryegrass. *J. Agric. Sci. Camb.*, **65**, 213–21.

Davies, A. (1966) The regrowth of swards of S.24 perennial ryegrass subjected to different pretreatments. *J. Agric. Sci., Camb.*, **67**, 139–44.

Davies, A. (1971) Changes in growth rate and morphology of perennial ryegrass swards at high and low nitrogen levels. *J. Agric. Sci., Camb.*, **77**, 123–34.

Davies, A. (1974) Leaf tissue remaining after cutting and regrowth in perennial ryegrass. *J. Agric. Sci., Camb.*, **82**, 165–72.

Davies, A. (1977) Structure of the grass sward, in *Proc. Int. Mtg Animal Production from Temperate Grassland* (ed. B. Gilsenan), An Foras Taluntais, Dublin, pp. 36–44.

Davies, A. (1981) Tissue turnover in the sward, in *Sward Measurement Handbook* (eds J. Hodgson, R. D. Baker, A. Davies, A. S. Laidlaw and J. D. Leaver), British Grassland Society, Hurley, pp. 179–208.

Davies, A. and Calder, D. M. (1969) Patterns of spring growth in swards of different grass varieties. *J. Br. Grassland Soc.*, **24**, 215–25.

Davies, A. and Evans, M. E. (1982) Tillering and regrowth of perennial ryegrass in the post flowering phase. *Rep. Welsh Pl. Breeding Stn. for 1981*, pp. 126–7.

Davies, A. and Simons, R. G. (1979) Effect of autumn cutting regime on developmental morphology and spring growth of perennial ryegrass. *J. Agric. Sci., Camb.*, **92**, 457–69.

Davies, A. and Thomas, H. (1983) Rates of leaf and tiller production in young spaced perennial ryegrass plants in relation to soil temperature and solar radiation. *Ann. Bot.*, **57**, 591–7.

Davies, A. G. and Troughton, A. (1971) Root volume and regrowth. *Rep. Welsh Pl. Breeding Stn for 1970*, p. 20.

Davies, A., Evans, M. E. and Sant, F. I. (1981) Changes in origin, type and rate of production of ryegrass tillers in the post-flowering period in relation to seasonal growth, in *Plant Physiology and Hebage Production* (ed. C. E. Wright), Occasional Symposium, No. 13, British Grassland Society, Hurley, pp. 73–6.

Davies, A., Evans, M. E. and Exley, J. K. (1983) Regrowth of perennial ryegrass as affected by simulated leaf sheaths. *J. Agric. Sci., Camb.*, **101**, 131–7.

Davies, I. (1956) Preliminary observations on the differential effect of time of cutting in spring on the relative herbage yields of two strains of timothy. *J. Agric. Soc., Univ. Coll. Wales, Aberystwyth*, **37**, 1–6.

Davies, I. (1969) The influence of management on tiller development and herbage growth. *Welsh Pl. Breeding Stn Tech. Bull. No. 3*.

Davies, I. (1973a) Regrowth characteristics of an S.23 perennial ryegrass sward defoliated at early stages of reproductive development. *J. Agric. Sci. Camb.*, **80**, 1–10.

Davies, I. (1973b) Factors affecting leafiness in herbage grasses, in *Evaluation of Breeding Material in Herbage Crops* (eds M. A. do Valle Ribeiro and P. O'Donnell), *Rep. Eucarpia Fodder Crops Mtg*, Dublin, September 1972, pp. 55–73.

Davies, I. (1975) Italian ryegrass varieties for successive conservation cuts. *Rep. Welsh Pl. Breeding Stn for 1974*, pp. 43–5.

Davies, I. (1976) Vegetative tiller development in cocksfoot. *Rep. Welsh Pl. Breeding Stn for 1975*, p. 24.

References

Davies, I., Davies, A., Troughton, A. and Cooper, J. P. (1972) Regrowth in grasses. *Rep. Welsh Pl. Breeding Stn for 1971*, pp. 79–94.

Dawson, K. P., Jewiss, O. R. and Morrison, J. (1983) The effects of strategic application of nitrogen on the improvement of seasonal yield from perennial ryegrass swards and its interaction with available water, in *Efficient Grassland Farming* (ed. A. J. Corrall), Occasional Symposium, No. 14, British Grassland Society, Hurley, pp. 171–4.

Deregibus, V. A., Sanchez, R. A. and Casal, J. J. (1983) Effects of light quality on tiller production in *Lolium* spp. *Pl. Physiol.*, **72**, 900–2.

Doak, B. W. (1952) Some chemical changes in the nitrogenous constituents of urine when voided on pasture. *J. Agric. Sci., Camb.*, **42**, 162–71.

Dovrat, A., Dayan, E. and Keulen, H. van (1980) Regrowth potential of shoots and roots of Rhodes grass (*Chloris gayana* Kunth.) after defoliation. *Neth. J. Agric. Sci.*, **28**, 185–99.

Edmond, D. B. (1964) Some effects of sheep treading on the growth of 10 pasture species. *N.Z. J. Agric. Res.*, **7**, 1–16.

Edmond, D. B. (1970) Effects of treading on pasture, using different animals and soils. *Proc. XIth Int. Grassland Congr.*, Queensland, pp. 604–8.

Evans, T. A. (1954) The effect of nitrogen applications at different dates on the seed yield of pedigree grasses. *J. Br. Grassland Soc.*, **9**, 53–60.

Forbes, T. D. A. and Hodgson, J. (1985) Comparative studies of the influence of sward conditions on the ingestive behaviour of cows and sheep. *Grass, Forage Sci.*, **40**, 69–77.

Frame, J. (1976) A comparison of herbage production under cutting and grazing (including comments on deleterious factors such as treading), in *Pasture Utilization by the Grazing Animal*, (eds J. Hodgson and D. K. Jackson), Occasional Symposium, No. 8, British Grassland Society, Hurley, pp. 39–40.

Frame, J. and Hunt, I. V. (1971) The effects of cutting and grazing systems on herbage production from grass swards. *J. Br. Grassland Soc.*, **26**, 163–71.

Garwood, E. A. (1969) Seasonal tiller populations of grass and grass/clover swards with and without irrigation. *J. Br. Grassland Soc.*, **24**, 333–4.

Gifford, R. M. and Marshall, C. (1973) Photosynthesis and assimilate distribution in *Lolium multiflorum* Lam. following differential tiller defoliation. *Austral. J. Biol. Sci.*, **26**, 517–26.

Graber, L. F., Nelson, N. T., Luekel, W. A. and Albert, W. B. (1928) Organic food reserves in relation to the growth of alfalfa and other perennial herbaceous plants. *Exp. Stn Rec.*, **58**, 326–7.

Grant, S. A., Barthram, G. T. and Torvell, L. (1981a) Components of regrowth in grazed and cut *Lolium perenne* swards. *Grass, Forage Sci.*, **36**, 155–68.

Grant, S. A., King, J., Barthram, G. T. and Torvell, L. (1981b) Responses of tiller populations to variation in grazing management in continuously stocked swards as affected by time of year, in *Plant Physiology and Herbage Production* (ed. C. E. Wright), Occasional Symposium, No. 13, British Grassland Society, Hurley, pp. 81–4.

Grant, S. A., Barthram, G.T., Torvell, L., King, J. and Smith, H.K. (1983a) Sward management, lamina turnover and tiller population density in continuously stocked *Lolium perenne*–dominated swards. *Grass, Forage Sci.*, **38**, 333–44.

Grant, S. A., King, J., Hodgson, J. and Bircham, J. S. (1983b) The efficiency of herbage utilisation in grazed swards, in *Efficient Grassland Farming* (ed. A. J. Corrall), Occasional Symposium, No. 14, British Grassland Society, Hurley, p. 308.

Greenhalgh, J. F. D. and Reid, G. W. (1969) The effects of grazing intensity on herbage consumption and animal production. 3. Dairy cows grazed at two intensities on clean or contaminated pasture. *J. Agric. Sci., Camb.*, **72**, 223–8.

Greenhalgh, J. F. D., Reid, G. W., Aitken, J. N. and Florence, E. (1966) The effects of grazing intensity on herbage consumption and animal production. 1. Short-term effects in strip-grazed dairy cows. *J. Agric. Sci. Camb.*, **67**, 13–23.

Greenwood, E. A. N. and Arnold, G. W. (1968) The quantity and frequency of removal of herbage from an emerging annual grass by sheep in a set-stocked system of grazing. *J. Br. Grassland Soc.*, **23**, 144–8.

Hamilton, B. A., Hutchinson, K. J., Annis, P. C. and Donnelly, J. B. (1973) Relationships between the diet selected by grazing sheep and the herbage on offer. *Austral. J. Agric. Res.*, **24**, 271–7.

Harris, W. (1973) Ryegrass genotype-environment interactions in response to density, cutting height, and competition with white clover. *N.Z. J. Agric. Res.*, **16**, 207–22.

Hebblethwaite, P. D., Wright, D. and Noble, A. (1980) Some physiological aspects of seed yield in *Lolium perenne* L, in *Seed production* (ed. P. D. Hebblethwaite), Butterworths, London, pp. 71–90.

Herriott, J. B. D. and Wells, D. A. (1963) The grazing animal and sward productivity. *J. Agric. Sci. Camb.*, **61**, 89–99.

Hodgson, J. (1966) The frequency of defoliation of individual tillers in a set-stocked sward. *J. Br. Grassland Soc.*, **21**, 258–63.

Hodgson, J. (1982) Influence of sward characteristics on diet selection and herbage intake by the grazing animal, in *Nutritional Limits to Animal Production from Pastures* (ed. J. B. Hacker) *Proc. Int. Symp.*, Queensland, 1981, Commonwealth Agricultural Bureaux, Farnham Royal.

Hodgson, J. and Ollerenshaw, J. H. (1969) The frequency and severity of defoliation of individual tillers in set-stocked swards. *J. Br. Grassland Soc.*, **24**, 226–34.

Hodgson, J., Bircham, J. S., Grant, S. A. and King, J. (1981) The influence of cutting and grazing management on herbage growth and utilisation, in *Plant Physiology and Herbage Production* (ed. C. E. Wright) Occasional Symposium, No. 13, British Grassland Society, Hurley, pp. 51–62.

Hunt, L. A. (1965) Some implications of death and decay in pasture production. *J. Br. Grassland Soc.*, **20**, 27–31.

Huokuna, E. (1960) The effect of differential cutting on the growth of cocksfoot (*Dactylis glomerata*). *Proc. VIII Int. Grassland Congr.*, Reading, pp. 429–33.

Hyder, D. N. (1972) Defoliation in relation to vegetative growth, in *The Biology and Utilisation of Grasses* (eds V. B. Youngner and C. M. McKell), Academic Press, New York, pp. 304–17.

Hyder, D. N. and Sneva, F. A. (1963) Morphological and physiological factors affecting the grazing management of crested wheatgrass. *Crop Sci.*, **3**, 267–71.

Ivins, J. D. (1954) Grazing without waste. *Agriculture, Lond.*, **60**, 561–3.

Jackson, D. K. (1974) Some aspects of production and persistency in relation to height of defoliation of *Lolium perenne* (var. S.23). *Proc. XIIth Int. Grassland Congr.*, Moscow, Vol. III, pp. 202–14.

Jackson, D. K. (1976) The influence of patterns of defoliation on sward morphology, in

References

Pasture Utilization by the Grazing Animal (eds J. Hodgson and D. K. Jackson), Occasional Symposium, No. 8, British Grassland Society, Hurley, pp. 51–60.

Jacques, W. A. and Edmond, D. B. (1952) Root development in some common New Zealand pasture plants. V. The effect of defoliation and root pruning on cocksfoot (*Dactylis glomerata*) and perennial ryegrass (*Lolium perenne*). *N.Z. J. Sci. Technol., Section A,* **3**, 231–48.

Ješko, T. and Troughton, A. (1978) Root tips removal in relation to net photosynthesis and growth in *Lolium perenne* L. *Biologia, Bratislava,* **33**, 65–71.

Jewiss, O. R. and Powell, C. E. (1966) The growth of S.48 timothy swards after cutting in relation to carbohydrate reserves and leaf area. *A. Rep. Grassland Res. Inst. for 1965*, pp. 67–72.

Jewiss, O. R., Robson, M. J., Woledge, J, Powell, C. E., Deriaz, R. E. and Williams, R. D. (1969) Effect of defoliation on single plants. *A. Rep. Grassland Res. Inst. for 1968*, pp. 53–5.

Jones, E. L. (1983) The production and persistency of different grass species cut at different heights. *Grass, Forage Sci.,* **38**, 79–87.

Jones, M. B. (1981) A comparison of sward development under cutting and continuous grazing management. in *Plant Physiology and Herbage Production* (ed. C. E. Wright), Occasional Symposium, No. 13, British Grassland Society, Hurley, pp. 63–7.

Jones, M. B., Collett, B. and Brown, S. (1982) Sward growth under cutting and continuous stocking managements: sward canopy structure, tiller density and leaf turnover. *Grass, Forage Sci.,* **37**, 67–73.

Kays, S. and Harper, J. L. (1974) The regulation of plant and tiller density in a grass sward. *J. Ecol.,* **62**, 97–105.

King, J., Lamb, W. I. C. and McGregor, M. T. (1979) Regrowth of ryegrass swards subject to different cutting regimes and stocking densities. *Grass, Forage Sci.,* **34**, 107–18.

Kley, F. K. van der and Ploeg, H. van der (1955) Grazing behaviour in relation to feed intake by rotationally grazed monozygotic dairy cattle twins in the Netherlands. *Landbouwk. Tijdschr., Wageningen,* **67**, 609–27.

Laidlaw, A. S. and Berrie, A. M. M. (1974) The influence of expanding leaves and the reproductive stem apex on apical dominance in *Lolium multiflorum*. *Ann. Appl. Biol.,* **78**, 75–82.

Lambert, D. A. (1963) The influence of density and nitrogen in seed production stands of S37 cocksfoot (*Dactylis glomerata* L.). *J. Agric. Sci., Camb.,* **61**, 361–73.

Lambert, D. A. (1964) The influence of density and nitrogen in seed production stands of S48 timothy (*Phleum pratense* L.) and S215 meadow fescue (*Festuca pratensis* L.). *J. Agric. Sci., Camb.,* **63**, 35–42.

Langer, R. H. M. (1958) A study of growth in swards of timothy and meadow fescue. I. Uninterrupted growth. *J. Agric. Sci., Camb.,* **51**, 347–52.

Langer, R. H. M. (1959) A study of growth in swards of timothy and meadow fescue. II. The effects of cutting treatments. *J. Agric. Sci., Camb.,* **52**, 273–81.

Laude, H. M., Riveros, G., Murphy, A. H. and Fox, R. E. (1968) Tillering at the reproductive stage in Hardinggrass. *J. Range Mgmt,* **21**, 148–51.

Lawrence, T. and Ashford, R. (1969) Effect of stage and height of cutting on the dry matter yield and persistence of intermediate wheatgrass, bromegrass and reed canarygrass. *Can. J. Pl. Sci.,* **49**, 321–32.

Lewis, J. (1968) Fertile tiller production and seed yield in meadow fescue (*Festuca pratensis* L.). 2. Drill spacing and date of nitrogen manuring. *J. Br. Grassland Soc.*, **23**, 240–6.

Lewis, J. (1969) Fertile tiller production and seed yield in meadow fescue (*Festuca pratensis* L.). 3. Date of spring defoliation and nitrogen application. *J. Br. Grassland Soc.*, **24**, 50–8.

Lucia Silva, G. R. de (1974) A study of variation in the defoliation and regrowth of individual tillers in swards of *Lolium perenne* L. grazed by sheep. Ph.D. Thesis, University of Reading.

MAFF (1969) *Rep. No. 5, Rosemaund Exp. Husbandry Farm 1969*, pp. 28–30, Ministry of Agriculture, Fisheries and Food National Agricultural Advisory Service.

Marsh, R. and Campling, R. C. (1970) Fouling of pastures by dung. *Herbage Abstr.*, **40**, 123–30.

Marshall, C. and Sagar, G. R. (1965) The influence of defoliation on the distribution of assimilates in *Lolium multiflorum* Lam. *Ann. Bot.*, **29**, 365–70.

Marshall, C. and Sagar, G. R. (1968) The distribution of assimilates in *Lolium multiflorum* Lam. following differential defoliation. *Ann. Bot.*, **32**, 715–9.

Marten, G. C. and Donker, J. D. (1966) Animal excrement as a factor influencing acceptability of grazed forage. *Proc. Xth Int. Grassland Congr.*, Helsinki, pp. 359–63.

May, L. H. (1960) The utilisation of carbohydrate reserves in pasture plants after defoliation. *Herbage Abstr.*, **30**, 239–45.

McCarty, E. and Price, R. (1942) Growth and carbohydrate content of important mountain forage plants in central Utah as affected by clipping and grazing. *USDA Tech. Bull. No. 818*.

McDiarmid, B. N. and Watkin, B. R. (1971) The cattle dung patch. 1. Effect of dung patches on yield and botanical composition of surrounding and underlying pasture. *J. Br. Grassland Soc.*, **26**, 239–45.

McIlvanie, S. K. (1942) Carbohydrate and nitrogen trends in bluebunch wheatgrass, *Agropyron spicatum*, with special reference to grazing influences. *Pl. Physiol.*, **17**, 540–57.

McIvor, P. J. and Watkin, B. R. (1973) The pattern of defoliation of cocksfoot by grazing sheep. *Proc. N.Z. Grassland Assoc.*, **34**, 225–35.

MacLusky, D. S. (1960) Some estimates of the areas of pasture fouled by the excreta of dairy cows. *J. Br. Grassland Soc.*, **15**, 181–8.

Mitchell, K. J. and Coles, S. T. J. (1955) Effects of defoliation and shading on short-rotation ryegrass. *N.Z. J. Sci. Technol., Section A*, **36**, 586–604.

Mittra, M. K. and Wright, M. J. (1966) Effects of blade removal and nitrogen level on growth and nitrate content of shoots of reed canarygrass. *Agron. J.*, **58**, 193–5.

Morris, R. M. (1969) The pattern of grazing in 'continuously' grazed swards. *J. Br. Grassland Soc.*, **24**, 65–70.

Neiland, B. M. and Curtis, J. T. (1956) Differential responses to clipping of six prairie grasses in Wisconsin. *Ecology*, **37**, 355–65.

Nolan, T. (1983) Meat production. in *Efficient Grassland Farming* (ed. A. J. Corrall), Occasional Symposium, No. 14, British Grassland Society, Hurley, pp. 77–83.

Nordestgaard, A. (1980) The effects of quantity of nitrogen, date of application and the influence of autumn treatment on the seed yield of grasses. in *Seed Production* (ed. P. D. Hebblethwaite), Butterworths, London, pp. 137–49.

References

Norman, M. J. T. and Green, J. O. (1958) The local influence of cattle dung and urine upon the yield and botanical composition of permanent pasture. *J. Br. Grassland Soc.*, **13**, 39–45.

Nyirenda, H. E. (1982) Factors influencing regrowth and persistency in Italian ryegrass populations. M.Sc. Thesis, University College of Wales, Aberystwyth.

Ollerenshaw, J. H. and Hodgson, D. R. (1977) The effects of constant and varying heights of cut on the yield of Italian ryegrass (*Lolium multiflorum* Lam.) and perennial ryegrass (*Lolium perenne* L.). *J. Agric. Sci., Camb.*, **89**, 425–35.

Ong, C. K. (1978) The physiology of tiller death in grasses. 1. The influence of tiller age, size and position. *J. Br. Grassland Soc.*, **33**, 197–203.

Ong, C. K., Marshall, C. and Sagar, G. R. (1978) The phsiology of tiller death in grasses. 2. Causes of tiller death in a grass sward. *J. Br. Grassland Soc.*, **33**, 205–11.

Oswalt, D. L., Bertrand, A. R. and Teel, M. R. (1959) Influence of nitrogen fertilization and clipping on grass roots. *Soil Sci. Soc. Am., Proc.*, **23**, 228–30.

Parsons, A. J., Leafe, E. L., Collett, B. and Stiles, W. (1983a) The physiology of grass production under grazing. I. Characteristics of leaf and canopy photosynthesis of continuously-grazed swards. *J. Appl. Ecol.*, **20**, 117–26.

Parsons, A. J., Leafe, E. L., Collett, B., Penning, P. D. and Lewis, J. (1983b) The physiology of grass production under grazing. II. Photosynthesis, crop growth and animal intake of continuously-grazed swards. *J. Appl. Ecol.*, **20**, 127–39.

Plancquaert, P. (1967) Etude sur l'exploitation des graminées fourragères. *Herbage Abstr.*, **37**, 90.

Reid, D. (1959) Studies on the cutting management of grass–clover swards. I. The effect of varying the closeness of cutting on the yields from an established grass–clover sward. *J. Agric. Sci., Camb.*, **53**, 299–312.

Reid, D. (1962) Studies on the cutting management of grass–clover swards. III. The effects of prolonged close and lax cutting on herbage yields and quality. *J. Agric. Sci., Camb.*, **59**, 359–68.

Reid, D. and MacLusky, D. S. (1960) Studies on the cutting management of grass–clover swards. II. The effects of close cutting with either a gang mower or a reciprocating-knife mower on the yields from an established grass–clover sward. *J. Agric. Sci., Camb.*, **54**, 158–65.

Roberts, H. M. (1965) The effect of defoliation on the seed producing capacity of bred strains of grasses. III. Varieties of perennial ryegrass, cocksfoot, meadow fescue and timothy. *J. Br. Grassland Soc.*, **20**, 283–9.

Ryle, G. J. A. (1961) Effects of light intensity on reproduction in S.48 timothy (*Phleum pratense* L.). *Nature, Lond.*, **191**, 196–7.

Ryle, G. J. A. (1965) Physiological aspects of seed yield in grasses, in *The Growth of Cereals and Grasses* (eds F. L. Milthorpe and J. D. Ivins), *Proc. 12th Easter School Agric. Sci., Univ. Nottingham*, pp. 106–18.

Ryle, G. J. A. (1970) Partition of assimilates in an annual and a perennial grass. *J. Appl. Ecol.*, **7**, 217–27.

Ryle, G. J. A. and Powell, C. E. (1975) Defoliation and regrowth in the graminaceous plant: the role of current assimilate. *Ann. Bot.*, **39**, 297–310.

Ryle, G. J. A. and Powell, C. E. (1976) Effect of rate of photosynthesis on the pattern of assimilate distribution in the graminaceous plant. *J. Exp. Bot.*, **27**, 189–99.

Sears, P. D. and Thurston, W. G. (1953) Effect of sheep droppings on yield, botanical composition, and chemical composition of pasture. III. Results of field trial at

Lincoln, Canterbury for the years 1944–1947. *N.Z. J. Sci. Technol, Section A,* **34,** 445–59.

Sears, P. D., Goodall, V. C. and Newbold, R. P. (1948) The effect of sheep droppings on yield, botanical composition, and chemical composition of pasture. II. Results for the years 1942–1944 and final summary of the trial. *N.Z. J. Sci. Technol, Section A,* **30,** 231–50.

Simons, R. G., Davies, A. and Troughton, A. (1972) The effect of the height of defoliation on two clones of perennial ryegrass. *J. Agric. Sci., Camb.,* **79,** 509–14.

Simons, R. G., Davies, A. and Troughton, A. (1974) The effect of cutting height and mulching on aerial tillering in two contrasting genotypes of perennial ryegrass. *J. Agric. Sci., Camb.,* **83,** 267–73.

Smith, A. (1968) Sward growth in relation to pattern of defoliation. *J. Br. Grassland Soc.,* **23,** 294–8.

Spiertz, J. H. J. and Ellen, J. (1972) The effect of light intensity on some morphological and physiological aspects of the crop perennial ryegrass (*Lolium perenne* L. var. 'Cropper') and its effect on seed production. *Neth. J. Agric. Sci.,* **20,** 232–46.

Sprague, V. G. and Sullivan, J. T. (1950) Reserve carbohydrates in orchard grass clipped periodically. *Pl. Physiol.,* **25,** 92–102.

Sullivan, J. T. and Sprague, V. G. (1943) Composition of the roots and stubble of perennial ryegrass following partial defoliation. *Pl. Physiol.,* **18,** 656–70.

Tainton, N. M. (1974) Effects of different grazing rotations on pasture production. *J. Br. Grassland Soc.,* **29,** 191–202.

Tallowin, J. R. B. (1981) An interpretation of tiller number changes under grazing, in *Plant Physiology and Herbage Production* (ed. C. E. Wright), Occasional Symposium, No. 13, British Grassland Society, Hurley, pp. 77–80.

Tayler, J. C. and Large, R. V. (1955) The comparative output of two seeds mixtures. *J. Br. Grassland Soc.,* **10,** 341–51.

Tayler, J. C. and Rudman, J. E. (1966) The distribution of herbage at different heights in 'grazed' and 'dung patch' areas of a sward under two methods of grazing management. *J. Agric. Sci., Camb.,* **66,** 29–39.

Thomas, H. and Davies, A. (1978) Effect of shading on the regrowth of *Lolium perenne* swards in the field. *Ann. Bot.,* **42,** 705–15.

Thomson, N. A. (1978) Factors affecting animal production: intake and utilisation by ewes grazing grass/clover and lucerne pastures. *Proc. N.Z. Grassland Assoc., 1977,* **39,** 86–97.

Thorne, G. N. (1959) Photosynthesis of lamina and sheath of barley leaves. *Ann. Bot.,* **23,** 365–70.

Vine, D. A. (1977) Development of a pasture model for grazing studies. Ph.D. Thesis, University of Edinburgh.

Wade, M. H. (1979) The effect of severity of grazing by dairy cows given three levels of herbage allowance on the dynamics of leaves and tillers in swards of *Lolium perenne*. M.Phil. Thesis, University of Reading.

Wade, M. H. and Baker, T. D. (1979) Defoliation in set-stocked grazing systems. *Grass, Forage Sci.,* **34,** 73–4.

Ward, C. Y. and Blaser, R. E. (1961) Carbohydrate food reserves and leaf area in regrowth of orchardgrass. *Crop Sci.,* **1,** 366–70.

Watkin, B. R. (1957) The effect of dung and urine and its interactions with applied

References

nitrogen, phosphorus and potassium on the chemical composition of pasture. *J. Br. Grassland Soc.*, **12**, 264–77.

Watkin, B. R. and Clements, R. J. (1978) The effects of grazing animals on pastures, in *Plant Relations in Pastures* (ed. J. R. Wilson), CSIRO, East Melbourne, pp. 273–89.

Weeda, W. C. (1965) The effects of frequency and severity of grazing by cattle on the yield of irrigated pasture. *N.Z. J. Agric. Res.*, **8**, 1060–9.

Weeda, W. C. (1967) The effect of cattle dung patches on pasture growth, botanical composition and pasture utilisation. *N.Z. J. Agric. Res.*, **10**, 150–9.

Weinmann, H. (1948). Underground development and reserves of grasses. *J. Br. Grassland Soc.*, **3**, 115–40.

Wheeler, J. L. (1958) The effect of sheep excreta and nitrogenous fertilizer on the botanical composition and production of a ley. *J. Br. Grassland Soc.*, **13**, 196–202.

White, J. and Harper, J. L. (1970) Correlated changes in plant size and number in plant populations. *J. Ecol.*, **58**, 467–85.

Wilson, D. and Robson, M. J. (1981) Varietal improvement by selection for reduced dark respiration rate in perennial ryegrass, in *Plant Physiology and Herbage Production* (ed. C. E. Wright), Occasional Symposium, No. 13, British Grassland Society, Hurley, pp. 209–21.

Yoda, K., Kira, T., Ogawa, H. and Hozumi, H. (1963) Self-thinning in overcrowded pure stands under cultivated and natural conditions. *J. Inst. Polytechnics, Osaka City Univ., Series D*, **14**, 107–29.

Younger, V. B. (1972) Physiology of defoliation and regrowth. in *The Biology and Utilization of Grasses* (eds V. B. Youngner and C. M. McKell), Academic Press, New York, pp. 293–303.

CHAPTER 4
The effects of season and management on the growth of grass swards

A. J. Parsons

4.1 INTRODUCTION

Despite its significance as an inexpensive feedstuff for ruminant production, in many temperate regions the grass crop is poorly utilized. Whether the aim is to increase total yield or to improve the economic efficiency of production, attempts to make better use of grass demand a detailed understanding of the limitations to production imposed by the environment, the seasonal development and the defoliation management of this crop.

A knowledge of the seasonal pattern of production of grass is of considerable practical value in agriculture. Such a knowledge enables farmers to match the feeding requirements of livestock to seasonal variations in the growth of grass (Owen, 1976; Holmes, 1980a). It enables them to decide when they can increase stocking rates without fear of loss of production or, alternatively, when any excess of production can be harvested and conserved (Doyle *et al.*, 1983). Moreover, an appreciation of the factors which limit grass growth from season to season is a first essential step in attempts to overcome these limitations by breeding or by management. In this way we can aim to extend the length of the growing season or otherwise modify the seasonal pattern of production to advantage (Borill, 1961; Cooper, 1960, 1964; Cooper and Breese, 1971; Robson, 1980).

In a sward provided with ample water and nutrients, seasonal changes in light energy and temperature play a major role in the production of grass, in that they ultimately define the potential of the environment to support grass growth (Cooper, 1970). Studies in controlled environments have proved invaluable in analysing the effects of individual environmental parameters and the fundamental principles of the response of plants to light and temperature (see Chapter 2). In the field, however, the grass crop experiences continuously changing combinations of several major environmental parameters. The first section of this chapter considers, briefly, the characteristics of the environment and of seasonal changes in the environment to which the temperate grass sward is exposed and to which it responds. The effect of seasonal changes in the environment on the growth of the grass sward is complicated by the progression from vegetative to reproductive development. This development is associated with marked seasonal differences in the way plants respond to their environment, which have important consequences on dry matter production. The second section of this chapter considers the seasonal changes in the physiology of the grass crop, which underlie seasonal differences in production.

Management also has a considerable effect on the amount of grass grown and, of more significance to the farmer, has a profound effect on the degree to which the amount grown is harvested. The final sections of this chapter describe the principal effects of management on the growth and utilization of the grass crop and the physiological basis for optimizing yield.

4.2 SEASONAL CHANGES IN THE ENVIRONMENT

4.2.1 The light environment

The seasonal pattern of radiant energy incident at a latitude of 52°N is shown in Fig. 4.1. Approximately 50% of the total radiation is photosynthetically active radiation (PAR) at wavelengths between 400 and 700 nm (McCree, 1972;

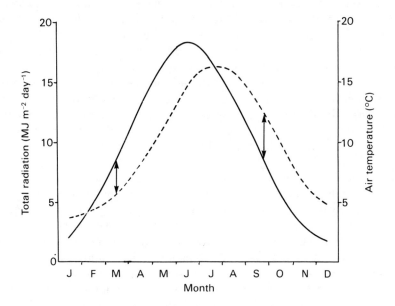

Figure 4.1 Seasonal patterns of radiation receipt (——) and air temperature (---); means of 10 years at Hurley England (52°N). Arrows highlight the difference in air temperature at the same radiation receipt in spring and autumn.

Seasonal changes in the environment

Monteith, 1973). The seasonal pattern of radiation is almost symmetrical about the midsummer solstice (22 June in the northern hemisphere) so that the total amount of radiation received during a period close to the equinox in spring (21 March) differs little from the amount received during an equivalent period close to the equinox in autumn (23 September). At 52°N daily totals of light energy receipt range from $1\,MJ\,m^{-2}\,day^{-1}$ (PAR) in midwinter to around $9\,MJ\,m^{-2}\,day^{-1}$ (PAR) in midsummer, as a result of changes in both light intensity and daylength (Woodward and Sheehy, 1983). In midwinter mean light intensities of just $35\,W\,m^{-2}$ (PAR) are experienced during a daylength of 8 h, whereas in midsummer mean intensities of $155\,W\,m^{-2}$ (PAR) are experienced during the course of a 16 h day. A description of seasonal changes in light intensity and daylength at other latitudes is presented by List (1966), Boucher (1975) and Charles-Edwards (1982).

Of even greater significance to grassland production than the amount of light *available* is the amount of light actually *intercepted* by the sward canopy. Light interception depends not only on the leaf area index of the crop, but on the optical properties of the leaves and the structure of the sward canopy – the way those leaves are disposed to the light (see Chapter 2). As we shall see, both the LAI and the structure of the sward canopy vary considerably during the course of the year as a result of management and state of development.

4.2.2 The temperature environment

Seasonal changes in the temperature of the sward environment result partly from global patterns of movement of large masses of air, but are determined locally by seasonal changes in the radiation receipt of the surfaces of the crop and soil. During periods of high radiation receipt in summer, these surfaces make a small net gain in energy whereas, in winter, a similar small net loss in energy is made (Milthorpe and Moorby, 1974; Woodward and Sheehy, 1983). Because of the thermal inertia of the large mass of soil, the seasonal changes in the *mean* temperature of the surface and the air close to that surface lag behind the seasonal changes in radiation receipt. At Hurley (52°N) the maximum light energy receipt occurs in June, whereas the period of maximum air temperature is some 4 weeks later, in July (Fig. 4.1). Similarly, minimum air temperatures occur not in December, but in January. As a result of this lag, temperatures in spring may be considerably lower than at a time of equal light energy receipt in autumn.

The lag between the seasonal pattern of radiation receipt and that of temperature increases with increasing depth beneath the soil surface, while the amplitude of seasonal variations in temperature is reduced (Fig. 4.2). As a result maximum temperatures below the soil surface occur later in the season than maximum temperatures at the surface (Russell, 1973).

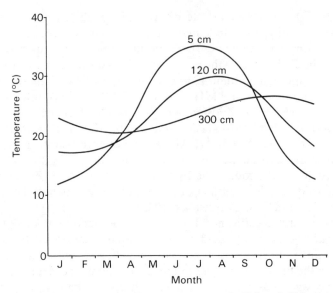

Figure 4.2 Seasonal patterns of soil temperature at 5, 120 and 300 cm beneath the soil surface. (After Russell, 1973.)

Variations in radiation receipt give rise not only to marked seasonal changes in temperature, but also lead to marked fluctuations in temperature during the course of a single day. During the daylight hours the temperature of radiation-intercepting surfaces rise more rapidly than that of the bulk of air and soil. At night the situation is reversed. This leads to complex profiles of temperature through the crop and soil (Fig. 4.3). The maximum variation in temperature occurs on bright sunny days in the region where most light is intercepted (Peacock, 1975a). Hence, the nature of the temperature profile depends not only on the diurnal patterns of light energy, but to a large extent on the LAI and structure of the crop – both factors which may be greatly modified by management. In a short grass sward (Fig. 4.3b) the maximum diurnal variation in temperature occurs close to ground level and there may be a substantial diurnal variation (approximately 10 °C) in the temperature of the upper layers of soil. In a tall grass sward (Fig. 4.3a) the maximum variation occurs at a greater height in the canopy, and the temperature of the soil surface varies little. The diurnal range of temperature within the crop canopy is striking. Maximum temperature variations of 35 °C in summer and 15 °C in winter are not uncommon, even in north temperate latitudes (Peacock, 1975a).

Clearly, at any one time different parts of a grass plant may be at widely different temperatures. This complicates attempts to identify the site of temperature perception for a given process (Monteith and Elston, 1971). Studies of the effect of contrasting temperatures between the leaves and roots have been carried out in perennial ryegrass by Peacock (1975b) and in maize by

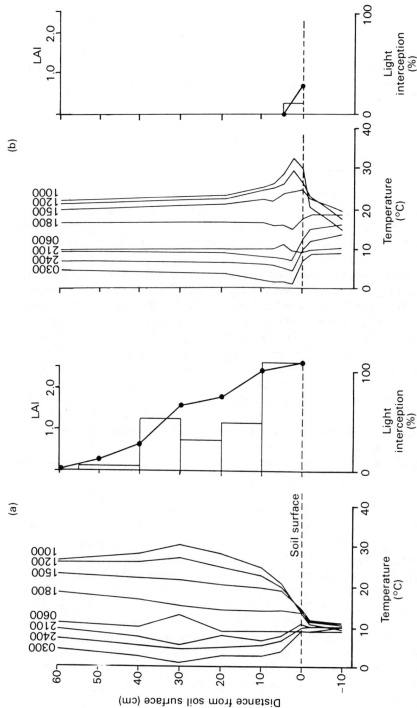

Figure 4.3 A series of soil and air temperature profiles observed at intervals during the course of a sunny day in (a) a tall and (b) a short perennial ryegrass crop. Vertical distributions of leaf area index and light interception are also shown. (After Peacock, 1975a.)

Kleinendorst and Brouwer (1970) and by Watts (1971). It is now well established that the rate of extension of grass leaves is most closely correlated with the mean temperature in the region of the shoot apex (Peacock, 1975b). However, the effect of a wide diurnal fluctuation in temperature in the light intercepting regions of the canopy on photosynthesis and the supply of assimilates for leaf growth is less certain.

4.3 SEASONAL PRODUCTION

4.3.1 The seasonal pattern of production

Many grasses of importance to temperate agriculture show a marked seasonal pattern of production, even when provided with ample nutrients and irrigated to maintain the soil close to field capacity. However, both the seasonal pattern of production and the total amount harvested are greatly modified by the defoliation management applied. 'Classic' accounts of the seasonal pattern of grass production are based on systems of infrequent severe cutting. This management reveals some important characteristics of the pattern of accumulation of dry matter, particularly when the sward is cut only once at the end of a single period of growth in late spring, and then cut on three or four occasions thereafter (Fig. 4.4a). Under these circumstances the sward continues to accumulate dry matter for longer, and achieves a greater ceiling yield during spring than during the successive regrowths in summer and autumn (Leafe et al., 1974). A series of regrowths from a number of swards cut at different times reveals a similar picture (Brougham, 1959a; Alberda and Sibma, 1968) (Fig. 4.4b). However, there are seasonal differences in the nature of what is harvested. Up to 80% of the total dry matter harvested in spring is made up of the elongated stems of reproductive tillers (Wilman et al., 1976; Osbourn, 1980), and these alone may contribute some 40% to a total annual production of 12–15 t ha^{-1} (Leafe et al., 1974). Because the digestibility of stem is less than that of grass leaves, the greater quantity harvested in spring is offset to some extent by a decrease in the digestibility of the herbage removed. However, recognized management practices can ensure that harvests of high digestibility are sustained throughout the year without substantial loss of total dry matter production (Minson et al., 1960; Green et al., 1971; Clarke et al., 1974).

A number of alternative cutting regimes have been proposed as a means to compare the seasonal pattern of production of a wide range of grass species and varieties (Anslow and Green, 1967; Morris and Thomas, 1972; Corrall and Fenlon, 1978). These regimes are based on infrequent cutting, but in each case a series of plots is harvested out of phase so that a mean growth rate is calculated for each week of the year. This approach emphasizes the high rates of dry matter accumulation in spring, the relative reduction or depression of growth rates in midsummer and the slower rates of growth in late summer and autumn (Fig. 4.4c).

Seasonal production

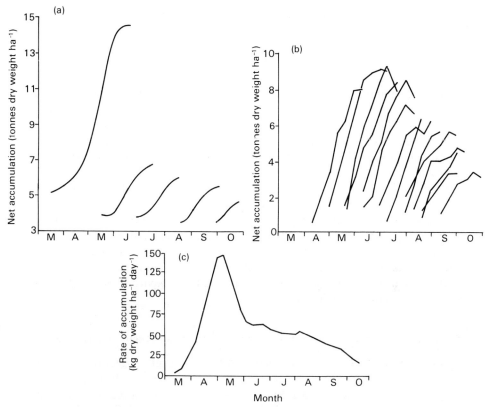

Figure 4.4 Seasonal patterns of production of perennial ryegrass under infrequent cutting managements. (a) Net accumulation above ground under a four-cut system. (Leafe et al., 1974.) (b) A succession of regrowths each following nitrogen application. (Alberda and Sibma, 1968.) (c) Mean rates of net accumulation above a cutting height of 4 cm from an overlapping series of cuts. (Corrall and Fenlon, 1978.)

Although frequently referred to as a 'classic' account of the seasonal pattern of production of grass, the pronounced variation in the rate of grass growth shown in Fig. 4.4c for an infrequently cut crop growing in a lowland environment is less evident under contrasting managements, particularly under continuous grazing. The marked enhancement of net accumulation rates in spring results, as we shall see, from the physiological advantages of having a large proportion of reproductive tillers in the crop. The depression of growth in midsummer results, in an irrigated crop, because regrowth at this time depends largely on the formation of a new generation of vegetative tillers – the apices of the existing reproductive tillers having been removed by cutting in early summer. Systems of management in which the grass crop is harvested early in spring, and in which reproductive development is suppressed, result in a more even distribution of production over the season (Korte et al., 1982; Johnson and Parsons, 1985b). Differences between the seasonal pattern of production of the same species have also been observed when grown at different altitudes (Morris and Thomas, 1972). With increasing altitude, the spring peak of production became less evident, particularly in the improved ryegrass variety S.23.

Observations of the seasonal pattern of accumulation of dry matter in cutting trials have a number of other important limitations. First, it is not possible from measurements of the net accumulation of herbage alone, to determine whether variations in 'growth' rate are due to changes in the amount of tissue produced (that is the rate of gross tissue production), or to changes in the degree to which the tissue produced is harvested. Second, the observations can take little account of the production and death of tissues below the cutting height, or in the roots. Thus, as indicated by Davies and Calder (1969), changes in the net accumulation of herbage can give the misleading impression that virtually no grass growth takes place during early spring at air temperatures below 5 °C.

A fundamental understanding of seasonal differences in production must be based on a knowledge of those processes involved in both the uptake and loss of matter in the whole grass crop – both shoots and roots – and of seasonal changes in the way in which the grass sward responds to its environment under contrasting defoliation managements.

4.3.2 Seasonal changes in the response of grass plants to their environment

In temperate zones the marked seasonal variations in light and temperature are a major cause of variations in the rate of growth of grass. In late autumn, overwinter and in early spring, for example, low temperatures and low light-energy receipt are both major limitations to production and define the length of the growing season (Cooper and Breese, 1971). However, these environmental parameters individually provide no immediately obvious basis for the asymmetric seasonal pattern of production seen in infrequently cut swards. Indeed, the greater rate of accumulation of dry matter in spring takes place even though the total amount of light energy received in spring is not consistently greater than in autumn, and despite the observation that mean temperatures in spring may be considerably lower than during a period of similar light energy receipt in autumn (Fig. 4.1). Thus, when compared at the same temperature, rates of accumulation of dry matter are greater in spring than in autumn (Anslow and Green, 1967). However, this change in the productivity of the grass crop coincides with the transition from vegetative to reproductive development and is associated with changes in several major physiological processes which alter the response of the grass crop to its environment.

(a) Seasonal changes in carbohydrate metabolism

Many temperate perennial grasses display a seasonal pattern of storage and utilization of carbohydrates. High molecular weight storage carbohydrates, typically fructosans (De Cugnac, 1931) begin to accumulate during late spring,

Seasonal production

towards the end of the period of reproductive development (Weinmann, 1952; Pollock and Jones, 1979). In reproductive tillers the accumulation of fructosans is greatest in the basal internodes (Smith, 1967). Following the senescence of reproductive tillers the accumulation of fructosans continues, in the new generation of vegetative tillers arising at the base of the reproductive stems, throughout summer and autumn into winter. There are marked differences between species and even between different selections within a variety (Breese, 1970) in the amount of fructosan that accumulates, and, to a lesser extent, in the site of accumulation. In general the accumulation is greater in the pseudostem (stubble) than in the leaf blade or in the root (Pollock and Jones, 1979; Baker and Garwood, 1961). Concentrations of up to 20% of dry weight are not uncommon in the stubble, for instance, of *Dactylis* (Baker and Garwood, 1961) and *Festuca, Phleum* and *Lolium* (Waite and Boyd, 1953; Pollock and Jones, 1979).

During early spring the strategy changes from one of storage to one of mobilization. As early as January in a number of grass species, there is a reduction in the level of high-molecular weight fructosans and a corresponding temporary accumulation of lower-molecular weight sugars, including sucrose (Pollock and Jones, 1979). It is uncertain what prompts this change in the metabolism of carbohydrates. The mobilization coincides with the very earliest stages of reproductive development – either vernalization or, in early 'flowering' varieties such as S.24 *L. perenne*, the initiation of reproductive development upon receipt of the critical daylength (Parsons and Robson, 1980). There is some evidence that a cold pretreatment, in the absence of subsequent reproductive development, may also lead to the mobilization of reserves (Behaeghe, 1978), but further research is necessary.

It has been suggested that the mobilization of reserves is evidence of a transition from a period of conservation to a period of growth (Behaeghe, 1978). However, the accumulation of low-molecular weight sugars may also play a role in the mechanism of cold-hardiness, and their consumption in growth may prejudice survival at low temperature. With few exceptions (Eagles and Othman, 1978a,b) varieties selected to show appreciable leaf growth in early spring survive less well at low temperatures (Cooper, 1964; Robson and Jewiss, 1968). In any case, the mobilization of reserves cannot contribute directly to an increase in the weight of the whole plant, only to a redistribution of weight within the plant.

(b) Seasonal changes in the response of leaf extension to temperature

Temperature is a major determinant of seasonal variation in both the rate of appearance and the rate of extension of leaves in the grass crop. In perennial ryegrass, the interval between the appearance of leaves decreases from 35 days in January to only 9 days in May (Table 4.1). The general effect of temperature on the rate of extension of grass leaves is also well established (see Chapter 2), but superimposed on this there is a change in the response of leaf extension to

TABLE 4.1 Seasonal changes in leaf appearance rates and the number of leaves per tiller in perennial ryegrass (after Davies, 1977)

	Jan.	Feb.	Mar.	Apr.	May	June	July	Aug.	Sept.	Oct.	Nov.	Dec.
Rate of leaf appearance (Days per leaf)	35.0	31.8	18.4	15.2	9.09	10.7	11.7	12.5	14.6	19.4	26.9	29.2
No. of live leaves per tiller	2.55	2.59	2.64	2.84	2.87	2.71	2.63	2.83	2.78	2.59	2.70	2.56

Seasonal production

temperature during the transition from vegetative to reproductive growth (Peacock, 1975c). Peacock observed that, when compared at the same temperature, leaves expanded faster in the spring than in the summer and autumn (Fig. 4.5). The difference in the rate of leaf extension was apparent over a wide range of temperatures, but was greatest when observed at about 15 °C. At this temperature leaves extended almost twice as quickly in spring as in autumn. Several studies have confirmed Peacock's observations (Thomas, 1977; Thomas and Norris, 1977), but the timing of the transition from the 'autumn' type to the 'spring' type response is less certain. It is apparent from Peacock's work that the transition takes place early in spring, when temperatures are low and leaf extension rates are slow, and that there is little significant difference between the autumn and spring responses. To study the timing of the transition, plants grown in and adapted to their natural environment have been transferred, in the short term, to a controlled environment at a standard temperature of 15 °C in order to demonstrate their potential for leaf extension (Parsons and Robson, 1980). In this study the transition from 'autumn' to 'spring' extension rates in *L. perenne* S.24 took place very early in spring (Fig. 4.6a). Rates of leaf extension at 15 °C in spring exceeded the autumn rates as early as February – before the formation of double ridges on

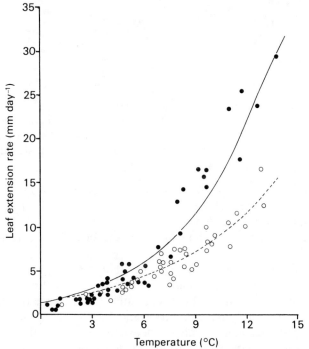

Figure 4.5 The different response of leaf extension to temperature in perennial ryegrass in Spring (●) and in autumn (○). Temperatures measured in the region of the stem apex. (Peacock, 1975c.)

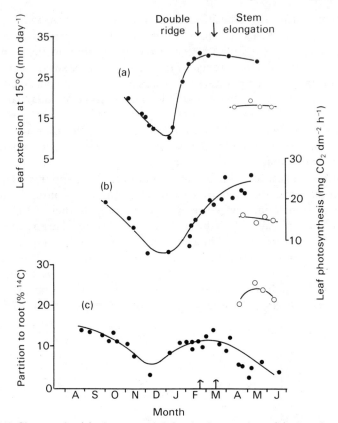

Figure 4.6 Changes in (a) the potential for leaf extension; (b) the photosynthetic potential of youngest fully expanded leaves and (c) the proportional partition of assimilates to root, during the transition from vegetative to reproductive growth in perennial ryegrass. Established swards (●); vegetative (unvernalized) swards (○). (Parsons and Robson, 1980, 1981a,b.)

the apex. The maximum potential for leaf extension was achieved in March – before the start of stem elongation – and was maintained throughout reproductive development. A high potential for leaf extension was not evident in the vegetative tillers that arose following the senescence of the reproductive tillers in May.

There is further circumstantial evidence that the change in the response of leaf extension to temperature is associated with the transition from vegetative to reproductive development. First, vegetative (unvernalized) swards growing in spring do not display an increased potential for leaf extension (Parsons and Robson, 1980). Second, variations between species in the timing of the transition from vegetative to reproductive development also correspond with differences in the timing of an increase in the potential for leaf extension (Peacock, 1976).

Seasonal production

It is tempting, but misleading, to consider the increase in the potential for leaf extension as synonymous with the greater rate of accumulation of dry matter in spring. Leaf extension is 'growth' in only one dimension. The role played by a change in the physiology of leaf extension in the seasonal pattern of dry matter production can only be understood in the context of changes in the carbon balance of the sward. Early in spring, low ambient temperatures limit the rate of extension of leaves, even though the potential for leaf extension may have begun to increase. Light energy receipt is also low and, at this time of low current assimilation, the extension of leaves may be provided for initially by the mobilization of reserve carbohydrates. Although it is a fascinating possibility, no causal relationship between the mobilization of reserves and the increased potential for leaf extension has yet been established.

Later in spring there is a marked increase in the supply of current assimilates for leaf growth. However, we have seen that temperatures in spring remain significantly lower than at the same light energy receipt in autumn – a factor which continues to limit the incorporation of these assimilates into leaf tissue. The increase in the response of leaf extension to temperature may play an important role in alleviating this limitation. As a result the sward maintains appreciable rates of leaf extension, despite the markedly lower temperature.

(c) Seasonal changes in the photosynthetic potential of leaves and canopy

Although the mobilization of reserve carbohydrate and the change in the response of leaf extension to temperature can be interpreted as a 'readiness' to grow in the reproductive crop in spring, the greater accumulation of dry matter at this time can only result from an increase in the uptake or retention of carbon.

The reproductive crop in spring has a far greater potential for photosynthesis than the vegetative crop in summer and autumn. When compared at the same high light intensity (400 W m^{-2} PAR), rates of 'gross' photosynthesis in a sward which intercepts virtually all incident light may exceed 10 g CO_2 m^{-2} h^{-1} in spring compared with only 6–7 g CO_2 m^{-2} h^{-1} in summer and autumn (Leafe et al., 1974; Leafe, 1978). One reason for the marked seasonal differences in the photosynthetic capacity of the canopy is a corresponding seasonal difference in the photosynthetic potential of individual leaves (see Chapter 2). In vegetative swards in summer and autumn the maximum rates of photosynthesis (P_m) of successive youngest fully expanded leaves decreases as the leaf area index of the sward is increased. By contrast, there is no such adverse relationship in swards which make the transition from vegetative to reproductive growth in spring. Indeed, the potential for photosynthesis of successive leaves increases throughout the spring (Fig. 4.6b), despite a considerable increase in LAI (Woledge and Leafe, 1976; Parsons and Robson, 1981a).

An extensive series of investigations by Woledge (1971, 1973, 1977, 1978, 1979) has demonstrated that seasonal changes in the photosynthetic potential of leaves can be explained largely in terms of how seasonal changes in the structure of the sward alter the extent to which young leaves suffer or avoid development in shade. However, the response to increasing LAI may be modified by seasonal changes in the environment. In late-summer and autumn, the detrimental effect of sward density in reducing the photosynthetic potential of individual leaves may be aggravated by the gradual reduction in light intensity at that time of year (Parsons and Robson, 1981a). During winter, low temperatures as well as low light intensity may influence the development of photosynthetic potential in leaves. However, in *Lolium perenne* the reponse to both light and temperature of leaves grown free from shade during winter differs little from that of leaves grown free from shade in summer and autumn (Woledge and Dennis, 1982).

The high photosynthetic potential of individual leaves in late spring is attributed to the elongation of stem internodes, which elevates growing points so that young leaves develop at a greater height in the canopy, relatively free from shade by older leaves (Woledge, 1977, 1978, 1979). However, the gradual increase in the photosynthetic potential of leaves which occurs before stem elongation in spring clearly cannot be attributed to the elevation of apices. One possible explanation is that increasing ambient temperatures, combined with a change in the response of leaf extension to temperature, lead to the production of leaves that extend more quickly than and are of greater final length than their predecessors. These leaves may penetrate the canopy of older, shorter leaves, formed in midwinter, and so complete their expansion and the development of their photosynthetic apparatus in an increasingly favourable light environment. Moreover, it is notable that leaves produced on reproductive tillers in early spring have a greater specific leaf area (the ratio of leaf area to leaf weight) than leaves on vegetative tillers in summer or on vegetative (unvernalized) plants growing in spring (Davies, 1971; Thomas and Norris, 1977; Behaeghe, 1978). Such economy in the investment of carbon in the expansion of photosynthetic surface may itself contribute to a greater acceleration of growth in spring.

Seasonal changes in the photosynthetic potential of individual leaves are not the only determinant of changes in the photosynthetic potential of the grass sward. There are also seasonal changes in the structure (or architecture) of the sward canopy. During a period of regrowth in a vegetative sward in summer and autumn, leaves become progressively more prostrate. The mean angle of leaves may decrease from 65 to 70° to less than 40° as the canopy increases in LAI (de Wit, 1965; Alberda, 1966; Davies, 1977). This tendency to a more prostrate arrangement of leaves increases in late season (Alberda, 1966; Sibma and Louwerse, 1977). The decline in leaf angle is most marked in the upper layers of the canopy where, at high LAI towards the end of a period of regrowth, leaf angle may decrease to less than 30° (Davies, 1977), and the canopy may become 'lodged'. It has also been suggested that as LAI increases during regrowth in summer and autumn, the prostrate layer that forms near the

Seasonal production

top of the canopy is made up of an increasing proportion of old leaves whose photosynthetic potential has declined with age (Leafe, 1972). Such an effect would be more pronounced in late autumn as successive new leaves become shorter than their predecessors. A combination of high LAI and a prostrate canopy leads to an inefficient use of light in canopy photosynthesis, particularly when the photosynthetic potential of those leaves which intercept the light is poor.

By contrast, during the course of the spring a considerable increase in LAI is associated with an erect canopy structure. This is partly the result of the maintenance of high leaf angle (de Wit, 1965; Alberda, 1966) and partly a result of stem elongation. Both these factors contribute to a more uniform distribution of light over the photosynthetic area of the canopy and so, at a time of high LAI, result in a more efficient use of light in canopy photosynthesis. Moreover, as young leaves expand near the top of the canopy during spring, the photosynthetic potential of the canopy continues to reflect the high photosynthetic potential of its youngest leaves (Parsons and Robson, 1981a).

The development of mathematical models of canopy photosynthesis (see Chapter 7), together with measurements of the photosynthesis of leaves and the canopy, has made it possible to estimate quantitatively the effect of leaf photosynthetic potential and sward structure on canopy photosynthesis. The models suggest that, in general, seasonal changes in the photosynthetic potential of those leaves which intercept the light are the major cause of seasonal changes in canopy photosynthesis, while observed changes in leaf angle have a relatively small effect (de Wit, 1965; Alberda and Sibma, 1968). In a recent study a change from a relatively prostrate to a more erect growth habit in a reproductive canopy during spring accounted for less than 10% of the observed increase in canopy photosynthesis (Parsons and Robson, 1981a). Measurements of canopy photosynthesis in vegetative swards 'lodged' artificially using a wire mesh (Leafe *et al.*, 1974; Sibma and Louwerse, 1977) suggest that a severe flattening of the upper layers of the sward may itself depress photosynthesis by 20–30%. The models confirm that such a marked deterioration in canopy structure could have a pronounced effect on canopy photosynthesis, should this occur naturally in late autumn.

(d) Seasonal changes in the carbon balance of the sward

The rate of net accumulation of dry matter in the field depends not only on the uptake of matter in gross photosynthesis and the corresponding rate of gross tissue production, but also on the simultaneous rates of loss of matter in respiration and tissue death. Consequently, an understanding of the carbon balance of the sward is needed to explain how the reproductive crop in spring can continue to accumulate dry matter for longer, and so achieve a greater ceiling yield, than the vegetative crop in summer and autumn.

The changes in the relationship between photosynthesis, respiration, gross tissue production and death which limit production in the simplified situation of a vegetative sward growing in a constant environment are described in Chapter 2, and are summarized in Fig. 4.7a. The rate of gross photosynthesis increases initially, as leaf area increases, up to the point where the crop intercepts virtually all incident light. Thereafter the rate of gross photosynthesis remains constant, or may even decrease as a result of a decline in the photosynthetic potential of young leaves. A substantial proportion of the gross photosynthetic uptake of carbon is lost from the crop as CO_2 during both 'synthetic' and 'maintenance' respiration, but the resulting pattern of gross tissue production broadly resembles that of gross photosynthesis. Because tissue clearly does not die the moment it is produced, there is a delay before an increase in the rate of gross tissue production gives rise to a corresponding increase in the rate of tissue death. So, while gross photosynthesis continues to increase, the current rate of death remains low relative to the current rate of gross tissue production and the sward continues to make a net gain in live weight. However, once there is little further increase in the rate of gross

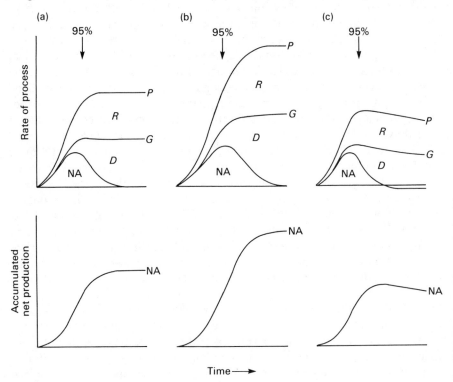

Figure 4.7 The relationship between the rates of gross photosynthesis (P), respiration (R), gross tissue production (G), net accumulation (NA) and death (D) in swards growing (a) in a constant environment; (b) in an increasing light environment as in spring and (c) in a decreasing light environment as in autumn. Lower diagrams show resulting patterns of net accumulation. Arrows show approximate time of 95% light interception.

Seasonal production

photosynthesis and gross tissue production, the rate of loss of matter by death 'catches up'. The rate of increase in live weight declines and a ceiling live weight is inevitably reached.

An extension of this simplified account can be used to explain the seasonal differences in the net accumulation of dry matter which are observed under infrequent cutting. Seasonal differences in the pattern of light energy receipt have a profound effect on the pattern of photosynthetic uptake of the sward in spring and autumn. As light energy increases during the spring, this leads to an increase in the photosynthetic uptake of carbon which may continue even after the sward has come to intercept virtually all incident light. This situation is aided by the sustained high photosynthetic potential of individual leaves. Because the photosynthetic uptake of the canopy continues to increase, the period when the gross rate of production of tissue exceeds the rate of loss of matter by death is extended. Thus, a high rate of net accumulation is sustained and the sward displays a longer 'linear' phase of growth and achieves a higher ceiling yield (Fig. 4.7b).

In autumn the photosynthetic potential of the canopy may already have begun to decrease by the time the sward has come to intercept virtually all incident light. The poor photosynthetic potential of the canopy, combined now with decreasing light energy receipt, leads to a reduction and eventual decline in canopy photosynthetic uptake. As a result the rate of death soon equals, and may even exceed, the rate of gross production of tissue, and the sward displays a shorter period of growth and a lower ceiling yield (Fig. 4.7c).

Mathematical models confirm that seasonal changes in the pattern of light energy receipt play a major role in the seasonal pattern of grass production (Parsons and Robson, 1982; Johnson and Thornley, 1983; Johnson and Parsons, 1985a).

During reproductive development there is a marked change in the rate of loss of tissues by death, which has a pronounced effect on the seasonal pattern of dry matter production. So great is the rate of turnover of tissue in a grass sward that a relatively small decrease in this rate greatly enhances the net accumulation of matter. The elongated stems of reproductive tillers do not exhibit the rapid turnover shown by the leaves of vegetative tillers. In the reproductive crop in spring, carbon invested in flag leaves, elongated stems and ears therefore remains longer on the plant than leaves in a vegetative sward would, and leads to an accumulation in the crop of dry matter that would otherwise have contributed to the rate of death. Moreover, the retention of this considerable weight of tissue within the crop does not incur a correspondingly large burden in maintenance respiration because the respiration of stems decreases markedly in late spring as their protein content declines (Jones et al., 1978).

(e) Seasonal changes in assimilate partition

Changes in the partition of assimilates between the major plant fractions – leaves, stems and roots – also contribute to seasonal differences in both the amount and the proportion of the crop that is available for harvest.

The seasonal pattern of root growth has been described for a number of grass species (Garwood, 1967a,b). Changes in the rate of production of new root tissue, combined with variations in the longevity of those tissues, result in large seasonal fluctuations in the active root mass (Garwood, 1967b; Behaeghe, 1978). By far the most active period of root production is the spring

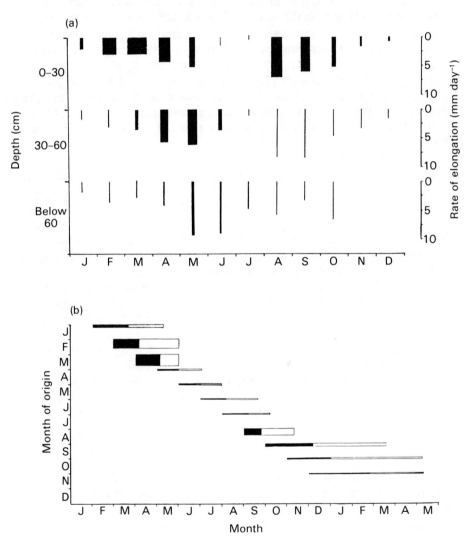

Figure 4.8 Seasonal patterns of root elongation and longevity. Diagrams show (a) the rate of elongation of roots in three soil horizons and (b) the period over which roots produced in 1 month continue to elongate (shaded) and the additional period over which those roots remain active (unshaded). In each case the thickness of the band is proportional to the number of roots involved. (Garwood, 1967b).

Seasonal production

(Fig. 4.8). In early spring there is an increase in the number of new adventitious roots produced in the horizon close to the soil surface. Later in spring the production of new adventitious roots declines, but considerable root growth continues in the form of an increase in the rate of elongation of existing root and in the number of roots elongating in deeper soil horizons. There is a marked decrease in the amount of root produced in midsummer, but the production of new adventitious roots resumes during the autumn. The roots produced in autumn play a major role in the perennation of the crop, forming the basis of the root system for the population of tillers that undergo reproductive development in the following spring. Differences between species in the seasonal pattern of root growth are small compared with this large seasonal fluctuation in root growth (Garwood, 1967a).

There is a broad similarity between the seasonal pattern of production of roots and of shoots. This demonstrates that much of the seasonal changes in the growth of both result from changes in the total uptake and retention of carbon by the crop, which determines the *amount* of assimlates to be partitioned, rather than from seasonal differences in the *proportion* of assimilates partitioned between the two fractions. Nevertheless, there has been much debate concerning the role played by seasonal changes in the proportional partition of assimilates between shoot and root in the seasonal pattern of production. It has been suggested that a large difference in assimilate partition between shoot and root in spring and in autumn might contribute directly to the considerable difference in yield (Alberda and Sibma, 1968; Leafe *et al.*, 1974). However, the most notable feature of recent measurements of assimilate partition in established canopies is that the proportion of assimilate partitioned to root is small (Parsons and Robson, 1981b; Colville and Marshall, 1981; Parsons *et al.*, 1983b), and is typically less than 15% (Fig. 4.6c). Root investment may be as great as 50% during the growth of seedlings but decreases (from 50 to 7%), even in vegetative plants, as the root system becomes established and as the crop increases in density (Ryle, 1970). This results because the active life of root tissue is greater than that of shoot (Garwood, 1967b; Troughton, 1981), so only a small proportion of current assimilation is necessary to sustain a comparatively large active root mass in an established crop. For these reasons, in a sward provided with ample water and nutrients, there appears to be only limited scope for a decline in the proportion of assimilates partitioned to root in late spring to lead directly to a greater production of shoot.

However, the indirect effects of even a small change in assimilate partition between shoot and root are profound. During periods of low LAI an increase in the proportion of assimilates retained in the shoot may increase leaf area and light interception and so, by compound interest, have a considerable effect in increasing production subsequently. In vegetative swards this effect is limited by the adverse consequences of increasing LAI on leaf photosynthetic development. However, in spring the distribution of carbon within the shoot to the meristems of elongating stem internodes is instrumental in sustaining the high photosynthetic potential of the reproductive canopy.

Finally, although all the grass shoot is potentially harvestable (Holmes, 1980b), in practice the harvest index is determined by what proportion of the standing crop exists above a given cutting (or grazing) height (Spedding, 1965a; Jackson, 1975). The partition of assimilates to elongating stems leads to a marked increase in the proportion of the standing crop that is more than 5 cm above ground. Such changes in structure contribute significantly to the potential yield that may be achieved under infrequent cutting, but the degree to which the grass crop is harvested in general depends on the defoliation management imposed.

4.4 MANAGEMENT

4.4.1 The place of physiology in understanding the effects of management on the production and utilization of grass

There is endless variation in the strategy for harvesting the grass sward. Although agronomic trials have provided a wealth of experience of the effect of management on yield, no single trial can aim to encompass more than a few of the possible treatments in any one year or at any one site. Our knowledge of the effects of management on the grass crop has advanced steadily, but it has been suggested that the past 50 years of grassland research 'was notable as much for controversy as for consensus' (Raymond, 1981). What this uncertainty has underlined is the need for an understanding of the fundamental (physiological) principles of just how the production and utilization of the grass crop are affected by management. In this way we may provide a rational basis by which to compare existing systems of management, an opportunity to identify those limitations to production which are specific to a given system of management, and ultimately a means by which to evaluate methods used to overcome them.

4.4.2 Principal effects of management on the flow of matter in the grass crop

There are two characteristics of the grass crop which are central to understanding the effects of management on production. First, because the grass sward displays a rapid turnover of tissues, any material that remains unharvested is soon lost to death (Fig. 4.9). This turnover is clearly the origin of a considerable potential loss to production. With a mean leaf appearance interval (April to September) of 11 days, an amount equivalent to the entire standing live weight of the crop may die and be replaced each month. So, in contrast with cereals, and the many crops which are harvested once, at the end of a single period of growth the grass crop must be harvested repeatedly.

Second, it is the photosynthetic tissue, the leaves, that are predominantly harvested. Over the season the repeated defoliation that is essential to harvest

Management

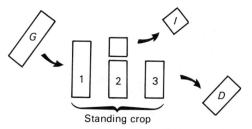

Figure 4.9 The flow of matter in a grazed grass sward. On each tiller the gross production of a new leaf (G) is countered by the loss of the oldest leaf (D) and the harvest (intake, I) of a proportion of each new leaf that appears. Increasing the proportion harvested reduces the proportion lost to death, but also reduces leaf area and the capacity to produce new leaves.

a proportion of the grass crop, inevitably reduces the leaf area and light interception of the canopy, interrupts canopy photosynthesis and so reduces the capacity for the production of new leaves (Fig. 4.9). Clearly, then, the way the sward is harvested on any one occasion has a profound effect on the amount grown, as well as on the degree to which the tissue produced is harvested. The objective of grassland management must be to strike a compromise between the conflicting demands of the grass plant, which needs to retain leaf area for photosynthesis, and the essential need to remove leaf tissue for harvest. This conflict leads to a dilemma that is central to grassland management.

The following sections consider the effect of two contrasting strategies for harvesting the grass sward – continuous and intermittent defoliation – on the balance between the uptake of matter in photosynthesis and the loss of matter in respiration and tissue death, and so on the amount of tissue that may be harvested over the season.

4.4.3 Continuous defoliation

The conflict between the needs of the grass plant and the needs of the harvester is most apparent in a system of continuous defoliation. By far the most common example of a system of management based on continuous defoliation is continuous grazing. In its simplest form continuous grazing describes a situation where animals maintain access to a single, often extensive area for a large proportion of the grazing season. Although the defoliation of the sward is not uniform (Harris, 1978), animals are continuously present and so have a continued effect on the uptake and loss of matter. Changes in leaf area index and in the rates of photosynthesis, tissue production, herbage intake and death are characteristically gradual.

Despite its prevalence as a means to harvest the grass crop, far less work has been carried out on the effects of continuous grazing on grass production than has been carried out on the effects of intermittent defoliation, whether by grazing or cutting. To a large extent this reflects the difficulty of measuring production under continuous grazing. Under this management, yield cannot be assessed from the net accumulation of herbage, as is conventional in agronomic studies, because a proportion of the tissues produced is continually removed by the animals. Indeed, both Spedding (1965b) and Morris (1969) describe a situation, as in Fig. 4.9, in which the gross rate of tissue production is equalled by the simultaneuous rate of removal of tissue by grazing and the loss of tissue to death and no net change in herbage mass takes place. Under these circumstances yield may only be assessed conventionally from the accumulation of animal products or from direct estimates of the rate of herbage intake by the animals (Baker, 1982; Le Du and Penning, 1982). Attempts have been made to estimate production under continuous grazing by using exclosure cages (t'Mannetje, 1978), but there is increasing evidence that the net accumulation of herbage within a 'cage' is an unreliable estimate of the amount of herbage consumed under continuous grazing (Vickery, 1975; Parsons *et al.*, 1984; Johnson and Parsons, 1985a).

A detailed understanding of the production and utilization of grass under continuous grazing has awaited the development of techniques to measure directly the flow of matter in the grass sward from the uptake and loss of carbon (Leafe, 1972; Vickery, 1975; Stiles, 1977) or from observations of the rate of tissue turnover (Grant *et al.*, 1983).

(a) The pattern and severity of defoliation under continuous grazing

The structure of the grass sward varies considerably in response to management. Swards maintained by continuous grazing at a low LAI, or low sward height, are characterized by a large number (40–60 000 m^{-2}) of small tillers. By contrast, swards maintained by continuous defoliation at a greater LAI, or greater sward height, are characterized by smaller numbers (10–25 000 m^{-2}) of larger tillers. These morphological adaptations have important consequences on the pattern and severity of defoliation experienced by individual plants in the sward. When a sward of high LAI, and a small number of large tillers, is defoliated to a height of just 3 cm, this results in the loss of a substantial proportion of the leaf tissue and in some plants, the expansion and restoration of leaf area depends on reserves (Davies, 1965; Davidson and Milthorpe, 1966a,b). However, a sward which is maintained long-term at a height of just 3 cm cannot depend continuously on reserves. Rather, in swards maintained at low LAI the structural adaptation of the crop decreases the relative severity of defoliation. As a result of the production of a large number of small tillers, each producing small leaves, a grass sward maintained at a height of less than 3 cm continues to experience on average only partial defoliation (Hodgson and Ollerenshaw, 1969; Morris, 1969; McIvor and Watkin, 1973). Thus, the structural adaptation enables the sward to sustain continuity of production from current assimilation.

Management

Although the adaptation to defoliation offers advantages in swards maintained at low LAI, this is not the case in swards maintained at high LAI. In these swards the elongation of stem internodes on reproductive tillers in spring and on the predominantly vegetative tillers in autumn, elevates leaves and even growing points into the grazed horizon (Hughes and Jackson, 1974; Jackson, 1975), and an increasing proportion of the leaf tissue is removed. Ultimately, these structural changes may lead to the mechanical collapse of the crop.

There is a further sense in which the continuously defoliated grass crop experiences only partial defoliation. Although the animals have access to the entire area, it is certain that not all of the sward is grazed at the same time. Rather, the sward consists of a population of tillers ranging from those that have most recently been defoliated to those that may have remained for a considerable period, intact. It is uncertain whether non-uniform defoliation leads to greater production (Smith, 1968; Harris, 1978). There is some evidence that the regrowth of defoliated tillers may be assisted by the translocation of assimilates from intact tillers on the same plant (Marshall and Sagar, 1965). However, in *Lolium multiflorum* this redistribution of assimilates is short-lived. To understand the consequences of continuous defoliation on production demands a knowledge of the area and the photosynthetic capacity of the leaves that remain in the crop, the gross rate of production of tissues and the degree to which the tissues produced are harvested.

(b) Characteristics of leaf and canopy photosynthesis in continuously grazed swards

The continuous removal of leaves by the grazing animal has a profound effect on the photosynthetic capacity of the remaining leaves and the canopy. In swards maintained at low LAI, young leaves continue to expand in high light, free from shade by older leaves, and so develop a high capacity for photosynthesis (Woledge, 1973, 1977, 1978). However, at the high stocking rates required to maintain a low LAI many leaves are defoliated while they are young (Morris, 1969, McIvor and Watkin, 1973). Thus, it is a proportion of the most photosynthetically efficient tissue that is removed. Nevertheless, the growing leaves (L1) and the remainder of the youngest fully expanded leaves (L2) contribute substantially to the total photosynthesis of the canopy. In a recent study in a sward maintained by continuous grazing at an LAI of 1.0, the growing leaves and the youngest fully expanded leaves together contributed some 75% to the total photosynthesis of the canopy (Table 4.2). Moreover, because young leaves expand in high light throughout the season, swards grazed continuously at a low LAI avoid the marked decrease in the photosynthetic potential of young leaves seen in the infrequently cut crop during periods of high LAI in summer and autumn (Parsons *et al.*, 1983a).

A further consequence of the continuous removal of leaf tissue is that a relatively large proportion of light energy is intercepted by the pseudostem or

TABLE 4.2 *Contribution of sward components to the area and the photosynthesis (radioactivity) of a hard-grazed sward as measured after short-term exposure to $^{14}CO_2$*

Category	Area (%)	Radioactivity (%)	Photosynthetic effectiveness ($g\,CO_2\,m^{-2}$ (leaf area) h^{-1})
L1	16.2	38.3	2.2
L2	26.4	38.9	1.4
L3	20.3	17.5	0.8
Sheath	37.1	4.5	0.1

sheath, as these tissues largely escape defoliation (de Lucia Silva, 1974). The photosynthetic potential of young sheath tissue is appreciable, albeit some 50% of that of young leaves (Thorne, 1959; Caldwell et al., 1981). However, in continuously grazed swards, successive leaves may be of similar length and their ligules coincide. Thus, most young sheath tissue is enveloped by a number of older leaf sheaths, with the oldest outermost. As this is of low photosynthetic capacity and may even be dead, the contribution of pseudostem to canopy photosynthesis is small. In the study described in Table 4.2 sheaths contributed less than 5% to the total photosynthesis of the canopy, even though the total area of sheath was as great as the area of young leaves. Thus, the good ground cover in an intensively grazed sward is misleading – much of the light is intercepted by tissues that contribute little to canopy photosynthesis.

Despite the high photosynthetic potential of individual leaves in a sward grazed to maintain a low LAI, canopy gross photosynthesis decreases progressively as the intensity of defoliation is increased (Parsons et al., 1983a; King et al., 1984). Clearly, the high photosynthetic potential of leaves and the structural adaptation of the crop compensate little for the overall reduction in leaf area.

Although the photosynthetic capacity of the grass sward is reduced by the continuous removal of photosynthetically efficient leaves, considerable photosynthetic assimilation may yet be achieved over the season as a result of the continuity of light interception (Fig. 4.10). Photosynthesis over the season in a sward maintained at an LAI of just 3.0 is close to that achieved in an infrequently cut crop – equivalent to some 60 t organic matter ha^{-1}. Moreover, continuous grazing avoids the marked decline in the photosynthesis of the canopy which arises following each cut – particularly that which arises in an infrequently cut crop at a time of high light energy in midsummer. As suggested by Donald and Black (1958), McMekan (1960) and Spedding (1965a), continuous grazing does offer an opportunity to maintain a sward at a LAI at which virtually all light is intercepted, and so offers the opportunity to maximize photosynthetic assimilation. However, as we shall see, maximum photosynthesis under continuous grazing is not synonymous with maximum yield.

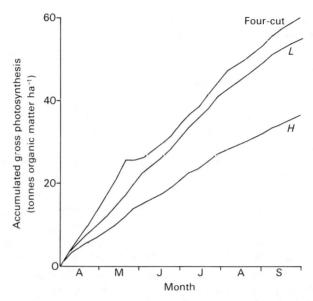

Figure 4.10 The effect of management on gross photosynthesis, accumulated over the grazing season, April to September, in swards maintained by lenient continuous grazing LAI = 3 (*L*); by hard continuous grazing LAI = 1 (*H*), and under infrequent severe cutting. (Parsons *et al.*, 1983a.)

(c) The effects of the intensity of continuous grazing on the growth and utilization of the sward

The intensity of continuous grazing not only affects the photosynthetic uptake of the sward, but also has a marked effect on both the rate of loss of matter during respiration and on the amount of assimilates partitioned to nonharvestable parts, and hence on the gross rate of production of shoot. In swards maintained at a high LAI, high rates of respiration result from the greater gross photosynthesis (and hence greater 'synthetic' respiration) and the greater weight of live tissue that there is to maintain (McCree, 1970). However, although the *amount* of matter lost in respiration is greater in swards of high rather than low LAI, in both cases a similar *proportion* – close to 45% – of gross photosynthesis is consumed by this route (Parsons *et al.*, 1983b). Similarly, measurements of the effect of the intensity of grazing on the partition of assimilates between the shoot and the non harvestable portions (predominantly the roots) indicate that the *amount* of carbon invested in roots is also greater in a sward of high rather than low LAI. However, once again, a similar *proportion* – close to 10% – of gross photosynthesis is involved. As a result the gross rate of production of shoot is greater in a sward maintained at a high LAI than in a sward maintained at a low LAI. This has been demonstrated both by measurements of the carbon balance of the crop (Parsons *et al.*, 1983b) and by direct estimates of the gross rate of production of leaves in the sward (Bircham and Hodgson, 1983; Grant *et al.*, 1983).

The most marked effect of the intensity of continuous grazing on the carbon balance of the grass crop is on the degree to which the tissues produced are harvested rather than lost to death. A general account of the effect of the intensity of defoliation on the relationship between the uptake and loss of matter in the continuously grazed sward is shown diagrammatically in Fig. 4.11 (Parsons *et al.*, 1983b; Johnson and Parsons, 1985a). In a sward maintained at high LAI photosynthesis and the gross rate of production of shoot are close to their maxima. However, in order to sustain the sward at high LAI it is necessary that only a small proportion of the tissue produced is harvested and a considerable proportion of leaf tissue must remain in the sward to contribute to the high rate of photosynthesis but also, inevitably, to contribute to a high rate of loss of matter to death. Direct measurements of the rate of death in continuously grazed swards confirm that the death rates increase almost linearly with the LAI maintained (Grant *et al.*, 1983). So, at low stocking rates, or in the total absence of grazing animals, the amount harvested is small.

As the intensity of defoliation is increased, and the sward is maintained at smaller LAI, a far greater proportion of the tissue produced is harvested, and a far smaller proportion remains in the sward to be lost to death. Initially, this increase in the overall utilization of the sward more than outweighs the detrimental effect of the intensity of defoliation in reducing photosynthesis and shoot growth, and the amount harvested per hectare is actually increased (Fig. 4.11). However, there are limitations to the extent to which increasing the overall utilization of the sward can continue to increase the amount harvested. Ultimately, the removal of too great a proportion of leaf tissue leads to a considerable reduction in light interception, and at very low LAI canopy photosynthesis and the amount harvested is reduced.

This analysis identifies a major limitation to production under continuous grazing. This is that high rates of photosynthesis and gross tissue production are not inevitably associated with high harvest yield. Maximum yield per hectare is achieved in a sward maintained at a LAI which is below the optimum for photosynthesis and shoot growth, but which provides the best compromise between plant growth and plant harvest (Fig. 4.11). This conclusion appears to conflict with the observation of Brougham (1956, 1957a, 1958) that maximum rates of net accumulation occur in a sward at an LAI at which 95% of light is intercepted, and with the suggestion that maximum production over the season should therefore be achieved in a sward maintained at close to full light interception (Brougham, 1958; Davidson and Donald, 1958). However, the principles described by Brougham were established in a sward undergoing a period of regrowth, that is a period of increasing LAI in the absence of defoliation, and not under a system of continuous grazing in which the LAI was maintained. As described previously, the high rates of net accumulation that occur during a period of regrowth depend not only on good light interception

Management

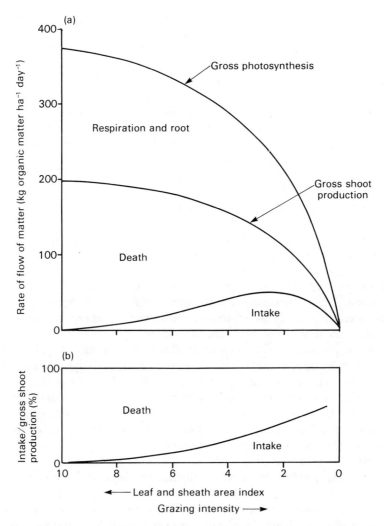

Figure 4.11 The effect of the intensity of continuous grazing on (a) the components of the production and utilization of grass in swards maintained by grazing at each of a wide range of LAI. The amount harvested as animal intake is replotted as a proportion of gross shoot production in (b) to describe the overall efficiency of utilization of herbage in each case. (Parsons *et al.*, 1983b.)

and high rates of photosynthesis, but to a large extent on advantages provided by the lag between a change in the rate of gross tissue production and a corresponding change in the rate of tissue death. In a sward maintained by continuous grazing the rate of gross tissue production varies little, and so there is no such advantageous relationship between tissue production and death (Parsons *et al.*, 1983b; Johnson and Parsons, 1985a).

A number of factors modify the general pattern of response to the intensity of defoliation shown in Fig. 4.11. First, the pattern of uptake and loss of tissues in relation to LAI is greatly influenced by season (Brougham, 1958). Seasonal changes in the photosynthetic potential of the canopy, combined with seasonal changes in light energy receipt, leads to a marked reduction in carbon assimilation and gross tissue production in a sward grazed to maintain a high LAI in autumn. As a result the number of animals and the total intake per hectare required to sustain a sward at a high LAI decreases over the season. Indeed, in late autumn it may be necessary to remove all the animals, and so reduce intake to zero, in an attempt to sustain a high leaf area. Structural changes in swards maintained by lenient grazing also contribute to a decrease in photosynthesis and tissue production in autumn as the proportion of photosynthetic to non-photosynthetic tissue declines (Hunt and Brougham, 1967).

Second, production per hectare also decreases in swards maintained by lenient grazing, as the distribution of LAI in the sward becomes less uniform as the season progresses (France et al., 1981). At the low stocking rates required to sustain a high LAI some areas of the sward are rejected by the animals. As the season progresses the situation arises where the animals harvest tissue from an area which is increasingly less than the total area of the sward. In the rejected areas the sward may approach a ceiling yield. This reduces estimates of production calculated over the sward as a whole.

Third, the amount harvested by animals from a continuously grazed sward depends heavily on the behavioural response of each animal to the morphological characteristics of the crop (Arnold, 1964; Williams et al., 1976) – a response which varies with the physiological status of the animals involved (Arnold and Dudzinski, 1967; Arnold, 1975). In swards maintained at low sward height the difficulties experienced by animals in removing tissue from the crop may become an overriding determinant of the amount harvested (Chacon et al., 1978; Penning et al., 1984; Black and Kenny, 1984).

It is clear that production under continuous grazing depends on an intimate interaction between the removal of tissue by the grazing animal and the effect of this on photosynthetic uptake, gross tissue production and the balance between intake and death. A greater appreciation of the physiology of production under continuous grazing in research and on the farm may best provide us with guidelines for management which enable us to strike an optimum balance between the growth and utilization of grass.

4.4.4 Intermittent defoliation

By far the greatest amount of work on the effects of management on grass production has been carried out in swards harvested by intermittent defoliation. Using this system of management, the sward characteristically undergoes a series of regrowths, each period of regrowth being terminated by a relatively short period of cutting or grazing.

Management

In contrast with continuous grazing, in which LAI and the components of the uptake and loss of matter change only gradually with time, intermittent defoliation is characterized by marked fluctuations in LAI and in the rate of processes involved in the uptake and loss of matter during each cycle of regrowth and defoliation. Moreover, not only does the rate of uptake and loss change during each cycle, but there are also marked changes in the relationship between these processes as the cycle proceeds.

Superficially, the estimation of yield under intermittent defoliation is simpler than under continuous defoliation. Consequently, the productivity of a grass sward is conventionally assessed from the amount of herbage that has accumulated above a chosen cutting or grazing height at the end of each successive period of regrowth. However, this apparent simplicity disguises the complex effect of variations in the severity and timing of defoliation on the balance between the uptake and loss of matter in the grass sward during regrowth and defoliation.

(a) Characteristics of leaf and canopy photosynthesis in intermittently defoliated swards

The photosynthetic capacity of the sward following defoliation depends on the amount of leaf area and the photosynthetic potential of the leaves remaining in the stubble. Immediately following defoliation the photosynthetic potential of the remaining leaves is a function of the light environment which the leaves experienced during their development in the sward before defoliation (Woledge, 1971, 1977). If the sward was previously vegetative and had attained a high LAI, then the leaves which remain following defoliation are poorly adapted for photosynthesis in the high light intensity to which they are suddenly exposed. Individual ryegrass leaves show some, but little, capacity for readaptation to environmental conditions (Woledge, 1971, 1973; Prioul *et al.*, 1980a,b). As a result of the poor photosynthetic potential of its young leaves, a sward defoliated from a high to a substantially lower LAI may initially display a rate of canopy photosynthesis per unit LAI less than that of a sward which has been maintained at that same low LAI. However, during the subsequent regrowth this situation may be reversed. Following severe defoliation new leaves expand in high light and develop a high capacity for photosynthesis. Moreover, during regrowth there is initially an increase in the area and the amount of light intercepted by the youngest, most photosynthetically efficient category of leaves – this being the only category in which regrowth can occur. Thus, during a period of regrowth canopy photosynthesis may become greater per unit LAI than in a sward maintained at the same LAI (Fig. 4.12).

Photosynthetic production *during* the period of defoliation is also important. In swards harvested by cutting, the removal of leaf area is virtually instantaneous, but in systems of rotational grazing the period of grazing down may be considerable. In a grazing system which has four paddocks in a strict rotation with a rest (regrowth) period of 21 days on each, the period of grazing down

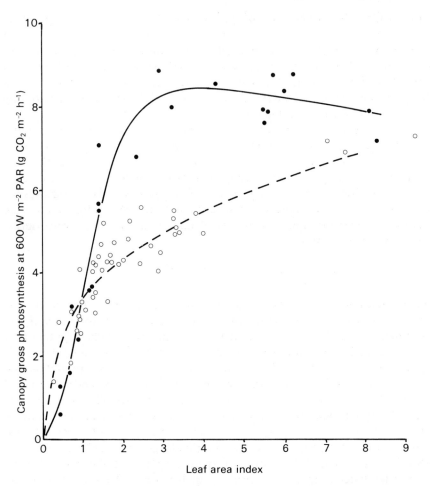

Figure 4.12 The relationship between canopy gross photosynthesis and LAI during a period of increasing leaf area during regrowth following rotational grazing (●), and in swards maintained at a wide range of leaf area (○) by continuous grazing. (Parsons, unpublished.)

must last for 7 days (Holmes, 1980a). Systems in which the grazing down period is as long as the period of regrowth are also not uncommon, and an appreciable proportion of the annual production is achieved actually during the period of grazing. During grazing down the photosynthetic potential of the canopy may be substantially less at a given LAI than during the period of regrowth (Deinum et al., 1981; King et al., 1984). However, it is uncertain if the low photosynthetic potential of the sward during the period of grazing down is aggravated by a disproportionately severe removal by the animal of the youngest leaves, or if

Management

there is simply a lack of the predominance of young leaves seen during the regrowth.

Clearly there can be no single, simple relationship between the LAI of the grass sward and the photosynthesis that is achieved. The photosynthetic potential of individual leaves and the structure of the canopy differs substantially depending on whether the sward is maintained at a given LAI, has been defoliated to that LAI, or has regrown to pass through that same LAI. This phenomenon greatly complicates the prediction of photosynthesis in models designed to assess the effect of management on yield (Noy-Meir, 1975; Johnson and Parsons, 1985a).

(b) The effect of the severity of defoliation on the supply of assimilates for growth

The principal effect of variations in the severity of defoliation is on the degree to which successive harvests interrupt light interception and photosynthesis, and so limit the supply of assimilates for growth. However, under severe defoliation, where a substantial proportion of the photosynthetic tissue is removed, the photosynthetic capacity of the sward may be insufficient to provide for the maintenance of surviving tissue and for the synthesis (and so restoration) of leaf area. Under these circumstances the production of new leaf tissue is supported initially from carbohydrate reserves. Because there is a respiratory loss of carbon associated with the synthesis of tissue, whether from current assimilates or from reserves, then, following severe defoliation, the sward as a whole displays a net loss in weight. The sward can only begin to make a net increase in weight when the leaf area produced from reserves increases the current photosynthetic assimilation of carbon sufficient to exceed the current loss of matter in respiration and death. The resulting delay is the basis of the 'lag' phase of growth which follows severe defoliation. The physiology of the effect of management on the mobilization and utilization of reserve carbohydrates and their relevance as a limitation to production in the grass crop during periods of regrowth is considered in detail in Chapter 3. It is evident that in many systems of temperate grassland management reserve carbohydrates play a critical, but quantitatively small role in production (Davidson and Milthorpe, 1966a,b; Harris, 1978).

The principles of the effect of variations in the severity of defoliation on the restoration of leaf area and light interception have been described by Brougham (1956, 1957b). In Brougham's experiments a sward grown to a height of 22 cm was subdivided and defoliated to three different heights: 12.5, 7.5 and 2.5 cm. The results showed that an increase in the severity of defoliation not only leads to a greater reduction in light interception immediately following defoliation, but also extends the time taken by the sward to regain full light interception. The swards defoliated to 12.5 cm took only 4 days to regain 95% light interception, whereas those defoliated to 7.5 and 2.5 cm took 16 and 24 days, respectively. This phenomenon is illustrated in Fig. 4.13a, using a

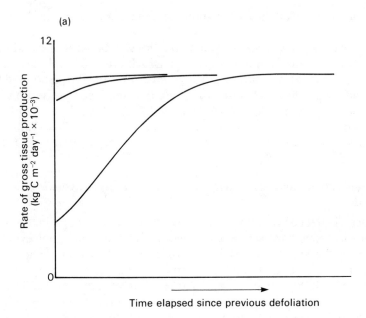

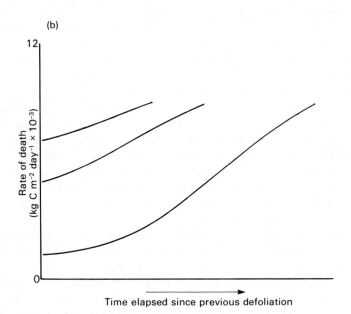

Figure 4.13 The effects of three severities of intermittent defoliation on (a) the rate of gross tissue production and (b) the rate of loss of tissue to death, as illustrated using a mechanistic model of grass production and senescence. The illustrations cover the period until the sward has reached 95% of its ceiling weight.

Management

mechanistic model of grass production and senescence (Johnson and Thornley, 1983; Johnson and Parsons, 1985a). It is clear that the effects of the severity of defoliation on photosynthesis and gross tissue production follow broadly the same patterns as those observed by Brougham for light interception. However, before we can consider the consequences of the severity of defoliation on growth we must first consider the effect of defoliation on losses by death.

(c) The effect of the severity of defoliation on losses through death

The effect of the severity of defoliation on losses by death in grass swards has received much less attention than the effect of the severity of defoliation on light interception and photosynthesis. Yet, as we have seen, a considerable proportion of the tissues present in the stubble or produced during regrowth is lost from the crop before harvest (Campbell, 1964; Morris, 1970; Smetham, 1975). There are considerable problems in measuring the rate of death of tissue in established grass swards (Davies, 1981). Rates of death cannot be estimated accurately from the accumulation of dead material because of the prodigious capacity of earthworms and other soil fauna to remove senescent material from the base of the sward. Nevertheless, from a combination of field and controlled environment studies there is an increasing body of information from which to construct an account of the effect of the severity of defoliation on the loss of matter by this route.

It is clear that the rate of loss of tissue by death depends on the size and the rate of turnover of the oldest category of tissue in the sward (Morris, 1970). An increase in the rate of loss by death has been observed during periods of regrowth in the field (Hunt, 1970; Grant *et al.*, 1981) and under controlled environment conditions where earthworms and fauna are absent (Robson, 1973a,b). This increase in the rate of death is predominantly the result of an increase in the size of leaves involved in the turnover of tissue (Hunt, 1965; Robson, 1973b). However, prolonged periods at a high LAI towards the end of a period of regrowth may lead to losses of tissue over and above those which result from the inevitable turnover of leaves and tillers. In dense swards the mutual shading of tillers may lead to an accelerated senescence of small tillers, or whole plants, as natural thinning occurs (Kays and Harper, 1974; Ong, 1978).

In addition to those losses that occur during each period of regrowth, further losses also occur during each period of defoliation. One major cause of losses during this period is that the rapid turnover of leaf tissue continues. However, where the period of defoliation is very short, tissue may also be destroyed and production lost as a result of mechanical damage by machinery, or by trampling, fouling and poaching by grazing animals (Watkin and Clements, 1978). Such damage may lead to the loss of whole plants or whole tillers from the sward, as well as the loss of single leaves. In wet weather, and particularly on poorly drained soils, these losses may become an overriding limitation to production. Although physical losses at the time of harvest are most evident in systems of intermittent defoliation – whether by cutting or rotational grazing – losses occur for similar reasons under continuous grazing. However, under continuous grazing the mechanical damage may be distributed more evenly over a greater area and this loss of production may be difficult to distinguish as separate from those factors affecting the overall utilization of the sward.

The principal effects of the severity of intermittent defoliation on the rate of loss of tissue by death are illustrated in Fig. 4.13b using the same mechanistic model of grass production and senescence as that used in Fig. 4.13a. The model demonstrates how an increase in the severity of defoliation leads to a decrease in the rate of death immediately following defoliation and also extends the time taken for the sward to regain maximum death rates – that is the time when death rate equals the gross rate of production of tissue and a ceiling yield is achieved. The broad similarity between the pattern of loss of tissue to death in Fig. 4.13b and the pattern of gross tissue production in Fig. 4.13a is more than coincidental. Regardless of the severity of defoliation, the rates of photosynthesis and death immediately following defoliation are both a reflection of the size of leaves remaining in the stubble, albeit in the case of photosynthesis and shoot growth it is the size of the youngest category of leaves that is most significant, whereas in the case of death it is the size of the oldest category of leaves that is important. Also regardless of the severity of defoliation, a sward which takes a long time to regain full light interception and the maximum rate

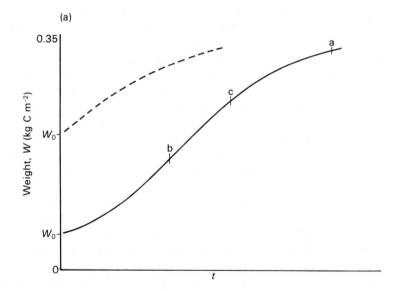

Figure 4.14 The characteristics of (a) the changes in the weight (W), (b) the instantaneous growth rate (dW/dt) and (c) the average growth rate ($W - W_0)/t$ during a period of regrowth following devere defoliation (———) or following lenient defoliation (– – –) where W_0 is W at $t = 0$ (see text for details.)

Management

of gross tissue production will also take a long time to regain maximum death rates. This is because changes in the rate of death must lag behind changes in the rate of gross tissue production. Clearly, these phenomena have important implications for the optimum timing of harvest.

(d) Severity of defoliation and the timing of harvest

The timing of successive harvests if often described in terms of a 'frequency' of defoliation, although this gives the misleading impression that the sward is

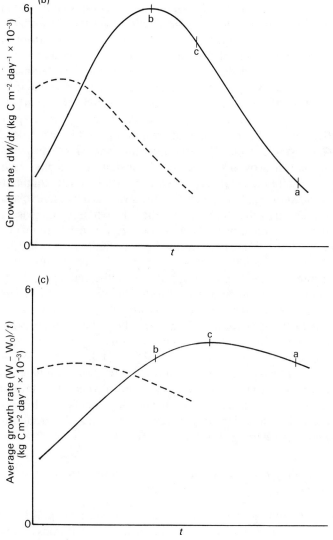

Figure 4.14 *continued*

necessarily harvested after periods of regrowth of similar duration. Indeed, fixed harvest intervals have long been criticized (McMeekan, 1960; Voisin and Lecomte, 1962). Clearly, in view of the marked seasonal variation in the physiological development of the grass sward and in the light and temperature environment, there is better opportunity to achieve an optimum balance between the uptake and loss of matter when the timing of harvest is based on physiological criteria.

One obvious feature of the events taking place during a period of regrowth is that, regardless of the severity of the previous defoliation, a ceiling to the yield is inevitably achieved if the sward remains unharvested. Failure to harvest the sward by this time can only decrease production over the season by extending the time passed without an increase in live weight. It is clear from Fig. 4.13a and b that following severe defoliation the effective length of the period of regrowth is increased and the sward continues for a considerable period to make a net gain in live weight. By contrast, in a leniently defoliated sward, the effective length of the period of regrowth is reduced. However, it is more complex to define the exact timing of harvest for optimum production. Harvesting at the ceiling yield (soon after 'a' in Fig. 4.14) clearly leads to the maximum harvestable yield $(W - W_0)$ within any single growth period, although the rate of growth (dW/dt) at this instant approaches zero. At 'b' (Fig. 4.14) the instantaneous rate of growth (dW/dt) is at a maximum, but what is required to maximize production is that the sward is harvested to achieve the maximum average growth rate, that is, at the time when the increase in weight $(W - W_0)$, divided by the time that has elapsed (t), is at a maximum (Watanabe and Takahashi, 1979). When the increase in the weight of the sward is sigmoid, as following severe defoliation, this can be seen to occur after the time of maximum instantaneous growth rate, but before the ceiling yield, as at 'c' in Fig. 4.14. Following lenient defoliation (dotted lines in Fig. 4.14) the maximum average growth rate may occur close to the time of the previous defoliation.

In addition to the general principles described above, there will be considerable seasonal variation in the effective length of a period of regrowth and in the LAI achieved before growth rates decline (Brougham, 1958, 1959a). Because the reproductive grass sward may continue to accumulate dry matter for longer and achieve a greater ceiling yield in spring than a vegetative sward in autumn, it has long been recognized that managements which involve less-frequent defoliation in spring may increase the annual net accumulation of matter, provided the herbage produced is harvested.

4.5 PHYSIOLOGICAL BASIS FOR OPTIMIZING PRODUCTION FROM GRASSLAND

As our understanding of the relationship between the uptake and loss of matter in the grass sward has advanced, so too have our notions of how best to defoliate the grass crop. As we have seen, under both continuous and inter-

Optimizing production from grassland

mittent defoliation, grassland management is necessarily a compromise between the need to retain leaf area for photosynthesis, and the need to remove leaf tissue before it dies and so achieve a yield.

Management must also take account of the specific objectives of the grass producer. One such objective is to maximize dry matter yield, although the seasonal distribution of that yield and the quality of the herbage removed must also be considered. A second objective is to optimize the individual performance of grazing animals. The economics of production of animal end-products (milk, meat, wool) may at times require a high level of intake per animal. However, if stocking rates are too low, in an attempt to ensure maximum intake per animal, this will result in reduced intake per hectare (Mott, 1960; Owen and Ridgman, 1968; Holmes, 1980a). An optimum combination of intake per head and per hectare must be struck. A third such objective is to maintain in the long term the structure of the grass sward so as to sustain high levels of production and to lessen the need for resowing. With these objectives and a knowledge of the physiology of the crop in mind, we can consider three examples of management that have been proposed to optimize production from grassland.

4.5.1 Infrequent severe cutting

It has become widely accepted that the maximum annual net accumulation of dry matter is achieved in swards managed by infrequent severe cutting (Anslow, 1967; Frame and Hunt, 1971; Smetham, 1975; Williams, 1980). This has been demonstrated in field trials in which swards were harvested at a range of regular intervals (Anslow, 1967) or at a variety of sward heights. Studies on the effect of cutting height on yield suggest between 2.5 and 5 cm throughout the season as an optimum (Reid, 1966; Harrington and Binnie, 1971). In the routine evaluation of grass varieties in the UK (NIAB, 1983/4a), swards cut four times over the season consistently outyield those cut nine times.

Although the merits of infrequent severe defoliation are apparent in swards cut at regular intervals, the greatest yields are achieved when the sward is defoliated in each growth period just as the ceiling yield of live tissue is approached – even though this may require that the sward is harvested after it has entered the period of declining growth rate (Watanabe and Takahashi, 1979). Seasonal changes in the physiology of production and in the environment dictate that this results in a single, extended period of growth in spring, and a series of shorter periods of regrowth in summer and autumn. In practice the timing of the harvest to achieve maximum accumulation is based, in the case of the vegetative canopy, on a time soon after virtually all incident light is intercepted. In the reproductive canopy the continued accumulation of herbage depends much on the development of reproductive tillers. Hence, the timing of harvest is associated with the emergence of close to 50% of the ears. This not only ensures substantial dry matter yield, but indicates the onset of a marked decline in the digestibility of the herbage to be removed (Green et al., 1971; NIAB, 1983/4b).

The success of this system of management depends much on the maximum expression of the physiological advantages shown by the reproductive crop in spring, in particular the substantial accumulation of reproductive stems. It is this dependence on the productivity of the reproductive canopy that results in the asymmetric seasonal pattern of production characteristic of this management.

Measurements of the carbon balance of the infrequently defoliated sward (Leafe et al., 1974; Parsons et al., 1983a) have demonstrated that the physiological advantages of the reproductive canopy and the value of harvesting severely to remove a large proportion of the standing crop more than outweigh the adverse effects of infrequent severe defoliation on the poor continuity of light interception and photosynthesis. Maximum yield is achieved under this system of management, even though the sward spends a considerable period at less than full light interception (Leafe, 1978; Robson, 1980). The total gross photosynthetic uptake of carbon over the season April to September may exceed the equivalent of $60 \, t \, ha^{-1}$ of organic matter. Some 40–45% of this production is lost through shoot respiration, and a further small proportion (10–15%) is accounted for in the growth and respiration of roots (Parsons and Robson, 1981b). Thus, of a gross shoot production of some $24–30 \, t \, ha^{-1}$, the amount harvested ranges between 12 and $15 \, t \, ha^{-1}$, so the overall efficiency of utilization is equivalent to some 25% of gross photosynthetic uptake, or 50% of the observed gross production of shoot. Thus, although infrequent severe defoliation results in maximum annual yields of dry matter, the amount lost to death may yet be as great as the amount harvested.

Infrequent severe cutting is effective in maintaining a sward structure appropriate for the continuation of this system of management. Infrequent defoliation in spring leads to the development of a high proportion of reproductive tillers, although throughout the year the total number of tillers remains low ($5–10\,000 \, m^{-2}$). Provided the sward is severely defoliated in late autumn, and so overwinters at low LAI, severe defoliation prevents any progressive long-term deterioration in sward structure.

4.5.2 Frequent lenient cutting or grazing

In the southern hemisphere an alternative system of management has been proposed to maximize the annual production of dry matter. Substantial annual yields which are as great as those reported above, have also been reported for a system of 'frequent lenient' defoliation, the development of which owes much to the work of Brougham (1956, 1957a,b, 1958, 1959b). The overall aim of this system is to use more lenient defoliation to avoid the periods of poor light interception following each harvest.

The hypothesis behind this system may be considered with reference to

Optimizing production from grassland

Fig. 4.7. It has been suggested that in swards managed by intermittent defoliation it is possible to define an upper limit to LAI, L1 in Fig. 4.7, beyond which the rate of net accumulation decreases substantially, and a lower limit (or residual leaf area), L2 in Fig. 4.7, above which the rate of net accumulation approaches its maximum value. Thus, it has been suggested that if a sward is defoliated leniently at L1, and the LAI is reduced to L2, then the sward will sustain close to maximum rates of net accumulation over a succession of periods of regrowth (Harris, 1978). It is important to note, however, that a leniently defoliated sward does not follow a part of the same sigmoid growth curve shown by a severely defoliated sward, as this hypothesis assumes. Nevertheless, high rates of net accumulation are achieved by harvesting the sward in this way. In practice the sward is defoliated either at, or 1–2 weeks beyond 95% light interception (Tainton, 1974). To gain the physiological advantages of the reproductive canopy, the period of regrowth is extended in spring.

There are significant disadvantages to this system of management, however, which result from the adverse long-term consequences of repeated lenient defoliation (Smetham, 1975). As was the case in swards managed by continuous lenient defoliation, swards maintained overall at a high LAI by frequent lenient defoliation also show high rates of loss of tissue to death and a progressive and long-term deterioration in sward structure. High rates of loss of tissue to death are acceptable in a situation where the system of management nevertheless results in high yields. However, the deterioration in sward structure is a serious disadvantage. It has been demonstrated that repeated lenient defoliation results in a progressive decrease in the proportion of green leaf to stem tissue in the herbage during successive regrowths (Hunt, 1965; Tainton, 1974; Korte et al., 1982). This phenomenon is particularly apparent at the base of the sward and results in a marked reduction in the leaf area, and so the photosynthetic capacity of the sward following each defoliation (Hunt and Brougham, 1967). As with continuous lenient defoliation, the regrowth of the sward is further prejudiced by the elongation of stem internodes during late spring and early summer, which elevates the photosynthetic tissue and some apices to a position above the height of defoliation (Hughes and Jackson, 1974; Jackson, 1975). As a result the sward does not remain leniently defoliated even when the sward is consistently harvested to a large height above ground level. Even when the severity of defoliation is based on a residual leaf area rather than height, a considerable reduction in tiller numbers is apparent, as is generally the case in swards that experience periods of high LAI (Kays and Harper, 1974). Altogether, these changes in morphology result in a situation where, although substantial production takes place in the spring months, this is followed by a progressive decrease in the rate of accumulation of tissue later in the season as well as a progressive decrease in the digestibility of the herbage removed.

In practice, in those parts of the world where frequent lenient defoliation was encouraged, a combination of lenient and severe defoliation is now generally advocated (Hunt and Brougham, 1967; Harris, 1978; Korte et al., 1982) to maintain sward structure and achieve a more uniform pattern of production. During spring a period of severe defoliation may be used to suppress the subsequent expression of reproductive development, to increase the proportion of vegetative tillers and so to minimise the delay in regrowth following the harvest of the reproductive crop. In autumn severe defoliation (an autumn 'clean-up') may again be used to increase the density of vegetative tillers and to maintain a high proportion of leaf to stem at the base of the sward.

4.5.3 Grazing – continuous and intermittent

There has been considerable controversy over the relative merits of continuous versus intermittent (rotational) grazing. The principal effects on dry matter production of the combinations of the severity and timing intermittent defoliation described in the examples above should apply whether the swards are cut or rotationally grazed. However, the effects of defoliation seen in cutting trials have not been consistently reflected in studies of animal production under rotational grazing (Brougham, 1970). In general, lower yields are achieved under rotational grazing than under cutting, largely reflecting the problems associated with using animals as effectively as cutting to remove a substantial proportion of the standing crop at each harvest. This is particularly apparent when the available herbage contains a large proportion of unpalatable, reproductive stem tissue (Rattray, 1978). Grazing animals may be encouraged to remove stem tissue, but this generally requires that their total diet is restricted. Also, defoliation by rotational grazing is generally more frequent than by cutting, and the accumulation of reproductive stems and the annual yield is consequently reduced (NIAB, 1983/4a). As a result yields per hectare under rotational grazing are closer to those achieved under continuous grazing, than to cutting, when the two grazing systems are compared at the same grazing intensity (McMeekan, 1960; Ernst et al., 1980).

Under continuous grazing, maximizing annual production per hectare requires that the sward be grazed to maintain a small LAI at which photosynthesis and the gross rate of production of shoot are reduced, but at which a large proportion of the tissue produced is harvested. Under these circumstances measurements of the carbon balance of continuously grazed swards indicate that the gross photosynthetic uptake over the season April to September may be reduced to $36 \, t \, ha^{-1}$. Losses in respiration and root growth account for some 45% of gross photosynthesis. Thus, gross tissue production is $18–20 \, t \, ha^{-1}$, of which $9–10 \, t \, ha^{-1}$ are harvested by the animals (Parsons et al., 1983b). The overall efficiency of production is therefore some 25% of gross photosynthesis, or 50% of the gross production of shoot. Clearly, continuous grazing is not intrinsically less efficient than infrequent severe cutting, as in both cases, only 50% of the gross production of tissue is harvested. Moreover, the amount lost to death under continuous grazing can be less than the amount lost under infrequent cutting.

To some extent the smaller amount harvested under continuous grazing, compared with cutting, is compensated for by a more uniform seasonal distribution of production which results from the supression of reproductive development during close grazing in sprint (Korte *et al.*, 1982; Johnson and Parsons, 1985b) and the sustained high digestibility of the herbage removed (Le Du *et al.*, 1981). Recent studies in the UK have demonstrated how it is possible to combine close to maximum yield per hectare with close to maximum yield per head (Penning *et al.*, 1984; Parsons, 1985). As a practical guideline to farmers, these studies suggest that this is achieved, for example, in a sward maintained at a height of between 4 and 6 cm by sheep. Moreover, swards maintained in this way retain a dense population of tillers, and show no progressive long-term deterioration in sward structure. Similar results have been obtained in swards maintained at a slightly increased height for dairy cattle (Le Du *et al.*, 1981).

4.5.4 The future

It has been a tradition in agricultural research to provide simple maxims as guidelines for management. Yet, in view of the complexity and plasticity of the response of the grass crop to defoliation, no single recipe for production can hold true under any but the most limited circumstances. In practice the need to sustain fluctuating livestock numbers against a background of seasonal differences in production requires a more flexible approach. An understanding of the physiological principles of the effects of defoliation on the uptake and loss of matter in the grass crop, supported by an appreciation of the importance of avoiding a marked deterioration in sward structure, will enable farmers to make their own management decisions as best suits their objectives and the changing conditions they experience. Such an understanding should increase the reliability and efficiency of production of the grass crop, and so increase confidence in the use of grass as an inexpensive feedstuff for ruminant production.

REFERENCES

Alberda, T. (1966) Responses of grasses to temperature and light, in *Proc. 12th Easter School in Agric. Sci., Univ. Nottingham, 1965* (eds F. L. Milthorpe and J. D. Ivins), Butterworths, London, pp. 200–12.

Alberda, T. and Sibma, L. (1968) Dry matter production and light interception of crop surfaces. III. Actual herbage production in different years as compared with potential values. *J. Br. Grassland Soc.*, **23**, 206–15.

Anslow, R. C. (1967) Frequency of cutting and sward production. *J. Agric. Sci., Camb.*, **68**, 377–84.

Anslow, R. C. and Green, J. O. (1967) The seasonal growth of pasture grasses. *J. Agric. Sci., Camb.,* **68,** 109–22.

Arnold, G. W. (1964) Responses of lambs to differing pasture conditions. *Proc. Austral. Soc. Animal Production,* **5,** 275–9.

Arnold, G. W. (1975) Herbage intake and grazing behaviour in ewes of four breeds at different physiological states. *Austral. J. Agric. Res.,* **26,** 1017–24.

Arnold, G. W. and Dudzinski, M. L. (1967) Studies on the diet of the grazing animal. II. The effect of physiological states in ewes and pasture availability on herbage intake. *Austral. J. Agric. Res.,* **18,** 349–59.

Baker, H. K. and Garwood, E. A. (1961) Studies on the root development of herbage plants. V. Seasonal changes in fructosan and soluble-sugar content of Cocksfoot herbage, stubble and roots under two cutting treatment. *J. Br. Grassland Soc.,* **16,** 263–8.

Baker, R. D. (1982) Estimating herbage intake from animal performance, in *Herbage Intake Handbook* (ed. J. D. Leaver), British Grassland Society, Hurley, pp. 76–94.

Behaeghe, T. J. (1978) The seasonal discrepancies between potential and actual grass growth. An essay of explanation by ecophysiological constraints. *Proc. 7th General Mtg Eur. Grassland Fedn,* Gent.

Bircham, J. S. and Hodgson, J. (1983) The influence of sward condition on rates of herbage growth and senescence in mixed swards under continuous stocking management. *Grass, Forage Sci.,* **38,** 323–31.

Black, J. L. and Kenny, P. A. (1984) Factors affecting diet selection by sheep. 2. Height and density of pasture. *Austral. J. Agric. Res.,* **35,** 565–78.

Borrill, M. (1961) Grass resources for out of season production. *Rep. Welsh Pl. Breeding Stn, 1960,* 107–11.

Boucher, K. (1975) *Global Climate,* English Universities Press, London.

Breese, E. L. (1970) Water-soluble carbohydrates in ryegrass, in *Jubilee Rep. Welsh Pl. Breeding Stn.,* 1919–1969.

Brougham, R. W. (1956) Effects of intensity of defoliation on regrowth of pasture. *Austral. J. Agric. Res.,* **7,** 377–87.

Brougham, R. W. (1957a) Some factors that influence the rate of growth of pasture. *Proc. N.Z. Grassland Assoc.,* Ruakura, pp. 109–16.

Brougham, R. W. (1957b) Pasture growth studies in relation to grazing management. *Proc. N.Z. Soc. Animal Production,* **17,** 46–55.

Brougham, R. W. (1958) Interception of light by the foliage of pure and mixed stands of pasture plants. *Austral. J. Agric. Res.,* **9,** 39–52.

Brougham, R. W. (1959a) The effect of season and weather on the growth rate of a ryegrass and clover pasture. *N.Z. J. Agric. Res.,* **2,** 283–96.

Brougham, R. W. (1959b) The effect of frequency and intensity of grazing on the productivity of a pasture of short-rotation ryegrass and red and white clover. *N.Z. J. Agric. Res.,* **2,** 1232–48.

Brougham, R. W. (1970) Frequency and intensity of grazing and their effects on pasture production. *Proc. N.Z. Grassland Assoc.,* **32,** 137–44.

Caldwell, M. M., Richards, J. H., Johnson, D. A., Nowak, R. S. and Dzurec, R. S. (1981) Coping with herbivory: Photosynthetic capacity and resource allocation in two semi-arid Agropyron bunch-grasses. *Oecologia,* **50,** 14–24.

Campbell, A. G. (1964) Grazed pasture parameters: dead herbage, net gain and utilization of pasture. *Proc. N.Z. Soc. Animal Production,* **24,** 17–28.

References

Chacon, E. A., Stobbs, T. H. and Dale, M. B. (1978) Influence of sward characteristics on grazing behaviour and growth of Hereford steers grazing tropical grass pastures. *Austral. J. Agric. Res.*, **29**, 89–102.

Charles-Edwards, D. A. (1982) *Physiological Determinants of Crop Growth*, Academic Press, London.

Clark, J., Kat, C. and Santhirasegaram, K. (1974) The effects of changes in heights of cutting and growth on the digestible organic matter production and botanical composition of perennial pasture. *J. Br. Grassland Soc.*, **29**, 269–73.

Colville, K. E. and Marshall, C. (1981) The pattern of growth assimilation of $^{14}CO_2$ and distribution of ^{14}C-assimilate within vegetative plants of *Lolium perenne* at low and high density. *Ann. Appl. Biol.*, **99**, 179–90.

Cooper, J. P. (1960) The use of controlled life-cycles in the forage grasses and legumes. *Herbage Abstr.*, **30**, 71–9.

Cooper, J. P. (1964) Climatic variation in forage grasses. I. Leaf development in climatic races of Lolium and Dactylis. *J. Appl. Ecol.*, **1**, 45–61.

Cooper, J. P. (1970) Potential production and energy conversion in temperate and tropical grasses. *Herbage Abstr.*, **40**, 1–15.

Cooper, J. P. and Breese, E. L. (1971) Plant breeding: forage grasses and legumes, in *Potential Crop Production – a case study* (eds J. P. Cooper and P. F. Wareing), Heinemann Educational, London, pp. 295–318.

Corrall, A. J. and Fenlon, J. S. (1978) A comparative method for describing the seasonal distribution of production from grasses. *J. Agric. Sci., Camb.*, **91**, 61–7.

Cugnac, A. De (1931) Recherches sur le glucides de graminées. *Ann. Sci. Naturelles (Bot.)*, **13**, 1–129.

Davidson, J. L. and Donald, C. M. (1958) The growth of swards of subterranean clover with particular reference to leaf area. *Austral. J. Agric. Res.*, **9**, 53–72.

Davidson, J. L. and Milthorpe, F. L. (1966a) Leaf growth in *Dactylis glomerata* following defoliation. *Ann. Bot.*, **30**, 173–84.

Davidson, J. L. and Milthorpe, F. L. (1966b) The effect of defoliation on the carbon balance in *Dactylis glomerata. Ann. Bot.*, **30**, 185–98.

Davies, A. (1965) Carbohydrate levels and regrowth in perennial ryegrass. *J. Agric. Sci., Camb.*, **65**, 213–21.

Davies, A. (1971) Growth rates and crop morphology in vernalized and non-vernalized swards of perennial ryegrass in spring. *J. Agric. Sci., Camb.*, **77**, 272–82.

Davies, A. (1977) Structure of the grass sward. *Proc. Int. Mtg Animal Production from Temperate Grassland,* Dublin, pp. 36–44.

Davies, A. (1981) Tissue turnover in the sward, in *Sward Measurement Handbook* (eds J. Hodgson, R. D. Baker, Alison Davies, A. S. Laidlaw and J. D. Leaver) British Grassland Society, Hurley, pp. 179–208.

Davies, A. and Calder, D. M. (1969) Patterns of spring growth in swards of different grass varieties. *J. Br. Grassland Soc.*, **24**, 215–25.

Deinum, B., t'Hart, M. L. and Lantinger, E. (1981) Photosynthesis of grass swards under rotational and continuous grazing. *Proc. 14th Int. Grassland Congr.,* Lexington, USA, pp. 407–10.

Donald, C. M. and Black, J. N. (1958) The significance of leaf area in pasture growth. *Herbage Abstr.*, **28**, 1–16.

Doyle, C. J., Corrall, A. J., Thomas, C., Le Du, Y.C. P. and Morrison, J. (1983) The integration of conservation with grazing for milk production: a computer simulation of the practical and economic implication. *Grass, Forage Sci.*, **38**, 261–72.

Eagles, C. F. and Othman, O. B. (1978a) Physiological studies of a hybrid between populations of *Dactylis glomerata* from contrasting climatic regions. I. Interpopulation differences. *Ann. Appl. Biol.*, **89**, 71–9.

Eagles, C. F. and Othman, O. B. (1978b) Physiological studies of a hybrid between populations of *Dactylis glomerata* from contrasting climatic regions. II. Intrapopulation differences. *Ann. Appl. Biol.*, **89**, 81–8.

Ernst, P., Le Du, Y, L. P. and Carlier, L. (1980) Animal and sward production under continuous grazing management – a critical appraisal. *Proc. Symp. Eur. Grassland Fed., Wageningen,* Pudoc, Wageningen, pp. 119–26.

Frame, J. and Hunt, I. V. (1971) The effect of cutting and grazing systems on herbage production from grass swards. *J. Br. Grassland Soc.*, **26**, 163–71.

France, J., Brockington, N. R. and Newton, J. E. (1981) Modelling grazed grassland systems: wether sheep grazing perennial ryegrass. *Appl. Geog.*, **1**, 133–50.

Garwood, E. A. (1967a) Seasonal variation in appearance and growth of grass roots. *J. Br. Grassland Soc.*, **22**, 121–30.

Garwood, E. A. (1967b) Studies on the roots of grasses. *Annual Rep. Grassland Res. Inst., Hurley,* 1966, pp. 72–9.

Grant, S. A., Barthram, G. T. and Torvill, L. (1981) Components of regrowth in grazed and cut *Lolium perenne* swards. *Grass, Forage Sci.*, **36**, 155–68.

Grant, S. A., Barthram, G. T., Torvill, L., King, J. and Smith, H. K. (1983) Sward management, lamina turnover and tiller population density in continuously stocked *Lolium perenne*-dominated swards. *Grass, Forage Sci.*, **38**, 333–44.

Green, J. O., Corral, A. J. and Terry, R. A. (1971) Grass species and varieties: relationships between stage of growth, yield and forage quality. *Tech. Rep. 8*, Grassland Research Institute, Hurley.

Harrington, F. J. and Binnie, R. C. (1971) The effect of height and frequency of cutting on grass production. *44th A. Rep. Agric. Res. Inst.,* Northern Ireland, pp. 17–24.

Harris, W. (1978) Defoliation as a determinant of the growth, persistance and composition of pasture, in *Plant Relations in Pastures* (ed. J. R. Wilson) CSIRO, Autralia, pp. 67–85.

Hodgson, J. and Ollerenshaw, J. H. (1969) The frequency and severity of individual tillers in set-stocked swards. *J. Br. Grassland Soc.*, **24**, 266–34.

Holmes, W. (1980a) Grazing Management, in *Grass – its Production and Utilization* (ed. W. Holmes), British Grassland Society, Hurley, pp. 125–73.

Holmes, W. (1980b) Chapter 1, Introduction, in *Grass – its Production and Utilization* (ed. W. Holmes), British Grassland Society, Hurley, pp. 1–5.

Hughes, R. and Jackson, D. K. (1974) Impact of grazing management on sward survival. *J. Br. Grassland Soc.*, **29**, 76.

Hunt, L. A. (1965) Some implications of death and decay in pasture production. *J. Br. Grassland Soc.*, **20**, 27–31.

Hunt, L. A. and Brougham, R. W. (1967) Some changes in the structure of a perennial ryegrass sward frequently but leniently defoliated during the summer. *N.Z. J. Agric. Res.*, **10**, 397–404.

Hunt, W. F. (1970) The influence of leaf death on the rate of accumulation of green herbage during pasture regrowth. *J. Appl. Ecol.*, **7**, 41–50.

Jackson, D. K. (1975) The influence of patterns of defoliation on sward morphology, in

References

Pasture Utilization by the Grazing Animal (eds J. Hodgson and D. K. Jackson), Occasional Symposium No. 8, British Grassland Society, pp. 51–60.

Jewiss, O. R. and Woledge, J. (1967) The effect of age on the rate of apparent photosynthesis in leaves of tall fescue (*Festuca arundinacea* Schreb.). *Ann. Bot.*, **31**, 661–71.

Johnson, I. R. and Parsons, A. J. (1985a) A physiological model of grass growth under grazing. *J. Theor. Biol.*, **112**, 345–67.

Johnson, I. R. and Parsons, A. J. (1985b) A physiological model of grass growth under continuous grazing managements on seasonal patterns of grass production. *Grass, Forage Sci.*, **40**, 449–58.

Johnson, I. R. and Thornley, J. H. M. (1983) Vegetative crop growth model incorporating leaf area expansion and senescence, and applied to grass. *Plant, Cell, Envir.*, **6**, 721–99.

Jones, M. B., Leafe, E. L., Stiles, W. and Collett, B. (1978) Patterns of respiration of a perennial ryegrass crop in the field. *Ann. Bot.*, **42**, 693–703.

Kays, S. and Harper, J. L. (1974) The regulation of plant and tiller density in a grass sward. *J. Ecol.*, **62**, 97–105.

King, J., Sim, E. M. and Grant, S. A. (1984) Photosynthetic rate and carbon balance of grazed ryegrass pastures. *Grass, Forage Sci.*, **39**, 81–92.

Kleinendorst, A. and Brouwer, R. (1970) The effect of temperature of the root medium and of the growing point of the shoot on growth, water content and sugar content of maize leaves. *Neth. J. Agric. Sci.*, **18**, 140–8.

Korte, C. J., Watkin, B. R. and Harris, W. (1982) Use of residual leaf area index and light interception as criteria for spring–grazing management of a ryegrass–dominant pasture. *N.Z. J. Agric. Res.*, **25**, 309–19.

Leafe, E. L. (1972) Micro-environment, carbon dioxide exchange and growth in grass swards, in *Crop Process in Controlled Environments* (eds A. R. Rees, K. E. Cockshull, D. W. Hand and R. G. Hurd), Academic Press, London, pp. 157–75.

Leafe, E. L. (1978) Physiological, environmental and management factors of importance to maximum yield of the grass crop, in *Maximising Yields of Crops*, ADAS/ARC Symposium, Harrogate, pp. 37–49.

Leafe, E. L., Stiles, W. and Dickinson, S. (1974) Physiological processes influencing the pattern of productivity of the intensively managed grass sward. *Proc. 12th Int. Grassland Congr., Moscow*, **1**, 442–57.

Le Du, Y, L. P., Baker, R. D. and Newberry, R. D. (1981) Herbage intake and milk production by grazing dairy cows. 3. The effect of grazing severity under continuous stocking. *Grass, Forage Sci.*, **36**, 307–18.

Le Du, Y. L. P. and Penning, P. D. (1982) Animal based techniques for estimating herbage intake. in *Herbage Intake Handbook* (ed. J. D. Leaver), British Grassland Society, Hurley, pp. 37–76.

List, R. J. (ed.) (1966) *Smithsonian Meteorological Tables*, Smithsonian Institution, Washington, D. C.

Lucia Silva, G. R. de (1974) A study of variation in the defoliation and regrowth of individual tillers in swards of *Lolium perenne* L. grazed by sheep. PhD. Thesis, University of Reading.

t'Mannetje, L. (1978) Measuring quantity of grassland vegetation in *Measurement of Grassland Vegetation and Animal Production* (ed. L. t'Mannetje), Commonwealth Bureau of Pastures and Field Crops, Farnham Royal, pp. 85–6.

Marshall, C. and Sagar, G. R. (1965) The influence of defoliation on the distribution of assimilates in *Lolium multiflorum* Lam. *Ann. Bot.*, **29**, 365–70.

McCree, K. J. (1970) An equation for the rate of respiration of white clover plants grown under controlled conditions, in *Production and Measurement of Photosynthetic Productivity* (ed. I. Malek), Proceedings IBP/PP Technical Meeting, Trebon, Purdoc, Wageningen, pp. 221–30.

McCree, K. J. (1972) The action spectrum, absorbtance and quantum yield of photosynthesis in crop plants. *Agric. Met.*, **9**, 191–216.

McIvor, P. J. and Watkin, B. R. (1973) The pattern of defoliation of Cocksfoot by grazing sheep. *Proc. N.Z. Grassland Assoc.*, **34**, 225–35.

McMeekan, C. P. (1960) Grazing management. *Proc. 8th Int. Grassland Congr.*, Reading, pp. 21–7.

Milthorpe, F. L. and Moorby, J. (1974) *An Introduction to Crop Physiology*, Cambridge University Press, London.

Minson, D. J., Raymond, W. F. and Harris, C. E. (1960) Study in the digestibility of S37 Cocksfoot, S23 Ryegrass and S24 Ryegrass. *J. Br. Grassland Soc.*, **15**, 174–80.

Monteith, J. L. (1973) *Principles of Environmental Physics*, Contemporary Biology Series, Edward Arnold, London.

Monteith, J. L. and Elston, J. F. (1971) Microclimatology and crop production in *Potential Crop Production* (eds P. F. Wareing and J. P. Cooper). Heinemann Educational, London, pp. 23–42.

Morris, R. M. (1969) The pattern of grazing in continuously grazed swards. *J. Br. Grassland Soc.*, **24**, 65–71.

Morris, R. M. (1970) The use of cutting treatments designed to simulate defoliation by sheep. *J. Br. Grassland Soc.*, **25**, 198–206.

Morris, R. M. and Thomas, J. G. (1972) The seasonal pattern of dry-matter production of grasses in the north pennines. *J. Br. Grassland Soc.*, **27**, 163–72.

Mott, G. O. (1960) Grazing pressure and the measurement of pasture production. *Proc. 8th Int. Grassland Congr.*, Reading, pp. 606–11.

National Institute of Agricultural Botany (1983/4a) *Recommended Varieties of Grasses*, Farmers Leaflet No. 16.

Noy-Meir, I. (1975) Stability in grazing systems: an application of predator–prey graphs. *J. Ecol.*, **63**, 459–81.

Ong, C. K. (1978) The physiology of tiller death in grasses. I. The influence of tiller age, size and position. *J. Br. Grassland Soc.*, **33**, 197–203.

Osbourn, D. F. (1980) The feeding value of grass and grass products, in *Grass – its Production and Utilization* (ed. W. Holmes), British Grassland Society, Hurley, pp. 70–124.

Owen, J. B. (1976) *Sheep Production*, Bailliere Tindall, London.

Owen, J. B. and Ridgman, W. J. (1968) The design and interpretation of experiment to study animal production from grazed pasture. *J. Agric. Sci., Camb.*, **71**, 327–35.

Parsons, A. J. (1985) New light on the grass sward and the grazing animal. *Span*, **28**, (2), pp. 47–49.

Parsons, A. J. and Robson, M. J. (1980) Seasonal changes in the physiology of S24 Perennial Ryegrass (*Lolium perenne* L.) 1. Response to leaf extension to temperature during the transition from vegetative to reproductive growth. *Ann. Bot.*, **46**, 435–44.

References

Parsons, A. J. and Robson, M. J. (1981a) Seasonal changes in the physiology of S24 Perennial Ryegrass (*Lolium perenne* L.) 2. Potential leaf and canopy photosynthesis during the transition from vegetative to reproductive growth. *Ann. Bot.*, **47**, 249–58.

Parsons, A. J. and Robson, M. J. (1981b) Seasonal changes in the physiology of S24 Perennial Ryegrass (*Lolium perenne* L.) 3. Partition of assimilates between root and shoot during the transition from vegetative to reproductive growth. *Ann. Bot.*, **48**, 733–44.

Parsons, A. J. and Robson, M. J. (1982) Seasonal changes in the physiology of S24 Perennial Ryegrass (*Lolium perenne* L.) 4. Comparison of the carbon balance of the reproductive crop in spring and the vegetative crop in autumn. *Ann. Bot.*, **50**, 167–77.

Parsons, A. J., Leafe, E. L., Collett, B. and Stiles, W. (1983a) The physiology of grass production under grazing. 1. Characteristics of leaf and canopy photosynthesis of continuously grazed swards. *J. Appl. Ecol.*, **20**, 117–36.

Parsons, A. J., Leafe, E. L., Collett, B., Penning, P. D. and Lewis, J. (1983b) The physiology of grass production under grazing. 2. Photosynthesis, crop growth and animal intake of continuously grazed swards. *J. Appl. Ecol.*, **20**, 127–39.

Parsons, A. J., Collett, B. and Lewis, J. (1984) Changes in the structure and physiology of a perennial ryegrass sward when released from a continuous stocking management: implications for the use of exclusion cages in continuously stocked swards. *Grass, Forage Sci.*, **39**, 1–9.

Peacock, J. M. (1975a) Temperature and leaf growth in *Lolium perenne*. I. The thermal microclimate, its measurement and relation to crop growth. *J. Appl. Ecol.*, **12**, 99–114.

Peacock, J. M. (1975b) Temperature and leaf growth in *Lolium perenne*. II. The site of temperature perception. *J. Appl. Ecol.*, **12**, 115–23.

Peacock, J. M. (1975c) Temperature and leaf growth in *Lolium perenne*. III. Factors affecting seasonal differences. *J. Appl. Ecol.*, **12**, 685–97.

Peacock, J. M. (1976) Temperature and leaf growth in four grass species. *J. Appl. Ecol.*, **13**, 225–32.

Penning, P. D., Steel, G. L. and Johnson, R. H. (1984) Further developments and use of an automatic recording system in sheep grazing studies. *Grass, Forage Sci.*, **39**, 345–51.

Pollock, C. J. and Jones, T. (1979) Seasonal patterns of fructosan metabolism in forage grasses. *New Phytol.*, **83**, 9–15.

Prioul, J. L., Brangeon, J. and Reyss, A. (1980a) Interaction between external and internal conditions in the development of photosynthetic features in a grass leaf. I. Regional responses along a leaf during and after low-light or high-light acclimation. *Pl. Physiol.*, **66**, 762–9.

Prioul, J. L., Brangeon, J. and Reyss, A. (1980b) Interaction between external and internal conditions in the development of photosynthetic features in a grass leaf. II. Reversibility of light-induced responses as a function of developmental stages. *Pl. Physiol.*, **66**, 770–4.

Rattray, P. V. (1978) Pasture constraints to sheep production. *Proc. Agron. Soc. N.Z.*, **8**, 103–8.

Raymond, W. F. (1981) Grassland research, in *Agricultural Research 1931–1981* (ed. G. W. Cooke), Agricultural Research Council, London, pp. 311–23.

Reid, D. (1966) Studies on cutting management of grass clover swards. 4. The effect of close and lax cutting on the yield of herbage from swards cut at different frequencies. *J. Agric. Sci., Camb.*, **66**, 101–6.

Robson, M. J. (1973a) The growth and development of simulated swards of perennial ryegrass. I. Leaf growth and dry weight changes as related to the ceiling yield of a seedling sward. *Ann. Bot.*, **37**, 487–500.

Robson, M. J. (1973b) The growth and development of simulated swards of perennial ryegrass. II. Carbon assimilation and respiration in a seedling sward. *Ann. Bot.*, **37**, 501–18.

Robson, M. J. (1980) A physiologist's approach to raising the potential yield of the grass crop through breeding, in *Opportunities for Increasing Crop Yields* (eds R. G. Hurd, P. V. Biscoe and C. Dennis), Association of Applied Biology, Pitman, London, pp. 35–50.

Robson, M. J. and Jewiss, O. R. (1968) A comparison of British and North African varieties of Tall Fescue (*Festuca arundinacea*). II. Growth during winter and survival at low temperatures. *J. Appl. Ecol.*, **5**, 179–90.

Russell, E. W. (1973) *Soil Conditions and Plant Growth*, 10th edn, Longman, London.

Ryle, G. J. A. (1970) Distribution patterns of assimilated ^{14}C in vegetative and reproductive shoots of *Lolium perenne* and *L. temulentum*. *Ann. Appl. Biol.*, **66**, 155–67.

Sibma, L. and Louwerse, W. (1977) The effect of changes in crop structure on photosynthesis, dry matter production and chemical composition of *Lolium perenne*. *Proc. 13th Int. Grassland Congr., Leipzig*, section 1, 2–12.

Smetham, M. L. (1975) The influence of herbage utilization on pasture production and animal performance. *Proc. N.Z. Grassland Assoc.*, 91–103.

Smith, A. (1968) Sward growth in relation to pattern of defoliation. *J. Br. Grassland Soc.*, **23**, 294–8.

Smith, D. (1967) Carbohydrates in grasses. II. Sugar and fructosan composition of the stem bases of Bromegrass and Timothy at several growth stages, and in different plant parts at anthesis. *Crop Sci.*, **7**, 62.

Spedding, C. R. W. (1965a) *Sheep Production and Grazing Management*, Bailliere, Tindall and Cox, London.

Spedding, C. R. W. (1965b) The physiological basis of grazing management. *J. Br. Grassland Soc.*, **20**, 7–14.

Stiles, W. (1977) Enclosure method for measuring photosynthesis, respiration and transpiration of crops in the field. *Tech. Rep. No. 18, Grassland Institute, Hurley*.

Tainton, N. M. (1974) Effect of different grazing rotations on pasture production. *J. Br. Grassland Soc.*, **29**, 191–202.

Thomas, H. (1977) The influence of autumn cutting regime on the response to temperature of leaf growth in perennial ryegrass. *J. Br. Grassland Soc.*, **32**, 227–30.

Thomas, H. and Norris, I. B. (1977) The growth responses of *Lolium perenne* to the weather during winter and spring at various altitudes in mid-Wales. *J. Appl. Ecol.*, **14**, 949–64.

Thorne, G. N. (1959) Photosynthesis of lamina and sheath of Barley leaves. *Ann. Bot.*, **23**, 365–70.

Troughton, A. (1981) Length of life of grass roots. *Grass, Forage Sci.*, **36**, 117–20.

Vickery, P. B. (1975) Grazing and net primary production of a temperate grassland. *J. Appl. Ecol.*, **9**, 307–14.

References

Voisin, A. and Lecomte, A. (1962) *Rational Grazing – the meeting of cows and grass*, Crosby Lockwood and Son, London.

Waite, R. and Boyd, J. (1953) The water-soluble carbohydrates of grasses. II. Grasses cut at grazing height several times during the growing season. *J. Sci. Fd. Agric.*, **4**, 257–61.

Watanabe, K. and Takahashi, Y. (1979) Effects of fertilization level on the regrowth of orchard grass. 1. Changes of yield and growth with time. *J. Jap. Soc. Grassland Sci.*, **25**, 195–202.

Watkin, B. R. and Clements, R. J. (1978) The effects of grazing animals on pastures, in *Plant Relations in Pastures* (ed. J. R. Wilson), CSIRO, Australia, pp. 273–89.

Watts, W. R. (1971) Role of temperature in the regulation of leaf extension in *Zea Mays*. *Nature*, **229**, 46–7.

Weinmann, H. (1952) Carbohydrate reserves in grasses. *Proc. 6th Int. Grassland Cong., Pennsylvania, Vol. 1*, 655–60.

Williams, C. M. J., Geytenbeck, P. E. and Allden, W. G. (1976) Relationships between pasture availability, milk supply, lamb intake and growth. *Proc. Austral. Soc. Animal Production*, **11**, 333–6.

Williams, T. E. (1980) Herbage production: grasses and leguminous forage crops, in *Grass – its Production and Utilisation* (ed. W. Holmes), British Grassland Society, Hurley, pp. 6–69.

Wilman, D., Ojuederie, B. M. and Asare, E. O. (1976) Nitrogen and Italian ryegrass. 3. Growth up to 14 weeks: yields, proportion, digestibilities and nitrogen content of crop fractions, and tiller populations. *J. Br. Grassland Soc.*, **31**, 73–9.

Wit, C. T. de (1965) Photosynthesis of leaf canopies. *Agric. Res. Rep. No. 663*, Pudoc, Wageningen, pp. 1–57.

Woledge, J. (1971) The effect of light intensity during growth on the subsequent rate of photosynthesis. *Ann. Bot.*, **35**, 311–22.

Woledge, J. (1973) The photosynthesis of ryegrass leaves grown in a simulated sward. *Ann. Bot. Appl. Bio.*, **73**, 229–37.

Woledge, J. (1977) The effects of shading and cutting treatments on the photosynthetic rate of ryegrass leaves. *Ann. Bot.*, **41**, 1279–86.

Woledge, J. (1978) The effect of shading during vegetative and reproductive growth on the photosynthetic capacity of leaves in a grass sward. *Ann. Bot.*, **42**, 1085–9.

Woledge, J. (1979) Effect of flowering on the photosynthetic capacity of ryegrass leaves grown with and without natural shading. *Ann. Bot.*, **44**, 197–207.

Woledge, J. and Dennis, W. D. (1982) The effect of temperature on photosynthesis of ryegrass and white clover leaves. *Ann. Bot.*, **50**, 25–35.

Woledge, J. and Leafe, E. L. (1976) Single leaf and canopy photosynthesis in a ryegrass sward. *Ann. Bot.*, **40**, 773–83.

Woodward, F. I. and Sheehy, J. E. (1983) *Principles and Measurements in Environmental Biology*, Butterworths, London.

CHAPTER 5
Mineral nutrients and the soil environment

D. W. Jeffrey

5.1 INTRODUCTION

This chapter is written from a broadly ecological viewpoint, with the aim of providing a concise guide that will enable a structured approach to thinking about grass and soil as a source of mineral nutrients. Disciplinary boundaries have led to an unfortunate fragmentation of the literature and research, and it is now appropriate to integrate across the fields of plant nutrition, community and ecosystem ecology, microbiology and soil science.

Four main ideas underlie the approach adopted.

(a) The prime object of study is the mycorrhizal grass plant and its rhizosphere (Clarkson, 1985). Recognition of this unit is long overdue in research strategies now that the era of low input intensive agriculture has arrived in the West. For agriculture in developing areas it always was, and will be, central to research (Chapin, 1980; Boddey and Dobereiner, 1984).

(b) The immediate context from a mineral nutrition viewpoint is the soil–plant continuum (Fig. 5.1) in which the supply of ions depends on solid phase–liquid phase relationships. The production of managed and natural grasslands may depend strongly on the 'availability' of nutrient ions and inorganic toxins. 'Availability' is not simple, but is based on a characteristic set of mechanisms involving the chemistry, microbiology and water relationships of the continuum. The interaction between changing availability and the dynamics of growth is also complex. The concept of 'essential' elements also needs review in moving from a physiological to an ecological context.

(c) In the long term, supply of ions or elements to vegetation depends on the continuity of function of grassland ecosystems as a whole. Grazing animals and decomposers are in many respects equivalent in maintaining a flow of ions (Fig. 5.2). This will be illustrated later by the nitrogen and phosphate cycles (Figs 5.8 and 5.9).

(d) Environmental constraints arising from adverse soil conditions give rise to responses which must be interpreted in an evolutionary context. Responses are expressed as inherited physiological mechanisms.

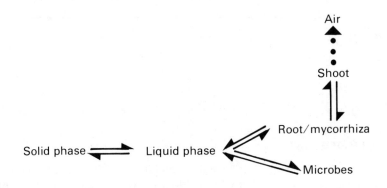

Figure 5.1 The soil–plant continuum is an essential basic study unit for investigating transfers of water and ions.

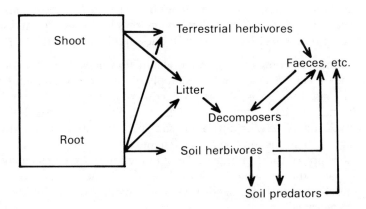

Figure 5.2 The flow of mineral elements in a grassland ecosystem. Herbivores and decomposers may be seen as equivalents where mineralization is being studied. Grasslands possess a web of relationships with organic matter as plant litter, faeces and other excreta as substrates for decomposers. Mineralization sustains the flow of 'available' mineral elements to the growing sward. See also Figures 5.8 and 5.9.

5.2 THE ROLE OF ASSOCIATED ORGANISMS

The association of grass plants with mycorrhizal fungi and rhizosphere bacteria must be considered to be the norm. While some negative records for mycorrhizal associations are known, e.g. *Poa annua* (Tutin, 1957) and *Chionochloa macra* (Crush, 1973), most grasses examined appear to be associated with vesicular arbuscular endophytes (see Saunders *et al.*, 1975). There are several important ways in which mycorrhizas and rhizospheres can influence the equilibrium between soil nutrients and the grass plant as follows.

(a) By physically extending the absorbing surface and thus exploring the larger volume of soil more effectively. This is the key role for mycorrhizas, and is of manifest importance in the uptake of phosphate. Absorption of sources of nitrogen, other minerals and water are also enhanced (Smith, 1980).

(b) By acting as sinks for nutrients. Bacterial films clearly absorb nutrients in parallel to and possibly in competition with, the supporting grass root. In this case it is not clear how bacterial sites may behave in terms of releasing nutrients to the root. Mycorrhizas are generally important as a site for phosphate absorption, which is stored as polyphosphate granules and then released to the carbohydrate-providing host (Cox *et al.*, 1975).

(c) By influencing chemically the solid–liquid phase equilibrium. All plants exude organic compounds from root surfaces, various studies agreeing that this carbon can amount to 20% of the total plant dry matter (Rovira, 1979). This organic material is the substrate for rhizosphere micro-organisms, and it and their metabolic products can influence the solubility equilibrium in various ways: for example, the H^+ ion concentration may rise and increase mineral weathering rates; or chelates, such as citrate, may be produced and increase mobility of metals. The chemical possibilities are numerous, and more careful experiments are required to elucidate particular situations. A particular case of an equilibrium change is the inhibition of nitrification by grass rhizosphere, an effect reducing the leaching loss of nitrate from Savannah ecosystems (Rice, 1984).

(d) By enzymic release of ions from solid-phase sources. The best illustrations concern the phosphatase activity of grass roots, demonstrated for upland grasses (Woolhouse, 1969) and cereals (Boero and Thien, 1979). Phosphatase activity could be of importance in soils with an otherwise untapped inositol phosphate pool. The inositol phosphates are probably the most important class of compounds in soil organic phosphate (Dalal, 1977). These compounds have great chemical stability, and enzymic hydrolysis appears to be the only possible means to release the orthophosphate ions (see Fig. 5.9, below). There is evidence for cereals that the enzyme is a metabolic product of the grass rather than the rhizosphere microorganisms (Boero and Thien, 1979). It is possible that mycorrhizal hyphae can release nitrogen by assimilating organic compounds for energy, but this possibility has not been explored with grasses.

(e) By fixing nitrogen. There is now substantial evidence that nitrogen-fixing bacteria form loose associations with the roots of grasses. These include tropical grasses (Dobereiner and Day, 1975; Boddey and Dobereiner, 1984) and seashore grasses – for example, the virtually cosmopolitan *Ammophila arenaria* in the UK (Hassouna and Wareing, 1964) and New Zealand (Line and Loutit, 1971), the salt-tolerant *Elymus farctus* (*Agropyon junceiforme*) in Ireland (Murphy, 1975) and the desert grass *Oryopsis hymenodes* (Wullstein et al., 1979).

On the other hand, a case is also known where the root exudates of a grass are inhibitory to both *Rhizobium* and the free-living nitrogen fixing bacteria. This species, *Aristidia oligantha*, is a widespread annual grass of semi-arid habitats in the USA. It has been studied in the context of old-field succession and its allelopathic action, together with associated herb species, is suspected of slowing nitrogen accretion (Rice, 1984). This would result in an attenuation of the the early stages of succession involving these particular species.

A key role of grass in natural and agro-ecosystems is the synthesis of protein following nitrogen assimilation; however, matching importance must be assigned to the rhizosphere N-fixation process. There are many grasslands in which the productivity of legume components is quite low or absent.

Yet nitrogen inputs or outputs may be observed or inferred. There is a major gap in our knowledge regarding these systems. The key questions to be answered are the following.

(a) Is rhizosphere-linked N-fixation a special or general phenomenon?
(b) How efficient is it?
(c) What are the optimizing environmental characteristics, especially in the light of what may be relevant in terms of management?
(d) How specific are the associations between grass and heterotrophic bacterial type? This may lead to possibilities for specific inocula, analogous to those available for *Rhizobium* and forage legumes.
(e) Does a mineralization step occur between the organic products of N-fixation and assimilation by root or mycorrhiza?
(f) If so, are other heterotrophs essential to the relationship?

Answering at least some of these questions may lead to improved strategies of applied research. First, this will place emphasis on what kind of management husbandry can best be employed to optimize the process in a given sward type. Secondly, it may prove profitable to investigate the selection and improvement of suitable heterotrophs as a means of increasing production or extending the environmental range of species (Boddey and Dobereiner, 1984). This could offer a better return than the transfer of genes for nitrogen fixation to grasses in the medium term.

Whilst it may prove possible in the long term to introduce the genes for nitrogen fixation into grasses, two shorter term strategies suggest themselves.

One is the improvement of knowledge of specificity of both grass and microbe with respect to the nature of interactions in the rhizosphere. Selection for improved tolerance to the presence of ammonium ions in soil solution, for example, would harmonize fertilizer use and biological nitrogen fixation. A search for cold-tolerant strains of effective tropical rhizosphere types also seems attractive.

A second strategy is to explore the genetic properties of the key groups *Azotobacter* and *Azopirillum* more intensively, as suggested by Elmerich (1984). Genetic manipulation is essentially complementary to selection, and the current knowledge of plasmids may lead to valuable returns.

5.3 THE CONTINUUM BETWEEN SOIL AND PLANT

The soil–plant–air continuum, which is the obvious unit of study in plant water relationships (see Chapter 6), is also relevant to ionic relationships; particularly mass transport of ions to roots and within the transpiration stream. However, a more elaborate unit must be utilized to understand fully and explain the ionic aspects of this organism–environment relationship. This may be simply defined as a solid phase–liquid phase–rhizosphere–root–shoot continuum. Within this system equilibria are dependent on physical characteristics (e.g. temperature and particle size), chemical characteristics (e.g. mineralogy, pH and redox state), biological characteristics (nature of soil organic matter and types and activity of microflora) and metabolic characteristics (referring especially to the state of the grass plant).

It is important to recognize these dimensions of complexity, even though some cannot be immediately addressed in any experimental sense. It is equally important to realize that a totally complete integration of all processes may never really be possible. The agronomist and ecologist will still have to rely on an essentially empirical view of some closed 'black boxes' in this continuum. The whole idea of mineral nutrient 'availability' is a key example of approaching the soil–plant continuum with a practical blend of theory and empiricism.

'Available' nutrients are those which are predictably accessible to the sward in the short term, usually the current growing season. In the case of the abundant monovalent cation potassium, it is widely agreed that the soil solution is in equilibrium with potassium ions in the cation exchange complex of the soil. Potassium ions absorbed by soil organisms are rapidly replaced from this source. In exhaustive cropping experiments (Barber, 1984), removing crops of grass from unreplenished soils reveals that the K^+ component of the cation exchange complex is diminished by each crop removal. In this situation slow replacement from 'non-available' mineral-K is observed. From this evidence, potassium availability may be confidently determined by estimating exchangeable potassium. This is a situation in which theoretical and practical concepts of availability correspond. A similar view may be taken of magnesium and calcium, even though solubility-product as well as ion-exchange considerations will play a role. Clearly, this view of availability is only possible because restoration of these particular elements from organic matter to the solid phase of the soil is also rapid and not unduly dependent on the biological activity.

For most other elements 'availability' is clouded, either because of the complex possibilities of the soil chemistry or the intense involvement of biological activity in the nutrient cycle. Because of this, estimates of nitrogen and phosphate availability in particular owe most to empiricism and least to detailed theory. Consequently, they will be dealt with as special cases because of their great agronomic and ecological importance as limiting factors to production.

5.4 A VIEW OF THE ESSENTIALITY OF ELEMENTS – DEFINING ESSENTIALITY IN PHYSIOLOGICAL AND ECOLOGICAL TERMS

Physiologists tell us that, according to the criteria of essentiality proposed by Arnon and Stout (1939), a number of elements are 'essential' for plant growth. However, if we investigate the mineral composition of grass-dominated herbage there are some striking differences between physiological essentiality and tissue composition (Fig. 5.3). These differences may only be explained in the ecosystem context, and the two cases of silicon and cobalt bear close inspection.

Whereas silicon is an abundant element in the composition of crustal rocks and soils, it is usually a minor constituent of plants compared with potassium or calcium. However, in some plant groups, notably the diatomaceous algae, horsetails (*Equisetum* spp.) and the grasses, it assumes major proportions. In many other angiosperms some silicification of particular structures such as spines or stinging hairs occurs. In all cases silicon, as silica (SiO_2), is confined to cell walls or siliceous cell inclusions called 'opaliths' (Iler, 1979). In strictly physiological terms, satisfactory growth can be demonstrated in silicon-less water culture. However, it is clear that in the face of fungal disease the silicon-less plant is much more vulnerable (Sanchez, 1976). It may also be inferred that silica-impregnated cell walls offer a substantial penalty to herbivores in terms of the physical forces required to shear cell walls. This is certainly a major factor in the tooth wear of grazing mammals and the open dentition of many groups. Silica might therefore be regarded as a first line of allelopathological protection for grasses.

At the other end of the concentration spectrum, cobalt is also regarded by the physiologist as non-essential for plant growth. In this case there appear to be no other subtle direct ecological benefits to the plant accruing from its cobalt

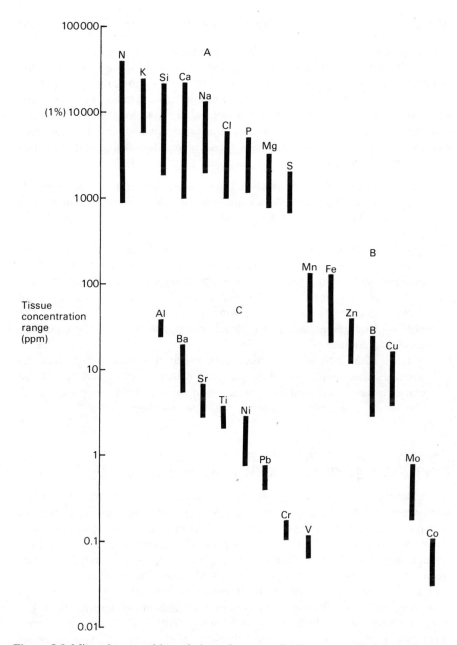

Figure 5.3 Mineral composition of plants from grassland sites compiled from data of Flemming (1973) and Tityanova and Bazilevich (1979). Group A are essential macronutrients plus silicon, sodium and chlorine. The concentrations for these last three elements do not represent abnormally high values. Group B are essential trace elements for herbage or their symbionts. Group C are non-essential elements for plants, and represent commonly encountered ranges. Selenium is missing from the analyses cited, and it is typically encountered in the low ppm range.

content. However, the plant may be clearly seen as the gateway to the ecosystem foodweb for this element. Cobalt is essential for nitrogen-fixing organisms, and hence for continued flow of fixed nitrogen to the ecosystem.

When considering the concept of essentiality, the working agronomist or ecologist needs a link between the biochemical or physiological characteristics and the complexities of ecosystem dynamics and environmental fluctuation. The approach adopted here is to classify ions (elements) of interest into a series of categories, as follows.

Category 1. Ion or element which is physiologically essential and supply is frequently an important direct limitation to the primary production of grassland. The elements nitrogen and phosphorus are the only universal cases.

Category 2. Essential elements for plant growth, which are usually supplied by soils at non-limiting rates: potassium, calcium, magnesium, iron, sulphur, manganese, copper, zinc, boron, molybdenum, sodium and chlorine.

Category 3. Not physiologically essential, but enhance survival of plants in an ecosystem context, e.g. silicon (selenium).

Category 4. Non-essential to plants, but supplied via plants to the foodweb: cobalt, selenium.

Category 5. Elements which, under certain circumstances, may lead to toxicities and hence growth constraints with ecological and agronomic consequences: aluminium, iron, manganese, lead, copper, zinc, chromium, magnesium, sodium, chlorine, sulphur.

5.5 SEASONAL CHANGES IN NUTRIENT ION DEPLOYMENT WITHIN THE PLANT

In both annual and perennial grasses great variations are to be expected in the deployment of the principle nutrient elements with season. The view of phosphate and nitrogen deployment in oats over a 20-week growing season, given by Williams (1955), seems to be a generally acceptable model for annual species (Fig. 5.4). Here the mobile elements are apparently moved tactically from organ to organ as vegetative growth and flowering proceeds. At senescence some 30% of total N and P resides in the seed.

In perennial grasses change in concentration in above-ground tissues is comparatively well studied. In Fig. 5.5, the systematic decline in tissue concentration is shown for three very different grass species. *Lolium perenne* shows a rapid decline, especially of N, from high concentration values (Flemming, 1973) *Agrostis stolonifera* in a Baltic seashore meadow, demonstrates a steady decline through its shorter growing season (Tyler, 1871), while the tropical grass *Hyparrhennia rufa* demonstrates the lowest absolute values, especially at the end of the growing season. The latter grass species is managed by burning, and has probably evolved in an African Savannah

Nutrient ion deployment

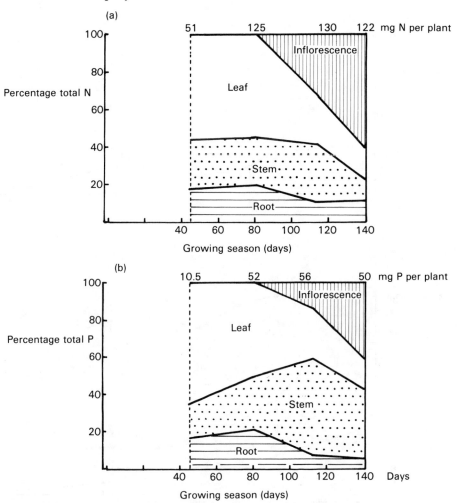

Figure 5.4 (a) Nitrogen and (b) phosphorus deployment in *Avena sativa* (oat) through a growing season to represent a likely pattern for annual grasses. (Data replotted in percentage terms from Williams, 1955.)

environment subject to natural burning (Daubenmire, 1972). The decline in N and P has several components: 'dilution' by tissue biomass growth, redeployment to underground storage and leaching losses to the soil. Not all nutrients behave in this manner, as the data for seasonal magnesium concentrations in *Lolium perenne* and *Agrostis stolonifera* indicate.

Other relatively immobile elements, including iron, manganese and lead, achieve highest concentrations at the end of the growing season. These elements are returned to the soil as leaves senesce. The curves from Tyler (1971) are shown in Fig. 5.6.

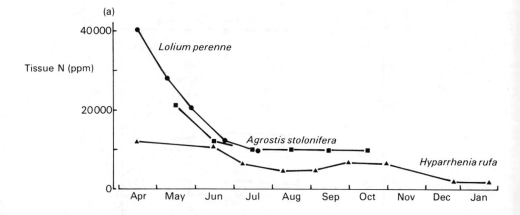
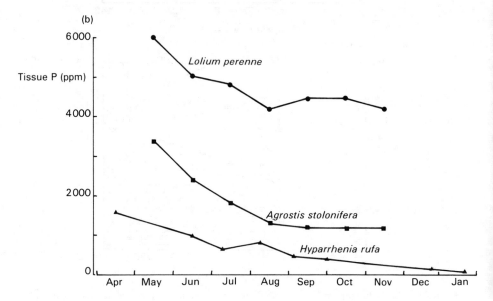
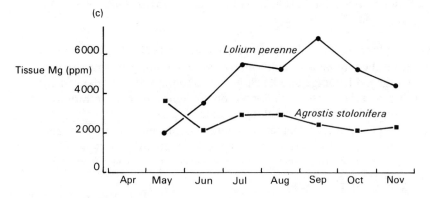

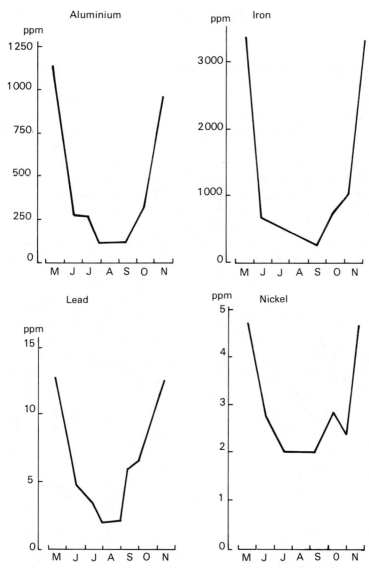

Figure 5.6 Seasonal change in concentrations of minor elements in shoots of *Agrostis stolonifera*, shore meadow, Denmark. Note differences in concentration scale. It is not clear how these patterns are generated, especially the enrichment of tissues at the end of the growing season. (Tyler, 1971.)

◀ **Figure 5.5** Seasonal patterns of elemental concentration of (a) nitrogen, (b) phosphorus and (c) magnesium in perennial grasses. *Lolium perenne*, fertilized pasture, Ireland (Flemming, 1973); *Agrostis stolonifera*, shore meadow, Denmark (Tyler, 1971); *Hyparrhenia rufa*, introduced savannah species, Costa Rica (Daubenmire, 1972). Phosphorus and nitrogen show a marked decline with season, a result of soil availability, 'dilution' by growth and possibly leaching. Magnesium demonstrates different patterns with species.

5.6 MACRONUTRIENT CYCLES

Presenting the nitrogen and phosphorus cycles with a commentary is a means of drawing attention to the view that ion supply is frequently a function of many interactions. This view is of utmost importance in grassland management where fertilizer input is minimal.

Fertilizer applications to grassland can be thought of as amplified natural flushes of availability. Agronomic practice seeks to time the application to optimize growth responses by: synchronizing with natural events in the soil; ensuring that the vegetation is capable of responding; and prolonging growth which may be terminated as the natural availablity pulse declines.

At a purely physiological level the effects of fertilizer–nutrient addition are sufficiently straightforward to be schematically represented (Fig. 5.7). Growth (production) responses are paralleled by a tissue concentration response. In some cases the phase following 'luxury' accumulation may be ion toxicity.

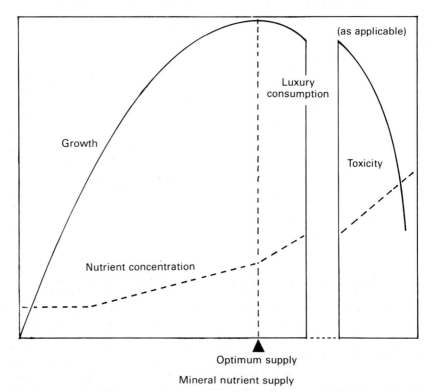

Figure 5.7 The response to increasing concentrations of an essential limiting nutrient may be seen in terms of plant growth and tissue concentration of the element. Growth requirements are satisfied when the growth response ceases. A phase of 'luxury' concentration increase then develops. These two phases may occur in the seasonal growth cycle. Finally, toxicity may intervene, with growth inhibition.

Macronutrient cycles

Ecosystem–process responses to fertilizer application may be more complex, and it must be realized that negative feedback is operating in some important systems. In particular, N and P nutrition becomes less efficient because mycorrhizal function is reduced under high fertility and all nitrogen-fixation processes are curtailed by free ammonia or nitrate. Addition of ammonia will stimulate nitrification, with a higher risk of nitrate loss. More studies are required of the 'mass-balance' of fertilizer–element recovery in agro-ecosystems to plan an optimum use of the energy (fossil fuel) and mineral resources that fertilizers represent.

5.6.1 The nitrogen cycle

Figure 5.8 has the plant at its centre, which is assigned the role of absorbing nitrate, or possibly ammonium, ions and synthesizing protein. Protein synthesis is followed in the cycle by its assimilation, either by the secondary consumers of the ecosystem or by decomposers following senescence. Digestion, catabolic metabolism, excretion and elimination of faeces by animals all have consequences for the cycle. Catabolic metabolism leads to the formation of nitrogen-rich excretory products including uric acid, urea and, in some cases, ammonia. Faeces may contain residual protein as well as a large population of voided micro-organisms. The net effect on the decomposer population of soil is to effect further catabolisms and for ammonium ion to be a principal product. It must be remembered that soil fungi may also immobilize nitrogen as the cell wall substance chitin, a glucosamine polymer. The fate of ammonium is generally directed towards nitrification by a part of the soil microbial population, the specialist pseudomonad chemo-autotrophs. Usually *Nitrosomonas* oxidizes ammonium to nitrite and *Nitrobacter* oxidizes nitrite to nitrate. Variable soil conditions such as wetting–drying or cooling–warming may cause pulses of ammonium or nitrate production (see Clarkson *et al.*, 1986; Runge, 1983).

In some grassland situations the nitrification process appears to be inhibited. Whilst there may be several causes for this, for example low bacterial populations or inhibition by soil moisture or other purely edaphic factors, one possibility is that nitrification inhibitors are secreted by grass roots – a so-called 'allelochemic' mechanism (Rice, 1984). Further investigation is required to determine if this is a widespread occurrence, as it elevates the importance of ammonium as a liquid-phase nitrogen source. This may be seen as having value in avoiding leaching losses.

The univalent ammonium cation may also be diverted to the cation-exchange complex of the soil. This is imagined to be readily interchangeable with ions of the liquid phase, and serves a buffering function for either direct absorption by the root systems or entry to the nitrification process.

Nitrate production is an Achilles' heel of the cycle, in that losses by leaching (commonplace) or denitrification (less common) can be of high order.

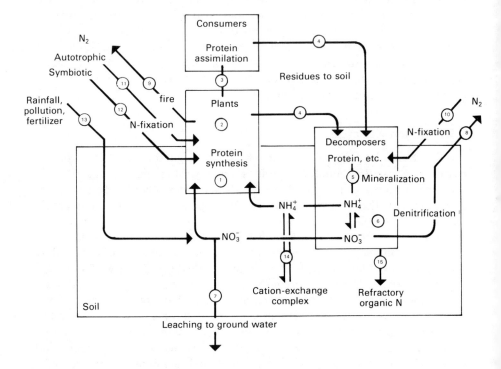

Figure 5.8 The nitrogen cycle. 1, Nitrogen assimilation entails nitrate reduction and the incorporation of ammonium into glutamine. The absorbtion step leads to the release of a counter ion, bicarbonate for nitrate and H^+ for ammonium. 2, Protein synthesis in ecosystems is predominantly a function of plants. 3, Predation takes many forms, but requires that amino acids be reconfigured. 4, Protein catabolism or senenescence and death releases a wide variety of substances to the soil surface. 5, Breakdown of excretory products, e.g. urea, or microbial catabolism of protein gives rise first to ammonium ions. 6, Oxidation of ammonium via nitrite to nitrate is accomplished by the pseudomonad heterotrophs *Nitrosomonas* and *Nitrobacter*. 7, Nitrate, unless absorbed, is vulnerable to leaching. 8, Denitrification is another possible fate. 9, Fire in grasslands can liberate nitrogen as volatile oxides and N_2, but mineralization to nitrate or ammonium may also occur. 10, Nitrogen fixation by heterotrophs may make an important contribution to low productivity systems. 11 and 12, Autotrophic and symbiotic fixation may not benefit grasses until the next season after senescence and mineralization. 13, Other inputs can be very variable. 14, The cation-exhange capacity of soil may contribute to the smoothing of seasonal fluctuations in mineralization. 15, Accumulation of organic N is variable and dependent on the activity of soil organisms.

Macronutrient cycles

Leaching is especially important during the winter months in humid climates.

Another regular loss in some grasslands is by burning, e.g. savannah grasslands (Daubenmire, 1972). The outcome of burning is uncertain because burnable vegetation has a low N content (Fig. 5.5), and although much of this is volatilized, mineralization may also be stimulated. In a semi-arid climate, fire-mineralized N may remain at the soil surface until growth recommences.

The final form of nitrogen removal is in the crop of grazing stock or cut forage. This may only represent a few percent of turnover in a low input system, and be readily replaced by natural means.

The most important inputs in natural grasslands are the biological nitrogen-fixation systems in which fixed carbon is sacrificed for energy to split the dinitrogen molecule. There are generally three systems for importance in grasslands.

(a) Symbiotic nodulated systems, which in grasslands usually means the *Rhizobium*–legume system. It should be remembered, however, that the nodulated legume is also mycorrhizal. For the grass component of a sward to benefit from symbiotic fixation, mineralization of legume tissue probably must always occur.
(b) Other autotrophic systems, blue-green algal crusts being frequently encountered in semi-arid grasslands. Their activity is characteristically intermittent, but mineralization of rewetted material may provide well-timed pulses of available nitrogen.
(c) Free-living heterotrophs, especially those already mentioned as associated with grass root surfaces, need re-evaluation. Measured rates are often low, but widespread uniform occurrence may offset the higher rates of more-thinly scattered symbiotic or autotrophic systems (Clark *et al.*, 1980). The occurrence of natural grasslands with few legumes and yet sustainable production indicates that this source of N input may be of great ecological and economic importance. Long-term experiments lasting one or more seasons indicate gains for grasslands of 5–90 kg N ha^{-1} year^{-1} (Dobereiner and De-Polli, 1980). This compares with several hundred kg N ha^{-1} year^{-1} for symbiotic systems (Sprent, 1979; Phillips, 1980).

5.6.2 The phosphorus cycle

The most difficult segment of this cycle (Fig. 5.9) to elucidate is that concerning the solid phase–liquid phase–plant surface complex. The quantity of phosphate in a given volume of soil is one of the most stable of the soil fertility factors. Losses of phosphate from soil in the temperate zone are low, with an estimated half-life of 20 000 years for losses of volcanic ash soils in New Zealand (Stevens and Walker, 1970). Total soil phosphorus may have a role in determining vegetation type (Jeffrey, 1987). The most oligotrophic grasslands tolerate soil total phosphate concentrations of between 75 and 150 mg P dm^{-3} soil volume,

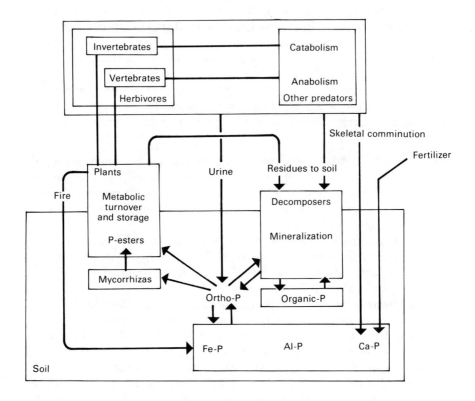

Figure 5.9 The phosphate cycle. Phosphate is assimilated by the plant-mycorrhizal system and rapidly esterified. Invertebrate predators may be of great importance in transferring P to the foodweb, whilst vertebrates accummulate P as skeletal apatite. Orthophosphate-rich urine may be produced by both types. Mineralization by decomposers leads also to the net production of orthophosphate. A separate outcome is the development of a soil organic P pool, which is only tapped through enzymic hydrolysis of stable P esters, especially inositol-phosphates. The orthophosphate pool is in equilibrium with the large pool of inorganic P. Within this pool shifts may occur between Ca–P, Al–P and Fe–P according to soil pH and redox potential or the presence of chelates. Fire may lead to a massive short circuit in the cycle.

whilst plagioclimax grasslands exist on unfertilized former woodland soils with 500 mg P dm^{-3} soil. In the former case a growth response to either phosphate addition or to N and P addition would be anticipated. Below 75 mg P dm^{-3} soil, grasses are replaced by sclerophyllous graminoids, especially of the Cyperacae.

In most cases, however, knowledge of total soil phosphate is not very meaningful in determining the transfer of phosphate from soil to plant. In principle this depends on (a) a set of complex solubility-product relationships between soil minerals and soil solution, (b) mineralization, including the possible enzymic hydrolysis of soil organic phosphate compounds, especially the inositol phosphates (Cosgrove, 1980; Woolhouse, 1969; Tate, 1984) and (c) the metabolic activities and physical penetration of the soil mass by root and mycorrhizal absorbing surfaces.

Many characteristics of a defined system are difficult to determine: for example, the chemistry of the lattice of phosphate minerals; the pH, organic chelate content and redox potential of soil liquid phase; the constitution of solid and dissolved organic phosphate; and the extent and activity of mycorrhizas. Thus, to determine the phosphate supply angronomists are reduced to devising empirical phosphate bioavailability estimations. These may have some bearing on the theory of phosphate supply, but do not mimic the processes described. The effectiveness of arbitrary or empirical procedures is assessed by correlating the phosphate fraction estimated with that absorbed by a particular crop. Procedures include extraction by reagents which relate to soil chemistry; equilibrium with ^{32}P-labelled solutions to determine the size of readily desorbable or 'labile' phosphate; and the use of anion-exchange resin as a sink for orthophosphate ions in soil solution. This last method offers much promise for field studies.

Metabolic esterification of phosphate ions leads effectively to the 'zero-sink' at the root or mycorrhiza surface. Hence, a diffusion pathway is created from any solid-phase sources. Within tissues phosphate is under strong metabolic control, being directed to priority sinks such as apex tissues, developing seeds or storage organs in vegetative tissues. Senescing tissue may contain very little phosphate. Vertebrate herbivores may play a significant part in the cycle, as catabolism leads to orthophosphate-rich urine in mature animals. Vertebrates also accumulate massive skeletons of hydroxyl-apatite, which appear to be dispersed by predators and decomposers either by fragmentation or dissolution in gastric fluids. Unlike the amino acids of proteins, organic phosphates are mineralized completely with each transfer within the foodweb.

5.7 THE COMPETITIVE STATUS OF GRASSES AND THEIR MACRONUTRIENT SUPPLY

The interspecific differences in nutrient concentration in grass tissue, illustrated in Fig. 5.5, imply that species have a particular elemental demand for optimum growth. This is well appreciated in agronomy, and it is clear that the capacity of a soil to support a particular species with a high nutrient demand is related to nutrient availability.

The balance amongst grass species and between grasses and herbs may be governed by macronutrient supply. In experiments carried out by Willis (1963), the application of nitrogen fertilizer to dune pasture (dry) and dune slacks (relatively moist) gave rise to large changes in species composition. *Festuca rubra* and *Poa pratensis* ssp. *subcaerulea* became dominant on fertilized dune pasture, eliminating annuals and smaller plants. The fertilized dune slack vegetation was dominated by *Agrostis stolonifera*, with the reduction or elimination of *Carex* species and a range of herbs.

In an experiment similar in concept, Jeffrey and Pigott (1973) applied N and P to a grassland in Upper Teesdale (UK) containing *Festuca ovina* and the relict arctic–alpine sedge *Kobresia simpliciuscula*. In this case the addition of phosphate released the grasses from their limitation, and the sward become dominated by *Festuca rubra*. This occurred within two seasons and the arctic–alpine species was virtually eliminated by the six-fold increase in cover. In contrast, an adjacent *Sesleria albicans*-dominated sward was virtually unaffected by the same nutrient addition.

The balance between grasses and nitrogen-fixing legumes is also linked to N and P relationships. Working with newly established grass–legume swards on reclaimed mine waste, Jeffrey and Maybury (1981) showed that low P/high N favours grasses, whereas high P/low N increases cover of legumes. Quantifying this balance (Fig. 5.10) is a useful method for managing such a system.

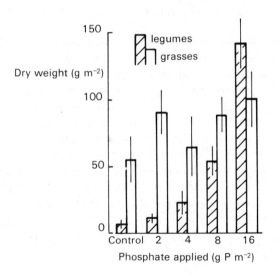

Figure 5.10 Yield of (□) grasses and (▨) legumes in a newly sown sward on rehabilitated mine wastes. Increasing applications of phosphate increase the proportion of legumes. For further details see Jeffrey and Maybury (1981).

5.8 ACCLIMATION OF GRASS PLANTS TO GROWTH-LIMITING SOIL CHARACTERISTICS

The most productive grasslands are on soils cleared from natural woodlands, and ecologically represent plagioclimaxes maintained by grazing or mowing. Limitations to growth arising from nutrient supply restrictions originate either from constraints to nutrient turnover or absolute nutrient depletion arising from cropping.

Nevertheless a significant proportion of world grassland is probably climax (Moore, 1964; Tieszen and Detling, 1983), with production constraints arising from climate, closely linked to low soil-water availability. Soil salinity, soil aeration or natural fire are additional primary limiting factors. Second-order soil phenomena associated with temperature and daylength constrain growth of grasses in alpine or high-latitude sites. In some regions of high rainfall, low base saturation is associated with high available aluminium combined with ion removal by leaching.

Environmental constraints may be viewed as driving forces in plant evolution, leading towards acclimation and survival. Reduced growth is obviously a penalty, and it may be questioned if the sought-after characteristics of yield 'close to theoretically achievable' improves the survival of any plant species. Agronomists and physiologists may use the term 'stress' to denote an unmanageable environmental factor depressing high growth. In an ecological context a better term is 'environmental constraint' to which all surviving species are axiomatically acclimated.

5.8.1 Accommodating to environmental constraints

In the majority of grasslands growth constraints are brought about by climate, soil moisture or macro-nutrient deficiency, especially nitrogen. It is accepted that species survive because of acclimation to these constraints. Three other kinds of environmental constraint are regularly, if infrequently, encountered: soil salinity, anaerobic soils and high concentrations of available toxic metals. It is worth briefly reviewing what is known about the mechanisms for acclimation.

(a) Salinity

There are many halophytic grasses, including the genus *Spartina* and species of *Festuca, Agrostis, Distichis, Elymus* and *Puccinellia*. If these are capable of surviving in substrates with solute potentials lower than -2.4 MPa (full sea water), then these halophytes must sustain a leaf-tissue water potential slightly more negative than this. Low water potential in tissue is largely achieved by vacuolar sap with high ionic concentration. However, cytoplasmic enzymes of halophytes are no less sensitive to ions such as sodium or chloride than non-halophytes (Stewart and Ahmad, 1983). These ions are generally excluded from cytoplasm, which raises the issue of generating a low solute potential in cytoplasm. The answer is that a series of organic solutes which are compatible with metabolism are being isolated from halophytes. It may be too early to comment on an overall physiological picture, but from the available work on grass halophytes, there are at least three profiles of compatible osmotic metabolite pattern (Briens and Larher, 1982).

(a) Species dependent mainly on soluble carbohydrates; i.e. sucrose and polyols such as sorbitol and mannitol, e.g. *Phragmites communis*.
(b) Species which accumulate both carbohydrate, as above, and nitrogenous compounds such as glycine betaine and proline, but with emphasis on carbohydrate, e.g. *Festuca rubra, Puccinellia maritima* and *Spartina townsendii*.
(c) Species in which nitrogenous solutes predominate, e.g. *Agropyron (Elymus) pungens* and *Distichlis spicata*. In these cases the compatible metabolite may account for a substantial percentage of total nitrogen in the plant.

In *Distichlis spicata* it has been demonstrated that rapid proline synthesis is the response of leaf tissue cultures exposed to increased sodium chloride concentration (Daines and Gould, 1985). In many saline environments substantial seasonal changes in soil salinity can be expected, and methods of plant osmoregulation are of interest.

An interesting speculation arising out of the distribution of compatible metabolites is the idea of a wider role for this set of metabolites. *In vitro* studies indicate that proline, betaine, sorbitol and manitol will stabilize enzymes against heat (Smirnoff and Stewart, 1985). Moreover, these metabolites accumulate in non-halophyte sand-dune species such as *Ammophila arenaria*. Their role in protecting against drought-induced heatload and possible desiccation deserves further study.

(b) Anaerobic Soils

Certain grass species are clearly able to sustain growth on permanently or regularly waterlogged soils. *Oryza sativa* (rice), *Molinia caerulea* (purple moorgrass), *Deschampsia caespitosa* (tufted hair grass), *Phragmites* (common reed) and *Echinochloa crus-galli* (barnyard grass) are well-known species from sites waterlogged with fresh water, whilst *Puccinellia maritima* and *Spartina* sp. are characteristic of anaerobic saline soils.

Damaging effects of anaerobic soils are due to enforced glycolysis of root tissue and high availability of Fe^{2+}, Mn^{2+} and S^{2-} (e.g. Kozlowski, 1984). The most frequently encountered protective response in grasses is morphological change to tissue structure. Lysis of cortical tissue provides for a gas-diffusion pathway from leaves to roots (aerenchyma). This not only provides for an internal aerobic environment, but also aerates the soil adjacent to the root. The

Acclimation of grass plants

need for glycolysis is avoided and ionic toxicity is prevented. There is also evidence, however, that species can adapt to the end-products of glycolysis (Crawford, 1978) and possess tolerance to toxic metals and sulphide ions (Singer and Havill, 1985; Havill et al., 1985). The view of Crawford (1978) that ethanol accumulation is avoided in flood-tolerant species can be regarded as a working hypothesis, but as Jackson and Drew (1984) point out, it does not seem to apply universally.

Flooding tolerance may not be absolute. If the first line of protection to waterlogging is aerenchyma, a second may be enzymic destruction of ethanol. The long duration and intensity of the redox effect may eventually overwhelm the defences of the plant. It is well known that even rice may eventually show symptoms of an iron toxicity syndrome.

(c) Accommodating to the toxicity of metals in soils

Three metal ions may frequently prove toxic to species under certain soil conditions. Trivalent aluminium (Al^{3+}) becomes increasingly available at a soil pH of less than 5.0. Divalent iron and manganese ions may dominate the soil solution under waterlogging conditions. Ordinarily, in aerated soils iron and manganese hydroxides and oxides have such low solubilities that the secretion of cheleting agents and the lowering of soil pH may be necessary to facilitate uptake in broad leaved species (Bienfait and Van der Mark, 1983). Mechanisms for iron and manganese efficiency have not been studied for grasses, although survival of extreme calcicole species (e.g. *Briza media*) implies an effective mobilizing system.

In geochemically uncommon circumstances, both natural and man-contrived, high concentrations of other toxic metals may occur in soils, especially lead, copper, zinc, cadmium, chromium and nickel. In all of these cases it is known that, on the one hand, species that are sensitive to the above metals fail to survive whilst, on the other hand, a limited number of species accommodate and succeed in completing their life-cycle. Grasses are included in both groups. The large literature on the availability of toxic metals, mechanisms of toxicity and tolerance was reviewed by Woolhouse (1983). Tolerance to excess metals, given their availability in the liquid phase, is seen as being due to immobilization at the cell wall or cell membrane surface, and to complexing within the cell, especially within the vacuole. A fascinating feature of the grasses are the large number of known metal-tolerant ecotypes, which make good models for studying both metal toxicity and tolerance. Tolerant ecotypes generally possess better metal binding capacity, which may lead to metal accumulation. At least two mechanisms for metal binding in grasses are proposed by recent literature.

In a study of zinc-tolerant and non-tolerant stains of *Deschampsia caespitosa* exposed to high zinc concentration, organic acids are responsible for zinc binding and tolerance (Godbold et al., 1984). Their work suggests that the tolerant strain complexes zinc in the cell vacuole with citrate. Malate may also be associated with zinc in the cytoplasm. Organic acids could complex a range of metal ions, including iron, nickel, aluminium and manganese, but evidence is so far lacking. An alternative metal-binding system is the metallothionein proteins, well known from animal systems. Here it binds copper, zinc and cadmium, and is known to be inducible, i.e. the protein quantity increases in response to the metal ion concentrations. Metallothioneins binding copper and cadmium with similar properties to those in animals have been isolated from *Agrostis gigantea* and *Zea mays* (Rauser, 1984a,b). These two distinct mechanisms are not mutually exclusive, and now that good analytical techniques are available, they may serve as ways to investigate tolerance and metal metabolism generally for Cu, Zn and Cd. An important topic is the transfer of metals from herbage to livestock in the two contexts of nutritional sufficiency and toxicity. The most enigmatic of the common metal tolerances is that of lead. It is difficult to separate availability phenomena from exclusion. Lead uptake is low even when toxicity is evident, and translocation from root to foliage is usually slight. There is no evidence at present that either of the two proposed metal-binding mechanisms operate in known lead-tolerant ecotypes.

The discovery of a series of protective metabolic systems which permit acclimation to extreme environmental conditons adds weight to the attitude that we are not dealing with 'stress'. In the author's view this word should be reserved for effects to the 'non-tolerant' form, unable to acclimate, and showing pathological symptoms leading towards death of the plant. Thus, 'stress' arising from soil conditions is due to mismanagement in agronomy and unusual catastrophe under natural or semi-natural conditions.

5.8.2 Conclusions

In managing grasslands for forage production we need to make the best use of all resources, including fertilizers, to achieve sustainable production. The relative price of fertilizer elements will increase as energy costs escalate, (nitrogen) and mineral reserves are depleted, and as the cost of working them and transportation increases. Hence, the relative economic benefits of injecting transient available nutrients into the soil–plant system will predictably decline.

Strategy for managing the minerals in a sward might be directed two ways.

(a) Making optimum use of applied nutrients.
(b) Seeking more-effective self-sustaining systems for nitrogen sufficiency and phosphate efficiency. In both cases the whole soil–plant system needs to be studied. This means very many possible combinations of experimental investigation.

A first objective might be to discover for each sward system minimum fertilizer regimes which permit the unimpeded function of natural nitrogen fixation and mycorrhizal activity. The profitablity of these regimes should be explored under a series of scenarios of different product values and fertilizer costs. A second objective could be to seek naturally occurring high-efficiency systems which might form a basis for further improvement. Heterotrophic nitrogen fixation and endotrophic mycorrhizas are virtually unexploited biological resources in genetic improvement terms.

The other application of grassland mineral nutrient studies is in nature conservation management. Here the objective is species richness, especially in terms of grasses, herbs and associated invertebrate and vertebrate grazers. The key variables for a particular equilibrium between species and soil environmental factors are seldom fully analysed. A key rule-of-thumb is to maintain the *status quo* at all costs. However, we must address the task of knowing the precise nature of the critical balance between mineral nutrient supply, constraining environmental factors and biological interactions. A management objective sharpens the perception of many otherwise 'pure' ecological problems. At present in Ireland the grasslands of the Burren, the seasonally flooded meadows of the Shannon valley and many ancient pastures are subjects for this approach.

Curiously, management for productivity and management for species diversity may have much of mutual interest. Research managers and conference organizers should aim to combine these two objectives more frequently.

REFERENCES

Arnon, D. I. and Stout, P. R. (1939) The essentiality of certain elements in minute quantity for plants with special reference to copper. *Pl. Physiol.*, **14**, 371–5.

Barber, S. A. (1984) *Soil Nutrient Bioavailability,* John Wiley and Sons, New York.

Boddey, R. M. and Dobereiner, J. (1984) Nitrogen fixation associated with grasses and cereals, in *Current Developments in Biological Nitrogen Fixation,* (ed. N. S. Subba Rao), Edward Arnold, London, pp. 277–314.

Boero, G. and Thien, S. (1979) Phosphatase activity and phosphorus availability in the Rhizosphere of corn roots, in *The Soil–Root Interface* (eds J. L. Harley and R. Scott Russel) Academic Press, London, pp. 231–242.

Bienfait, H. F. and Van der Mark, F. (1983) Phytoferritin and its role in iron metabolism, in *Metals and Micronutrients: Uptake and Ultilization by Plants* (eds D. A. Robb and W. S. Pierpoint) Academic Press, London, pp. 111–23.

Briens, M. and Larher, F. (1982) Osmoregulation in halophytic higher plants: a comparative study of soluble carbohydrates, polyols, betaines and free proline. *Plant, Cell, Envir.*, **5**, 287–92.

Chapin, F. S. (1980) The mineral nutrition of wild plants. *A. Rev. Ecol. Systematics,* **11**, 233–60.

Clark, F. E., Cole, C. V. and Bowman, R. A. (1980) Nutrient cycling in grasslands, systems analysis and man, in *International Biological Programme 19* (eds A. L. Breymeyer and G. M. Van Dyne) Cambridge University Press, Cambridge, pp. 659–712.

Clarkson, D. T. (1985) Factors affecting mineral acquisition by plants. *A. Rev. Pl. Physiol.*, **36**, 77–115.

Clarkson, D. T., Hooper, M. J. and Jones, L. H. P. (1986) The effect of root temperature on the uptake of nitrogen and the relative size of the root system in *Lolium perenne* 1. Solutions containing both NH_4^+ and NO_3^-. *Plant, Cell, Envir.*, **9**, 535–45.

Cosgrove, D. J. (1980) *Inositol Phosphates: Their Chemistry, Biochemistry and Physiology*, Elsevier, Amsterdam.

Cox, G., Saunders, F. E., Tinker, P. B. and Wild, J. A. (1975) Ultrastructural evidence relating to host–endophyte transfer in a vesicular–arbuscular mycorrhiza, in *Endomycorrhizas* (eds F. E. Saunders, B. Mosse and P. B. Tinker), Academic Press, London, pp. 297–312.

Crawford, R. M. M. (1978) Metabolic responses to anoxia, in *Plant Life in Anaerobic Environments* (eds D. D. Hook and R. M. M. Crawford) Science Publishers, Ann Arbor, Michigan, pp. 119–136.

Crush, J. R. (1973) Significance of endomycorrhizas in Otago tussock grasslands. *N. Z. J. Bot.*, **11**, 645–60.

Daines, R. J. and Gould, A. R. (1985) The cellular basis of salt tolerance studied with tissue cultures of the halophytic grass *Distichlis spicata*. *J. Pl. Physiol.*, **119**, 269–80.

Dalal, R. C. (1977) Soil organic phosphorus. *Adv. Agron.*, **29**, 83–117.

Daubenmire, R. (1972) Ecology of *Hyparrhenia rufa* in derived savannah in north-western Costa Rica. *J. Appl. Ecol.;* **9**, 11–23.

Dobereiner J. and Day, J. M. (1975) Associative symbioses in tropical grasses: characterisation of microorganisms and dinitrogen fixing sites, in *Proc. First Int. Symp. Nitrogen Fixation*, (eds W. E. Newton and C. J. Nyman), Washington State University Press, Pullman, Washington, pp. 518–38.

Dobereiner, J. and De-Polli, H. (1980) Diazotrophic rhizocoenoses, in *Nitrogen Fixation* (eds W. D. P. Stewart and J. R. Gallon), Annual Proceedings of the Phytochemical Society of Europe No. 18, Academic Press, London, pp. 301–34.

Elmerich, C. (1984) *Azotobacter* and *Azospirillum* genetics and molecular Biology, in *Current Developments in Biological Nitrogen Fixation* (ed. N. S. Subba Rao), Edward Arnold, London, pp. 315–46.

Flemming, G. A. (1973) Mineral composition of herbage, in *Chemistry and Biochemistry of Herbage*, Vol. 1 (eds G. W. Butler and R. W. Bailey), Academic Press, London, pp. 529–66.

Godbold, D. L., Horst, W. J., Collins, J. C., Thurman, D. A. and Marschner, H. (1984) Accumulation of zinc tolerant and non-tolerant ecotypes of *Deschampsia caespitosa*. *J. Pl. Physiol.*, **116**, 59–69.

Hassouna, M. G. and Wareing P. F. (1964) Possible role of rhizosphere bacteria in the nitrogen nutrition of *Ammophila arenaria*. *Nature*, **202**, 467–9.

Havill, D. C., Ingold, A. and Pearson, J. (1985) Sulphide tolerance in coastal halophytes. *Vegetatio*, **62**, 279–85.

Iler, R. K. (1979) *The Chemistry of Silica*, John Wiley, New York.

Jackson, M. J. and Drew, M. C. (1984) Effects of flooding on growth and metabolism of herbaceous plants. in *Flooding and Plant Growth* (ed. T. T. Kozlowski), Academic Press, Orlando, pp. 47–128.

Jeffrey, D. W. (1987) *Soil–Plant Relationships – an Ecological Approach*, Croom-Helm, London.

Jeffrey, D. W. and Maybury, M. M. (1981) Scientific studies of mine waste revegetation: The assessment and reduction of heavy metal toxicity in the revegetation of mining wastes. *Irish J. Envir. Sci.*, **1**, 49–56.

References

Jeffrey, D. W. and Pigott, C. D. (1973) The response of grasslands on sugar-limestone in Teesdale to application of phosphorus and nitrogen. *J. Ecol.*, **61**, 85–92.

Kozlowski, T. T. (ed.) (1984) *Flooding and Plant Growth*, Academic Press, Orlando.

Line, M. A. and Loutit, M. W. (1971) Non-symbiotic nitrogen-fixing organisms from some New Zealand tussock-grassland soils. *J. Gen. Microbiol.*, **66**, 309–18.

Moore, C. W. E. (1964) Distribution of grasslands, in *Grasses and Grasslands* (ed. C. Barnard), Macmillan, London, pp. 182–205.

Murphy, P. M. (1975) Non-symbiotic nitrogen fixing bacteria in Irish soils. *Proc. R. Irish Acad.*, **75**, 453–64.

Phillips, D. A. (1980) Efficiency of symbiotic nitrogen fixation in legumes. *A. Rev. Pl. Physiol.*, **31**, 29–49.

Rauser, W. E. (1984a) Estimating metallothionein in small root samples of *Agrostis gigantea* and *Zea mays* exposed to cadmium. *J. Pl. Physiol.*, **116**, 253–60.

Rauser, W. E. (1984b) Isolation and partial purification of cadmium-binding protein from roots of the grass *Agrostis gigantea*. *Pl. Physiol.*, **74**, 1025–9.

Rice, E. L. (1984) *Allelopathy*, 2nd edn, Academic Press, Orlando.

Rovira, A. D. (1979) Biology of the soil–root interface, in *The Soil–Root Interface* (eds J. L. Harley and R. S. Russell), Academic Press, London, pp. 145–60.

Runge, M. (1983) Physiology and ecology of nitrogen nutrition, in *Physiological Plant Ecology* Vol. IV (eds O. L. Lange, P. S. Nobel, C. B. Osmond and H. Ziegler), Springer, Berlin, pp. 163–200.

Sanchez, P. A. (1976) *Properties and Management of Soil in the Tropics*, John Wiley and Sons, New York.

Saunders, F. E., Mosse, B. and Tinker, P. B. (eds) (1975) *Endomycorrhizas*, Academic Press, London.

Singer, C. E. and Havill, D. C. (1985) Manganese as an ecological factor in salt marshes. *Vegetatio*, **62**, 287–92.

Smirnoff, N. and Stewart, G. R. (1985) Stress metabolites and their role in coastal plants. *Vegetatio*, **62**, 273–8.

Smith, S. E. (1980) Mycorrhizas of autotrophic higher plants. *Biol. Rev.*, **55**, 475–510.

Sprent, J. (1979) *The Biology of Nitrogen Fixing Organisms*, McGraw-Hill, London.

Stewart, G. R. and Ahmad, I. (1983) Adaptation to salinity in angiosperm halophytes, in *Metals and Micronutrients, Uptake and Utilization by Plants*, (eds D. A. Robb and N. S. Pierpoint), Academic Press, London, pp. 33–50.

Stevens, P. R. and Walker, T. W. (1970) The chronosequence concept and soil formation. *Q. Rev. Biol.*, **45**, 333–50.

Tate, K. R. (1984) The biological transformation of P in soil. *Pl. Soil*, **76**, 245–56.

Tieszen, L. L. and Detling, J. K. (1983) Productivity of grassland and tundra, in *Physiological Plant Ecology*, Vol. IV (eds O. L. Lange, P. S. Nobel, C. B. Osmond and H. Ziegler), Springer, Berlin, pp. 173–203.

Titlyanova, A. A. and Bazilevich, N. I. (1979) Ecosystem synthesis of meadows – nutrient cycling, in *Grassland Ecosystems of the World*, (ed. R. T. Coupland), Cambridge University Press, Cambridge, pp. 170–80.

Tutin, T. G. (1957) A contribution to the experimental taxonomy of *Poa annua* L. *Watsonia*, **4**, 1–10.

Tyler, G. (1971) Studies in the ecology of Baltic sea-shore meadows IV. Distribution and turnover of organic matter and minerals in a shore meadow ecosystem. *Oikos*, **22**, 265–91.

Williams, R. F. (1955) Redistribution of mineral elements during development. *A. Rev. Pl. Physiol.*, **6**, 25–42.

Willis, A. J. (1963) Braunton Burrows: the effects on the vegetation of the addition of mineral nutrients to the dune soils. *J. Ecol.*, **51**, 353–74.

Woolhouse, H. W. (1969) The acid phosphatases of plant roots, in *Ecological Aspects of the Mineral Nutrition of Plants* (eds I. H. Rorison, A. D. Bradshaw, M. J. Jennings and P. B. Tinker), Blackwell, Oxford, pp. 357–80.

Woolhouse H. W. (1983) Toxicity and tolerance in the responses of plants to metals, in *Physiological Plant Ecology* Vol. III (eds O. L. Lange, P. S. Nobel, C. B. Osmond and H. Ziegler), Springer, Berlin, pp. 245–300.

Wullstein, L. H., Bruening M. L. and Bollen, W. B. (1979) Nitrogen fixation associated with sandgrain root sheaths (rhizosheaths) of certain xeric grasses. *Physiol. Pl.*, **46**, 1–4.

CHAPTER 6
Water relations

M. B. Jones

6.1 INTRODUCTION

On a world scale it is arguable that shortage of water is the single most important factor limiting crop yields (Begg and Turner, 1976). However, in many situations there is an interaction between water deficits and nutrient deficiency (Garwood and Williams, 1967), which is due primarily to the non-availability of nutrients in dry soil horizons; this means that it is often difficult to separate these two areas of influence on plant growth. In addition, high temperatures and low humidities which have direct effects on plant processes are often associated with conditions of water deficit, so that the influence of climatic factors during periods of water shortage are complex. Water and nutrient availability are both subject to change by management but, in contrast with the extensive use of fertilizers, irrigation of agricultural grasslands is not widely practised, although there are some situations where it is economically viable, even in temperate regions (Doyle, 1981). Despite the limited scope for reducing water shortage, the understanding of how grasses respond when suffering water deficits may help to improve management and it may also help in the selection of new, more stress-tolerant varieties.

The growth of grasses can also be affected by excess of water, that is waterlogging or flooding. Flooding leads to the rapid depletion of soil oxygen and to changes in the physiological processes of plants that markedly influence their growth and survival (Kozlowski, 1984). While much has been published on the effects of drought on agricultural plants, relatively little attention has been paid to the response of these plants to flooding. This is presumably because flooding in temperate environments does not normally occur during the time when growth is most rapid. However, the differential sensitivity of grass species to flooding could have a marked effect on sward composition (Morrison and Idle, 1972). This is because certain grass species such as *Holcus lanatus* seem to be able to adapt to waterlogging by producing nodal and soil surface roots which can exploit the more aerated upper layers of the soil (Watt and Haggar, 1980). Due to the dearth of information on the effects of waterlogging on agricultural grasses the emphasis in this chapter will be on the effects of water deficits which induce a stress. The term 'stress' is difficult to define precisely, but it is used widely in plant biology to describe any factor that disturbs the normal functioning of the plant (Kramer, 1980).

The aim is to explain the importance of water to the grass plant and in particular to show how plants obtain and use water, and what effects shortage of water have. To help understand this it is important to appreciate that grassland vegetation is a component of a cycle of water from the soil to plant to atmosphere and back again to the soil; the 'hydrological cycle'. The components of this hydrological cycle in vegetation are illustrated in Fig. 6.1, where the cycle is considered in terms of inputs and losses of water to and from the soil and plant.

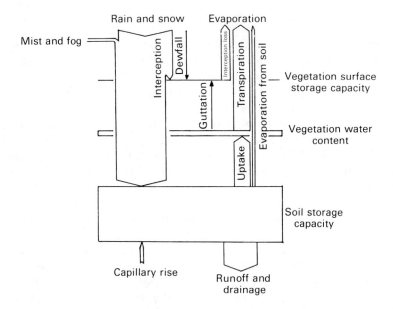

Figure 6.1 The hydrological cycle in a vegetation–soil system. Components approximately to scale for annual rainfall of 1000 mm and evapotranspiration of 500 mm. (From Rutter, 1975.)

6.2 THE HYDROLOGICAL CYCLE

Clearly, the magnitude of the components of the hydrological cycle vary with weather conditions, soil type and vegetation and the illustration used applies to a situation where annual rainfall is 1000 mm and evaporation 500 mm. There are two features of the cycle which are of particular importance to the plant's water relations. The first is that, as well as moving through the cycle, water is also stored in parts of the cycle. These storage sites are at the plant surface,

The hydrological cycle

within the plant and within the soil, and the orders of magnitude of these components are approximately 1, 10 and 100, respectively. This capacity of soil to store relatively large amounts of water is very important for the survival of plants, particularly when rainfall is intermittent.

The second point is that when the grass plant is actively growing there is an almost continuous flow of water through the plant from the soil to the atmosphere. Plants often transpire more than their own weight of water in a day, and this is an inevitable consequence of the uptake of CO_2 from the atmosphere in photosynthesis. The reason for this is that the major route for diffusion of CO_2 into the leaves is through the stomata and down a concentration gradient from the atmosphere to the chloroplasts, while at the same time a gradient in water vapour concentration exists in the opposite direction from the cell walls to the atmosphere, which drives the process of transpiration.

The pathway for water movement from the soil through plants to the air has been called the soil–plant–atmosphere continuum (Philip, 1966), and was most conveniently treated by van den Honert (1948) as a number of components where the resistances to water movement are in series, through both liquid and vapour phases, so that

$$E = \frac{\Psi_r - \Psi_s}{r_{s,r}} = \frac{\Psi_l - \Psi_r}{r_{r,l}} = \frac{\Psi_a - \Psi_l}{r_{l,a}} = \frac{\Psi_a - \Psi_s}{r_{\text{total}}} \tag{6.1}$$

where E is the water flux (transpiration) through the system; $r_{s,r}$, $r_{r,l}$ and $r_{l,a}$ are the resistances of the transfer pathways between soil (s), root (r), leaf (l) and atmosphere (a); and Ψ_s, Ψ_r, Ψ_l and Ψ_a are the water potentials along the pathway. This analysis is similar to Ohm's law describing the flow of electricity, and this treatment of water movement in liquid and gaseous phases is often termed the Ohm's law analogy. Clearly, using this analogy the resistance to water movement through the various parts of the system can be determined from a knowledge of the fall in water potential across them. Typical values of well-watered soils are close to -0.5 MPa while Ψ_l is around -1.5 MPa when transpiration is occurring, giving $\Delta\Psi$ in the liquid phase of 1.0 MPa. The water potential in the atmosphere is commonly lower than -50 MPa at typical levels of relative humidity, giving a $\Delta\Psi$ in the vapour phase more than an order of magnitude grater than in the liquid phase. It is therefore apparent that the major resistance in the continuum is associated with transpiration rather than uptake. Because of this conclusion, work on understanding the physiological control of water movement through plants has emphasized the importance of the major variable resistance between the intercellular spaces of the leaf and the air – the stomata. Although this interpretation highlights the importance of stomata it must be pointed out that there is a discontinuity at the leaf surface between the physics of liquid and vapour transfer (Davies, 1985) such that movement of liquid water in the plant and soil is proportional to the water-potential gradient while gaseous movement is proportional to the vapour-pressure difference. Because there is a logarithmic relationship between water potential and vapour pressure, it is clear that, strictly speaking, water vapour flux from the leaf cannot be proportional to the water-potential gradient. Calculations using $\Delta\Psi$ in the vapour phase in fact lead to a substantial overestimate of the magnitude of the relative resistance in the vapour phase (Kramer, 1983), although they do not invalidate the conclusion that the stomata have a central role in controlling transpiration.

Further complications in the use of the van den Honert analysis arise first because it assumes constant resistances in parts of the pathway such as the roots, whereas the resistances have in some cases been shown to decrease with flow rate (Boyer, 1974). Secondly, it assumes a steady state, whereas flow through different parts of the plant often occurs at different rates. For example, when transpiration increases rapidly after sunrise, absorption lags behind transpiration and there is a reduction in leaf and stem water content. The use of this stored water in transpiration draws upon the capacitance of the plant tissues for water, where capacitance is analogous to the property of some electrical circuits to store electricity in capacitors or condensers in such a manner that it can be released later into the circuit. Capacitance in plants is defined as the ratio of the change in tissue water content to the change in water potential (Powell and Thorpe, 1977). The water storage capacity of different plant parts are variable, but roots could be an important reserve, as they have been shown to shrink to as little as 60% of their turgid diameter when transpiration rates are high (Davies, 1985).

Clearly, equation (6.1) is an oversimplified model of the flow of water through the soil–plant–atmosphere continuum, and in some cases it may even be misleading (Davies, 1985). However, it is still useful in establishing the basic principles upon which the flow of water through plants can be understood.

6.2.1 Inputs of water

The major hydrological input to the plant–soil system is precipitation in the form of rainfall or snow, except where irrigation is practised. There is, of course, a very wide range of annual levels of precipitation in addition to the marked seasonal variations in this input. Agricultural grasses, like all other major crops, are termed aridopassive, because they carry no active photosynthetic tissue during prolonged dry periods and consequently they have a growth pattern which is often closely in phase with water input (Fischer and Turner, 1978).

In addition to rain and snow, water can reach the soil and vegetation in the form of mist, fog or dew. The meteorological conditions which produce mist and fog are very localized and their importance to water inputs are therefore variable, although in general of little significance compared with the input through precipitation. Dew forms on surfaces if night-time longwave radiation exchange causes the surface temperature to fall below the dewpoint of the surrounding air. Monteith (1973) has shown that the maximum rate of dew

The hydrological cycle

formation, which occurs when the air is saturated, is about 0.06–0.07 mm h^{-1}, and this corresponds well with the measured amounts of dew formation ranging between 0.2 and 0.4 mm night^{-1}. The source of water falling as dew can be either the atmosphere or the soil, but a net gain to the plant–soil system results only from atmospheric supply. Although dew is only a small proportion of the total water input, it can have a significant effect on the water relations of grasses. For example, Kerr and Beardsell (1975) showed that the leaf water potential (Ψ_l) of *Paspalum dilatatum* declined less rapidly between sunrise and midday when dew was present than when it was absent (Fig. 6.2). They suggested that because leaf expansion can be reduced by a small decline in Ψ_l (Boyer, 1968), the presence of dew on leaves after sunrise may prolong the period of growth.

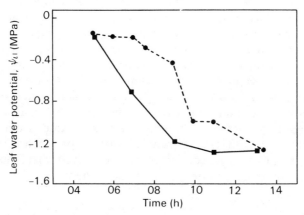

Figure 6.2 The effect of dew on the morning decline in leaf water potential (Ψ_l) in *Paspalum dilatatum*, ●--●, measurement on morning with dew, ■——■, measurements on morning without dew. (From Kerr and Beardsell, 1975.)

6.2.2 Canopy interception

When precipitation falls, a proportion of it is first intercepted by the canopy, the amount depending on how much vegetation is present. Some of the water is retained, particularly if the canopy is dry, but when its storage capacity is exceeded the rest falls or is directed to the soil by stemflow. Stemflow of precipitation may provide more water for some species and give them a competitive advantage in mixed communities (Ndawula-Senyemba *et al.*, 1971). The storage capacity of a grass sward depends on the leaf area over which a film of water or droplets can reside. Merriam (1961) obtained values of 0.5 and 2.8 mm for *Lolium perenne* with sward heights of 10 and 48 cm, respectively. The importance of this relatively small component of water storage to the hydrological cycle is that water evaporates from it directly to the atmosphere and is therefore not available to the plant. The ease of the loss by evaporation in relation to transpiration can be expressed as the 'relative transpiration rate' (Monteith, 1965), and as far as grass canopies are concerned it is very close to unity (McMillan and Burgy, 1960). Thus, interception causes no greater loss than would occur through transpiration and the Penman–Monteith formula (see below) used for calculating potential transpiration from a grass canopy is also a measure of potential evaporation from wet surfaces (Rutter, 1975).

6.2.3 Losses of water from soil and plant

The loss of water from the soil–plant system is essentially an energy-dependent evaporative process. Working on this principle, Penman (1948) was the first to derive physically based equations to describe evaporation from open water surfaces or from short green vegetation, particularly grassland, amply supplied with water. Penman assumed that in effect evaporation occurred on the outer surface of leaves rather than within their substomatal cavities. However, movement of water vapour from the leaf mesophyll cells through the stomata and into the bulk air above the canopy is a diffusion-limited process, and stomatal and aerodynamic conductances (the reciprocal of resistances) are therefore important factors controlling the loss of water from vegetation. Consequently, Monteith (1965) included surface parameters in terms of stomatal and boundary-layer conductances, and derived what is now known as the Penman–Monteith equation, which can be written as

$$\lambda E = \frac{sA + c_p \varrho D g_a}{s + \gamma(1 + g_a/g_c)} \tag{6.2}$$

where E is the rate of evapotranspiration, A is available energy (net radiation minus soil heat flux) (W m^{-2}), s is the rate of change of saturated vapour pressure with temperature (kPa °C^{-1}), λ is the latent heat of vaporization of water (2440 J g^{-1}), γ is the psychrometric constant (66 Pa °C^{-1}), ϱ is the density of air (kg m^{-3}), c_p is the specific heat of air (J g^{-1} °C^{-1}), D is the vapour deficit of the air (kPa) and g_c and g_a are the canopy stomatal and boundary layer conductances (m s^{-1}), respectively.

This equation is used extensively to estimate the water loss from grass swards, and four of the input variables required – available energy, vapour pressure deficit, air temperature and windspeed (used to calculate g_a) – are easily determined by an automatic weather station placed above the canopy. Simple corrections can be introduced to allow for the difference between surface and screen temperature (Gardner and Field, 1983). The determination of canopy stomatal conductance is more difficult, requiring the measurement of profiles of temperature, humidity and windspeed above the canopy (Monteith, 1965). Alternatively, it can be calculated from the product of leaf area index and individual leaf conductances measured with a diffusion

The hydrological cycle

porometer (Szeicz et al., 1973; Squire and Black, 1981). However, in this case a difficulty arises because of the variation in leaf conductance within the stand, so that canopy conductance is estimated more accurately by dividing the canopy into several distinct layers. Jarvis (1981) discusses in detail the problems associated with the determination of canopy conductance.

Interestingly, for grasses the problems of determining canopy conductance are not important for well-watered swards of uniform height which are completely shading the ground, because potential evapotranspiration can be predicted without resort to canopy stomatal conductance values (Penman, 1948; Monteith, 1965; Thom and Oliver, 1977; Al-Nakshabandi, 1983). In other situations, however, particularly where soil moisture limits water use and light interception by the sward is not complete, stomatal conductance and the leaf area from which transpiration takes place become very important in regulating the water loss from the canopy. The Penman–Monteith equation can be used to estimate the development of soil moisture deficits when the effect of increasing deficits on canopy stomatal conductance are taken into account. In Britain, MORECS is an acronym for the Meteorological Office Rainfall and Evaporation Calculation System, which has attempted to do this for 40-km grid squares, assuming a vegetation cover of close-cut or grazed grass (Thompson, 1981). An evaluation of the success of MORECS in estimating soil moisture deficits (Gardner and Field, 1983) found a definite bias towards overestimating true soil moisture deficits in nearly all years except in very dry summers, when underestimation occurred. The reasons for these errors have not been identified but changes in canopy conductance due to stomatal response to water stress (see p. 226) are probably important, although difficult to incorporate in the model in a predictive manner.

Direct evaporation of water from the soil surface, rather than after passage through the plant, can be a major route for water loss where vegetation does not completely cover the soil and the soil surface is wet. Under these circumstances soil evaporation depends largely upon the amount of radiation reaching the surface, and this is determined by the leaf area and orientation of leaves in the canopy above the soil. However, as the soil surface dries it is more likely that the moisture content and water conductivity characteristics of the soil will determine evaporation. Hence, under intermittent rainfall regimes where soil drying occurs, soil evaporation is least with coarse-textured soil surfaces, because they retain less water in the surface zone subject to most evaporative loss.

The amount of soil evaporation can be a significant proportion of total evapotranspiration, and van Keulen (1975) has shown that loss from fertilized annual pastures in Israel (mean annual rainfall 250 mm, fine sand surface texture soil) averaged 38% of total evapotranspiration over 3 years. Furthermore, the amount of evaporation depended strongly on rainfall distribution, with smaller losses over the growing season if the average seasonal rainfall was spread over fewer wet days.

6.2.4 Control of transpiration

As we have seen already, the loss of water vapour in transpiration is the inevitable consequence of the stomata being open in the light to allow the diffusion inwards of CO_2 for photosynthesis. The amount of water vapour lost in transpiration, expressed as a flux (E), is proportional to the difference in specific humidity Δq(kg kg^{-1}) between the leaf intercellular spaces and air, so (after Jarvis, 1981)

$$E = \Delta q \varrho g \tag{6.3}$$

where E is in kg m^{-2} s^{-1}, ϱ is the density of air (kg m^{-3}) and g is the leaf conductance (m s^{-1}). As most of the water vapour flux is through the stomata, leaf conductance is synonymous with stomatal conductance if the leaf boundary layer conductance is high compared with the stomatal conductance. This is normally the case with typically long and narrow grass leaves, as boundary layer conductances are always high when leaf dimensions are small, even at low windspeeds (Grace, 1983).

The stomata are clearly important in controlling transpiration, and those of most temperate grasses seem to be confined to the upper or adaxial surface of the leaf blade, where they are usually situated in rows at the bottom of ridges (Wilson, 1971; Silcock and Wilson, 1981). The number and size of the stomata largely determines leaf conductance (Wilson, 1975a), but the depth of the ridges can also influence the amount of transpiration because the boundary layer conductance decreases where there are deep ridges. However, despite this, Wilson (1975b) has shown that plants of *Lolium perenne* selected for deep ridging will lose more water than those selected for shallow ridging, particularly when water supply is restricted. Similar results were obtained by Silcock and Wilson (1981) using seven contrasting *Festuca* species with leaf surfaces ranging from very flat to highly convoluted. This rather surprising consequence is because the stomata of deep-ridged leaves tend to remain open longer as water stress develops, and they are therefore poor conservers of water. The reason why the stomata remain open longer in deep-ridged leaves is not clear, but it may be because of the more humid microenvironment of the epidermal surface at the bottom of the ridges (see p. 229) brought about by the relatively slow mixing of the air due to lower boundary layer conductances.

6.2.5 Water use efficiency

Because water loss in transpiration is an inevitable consequence of CO_2 uptake for photosynthesis, the amount of net CO_2 fixation per unit of water transpired can be considered as a measure of the relative efficiency of water use. Water use efficiency (WUE) is simply the ratio of net assimilation to water loss, but in practice the definitions of assimilation and water use vary. The agronomist may

The hydrological cycle

express assimilation in terms of dry matter production or yield, which may or may not ignore root dry matter, while the physiologist may express it in terms of net CO_2 exchange (P_n). The extrapolation of the latter from instantaneous values to days or longer is difficult because the proportion of P_n lost in respiration should be taken into account, although it is difficult to measure. Also, water loss may be expressed as transpiration or alternatively total evapotranspiration. Consequently, WUE can be determined at the leaf level, from gas-exchange measurements of illuminated leaves, or at the canopy level in terms of dry matter gain per unit of water lost. Of course, the latter is the more relevant for understanding crop performance, but it is easier to be precise about leaf gas-exchange measurements. Fortunately, despite the apparent anomalies, measurements made at the leaf and canopy level seem to be very similar (Fischer and Turner, 1978).

The water use efficiency of leaves can be equated to the rate of net CO_2 exchange (P_n) divided by the transpiration rate (E), so that following Fischer and Turner (1978)

$$\text{WUE} = \frac{P_n}{E} = \frac{\Delta_c(r_a + r_s)}{\Delta_e(r_a' + r_s' + r_i')} \tag{6.4}$$

where Δ_c and Δ_e are the leaf-to-air concentration differences for CO_2 and water vapour ($g\,m^{-3}$), respectively, and r_a', r_s' and r_i' are the boundary layer, stomatal and internal (or residual) resistances to CO_2 diffusion, while r_a and r_s are the boundary layer and stomatal resistances to water vapour ($s\,m^{-1}$) (resistance is the reciprocal of conductance). The internal resistance (r_i') applies to CO_2 diffusion only, and incorporates all intracellular factors. The molecular diffusion coefficient for CO_2 is somewhat smaller than the coefficient for water vapour, and their ratio is usually taken as 1.6. If, as is the common practice, the CO_2 concentration at the chloroplasts (site of carboxylation) is assumed to be zero, Δ_c equals the concentration of CO_2 in the atmosphere (0.58 $g\,m^{-3}$ at 25°C). Alternatively, the CO_2 concentration at the chloroplasts is taken to be equal to the CO_2 compensation point, the CO_2 concentration at which gross photosynthesis balances respiration (Jones, 1983). The CO_2 compensation is zero for C_4 species and around 0.08 $g\,m^{-3}$ for C_3 species. The highest WUE to be expected can then be calculated assuming that r_i' is zero (infinitely high photosynthetic affinity for CO_2) when at an air temperature of 25°C and air relative humidity of 50% (air saturation deficit of 12 $g\,m^{-3}$) the WUE would be 30 g $CO_2\,kg^{-1}\,H_2O$. This potential level of WUE is achieved only by succulent species, and all other plants have WUE values much lower than this. Clearly there is no unique value for the WUE of any species, as it can be seen from equation (6.4) that environmental factors, which are rarely constant in nature, can influence it. In particular, air saturation deficit will determine Δ_e and WUE will decrease as the saturation deficit increases. In addition, as Δ_e also depends on temperature, then air temperature will influence WUE, which will fall as temperature rises. In general WUE will be greater in cool humid climates than in warm arid climates.

The relative values of the various components of leaf resistance can also influence WUE. The main variable resistance is stomatal, and it can be shown that WUE should almost always increase as stomatal resistance increases. This is because if the CO_2 concentration at the chloroplasts is assumed constant, then WUE is proportional to $(r_a + r_s)/(r'_a + r'_s + r'_i)$ in a given environment, and as r_s increases so does this ratio. In fact this ratio often remains remarkably constant for a particular species as environmental conditions change, and this may explain why the WUE of some crops have been shown to be constant over the whole growing season if the saturation deficit of the air does not vary greatly (Day *et al.*, 1978).

The internal resistance can have a marked effect on WUE, as is illustrated by the difference between C_3 and C_4 species. In C_4 species the internal resistance is normally about half that in C_3 species, and also their stomatal resistance is generally higher than C_3 species (Körner *et al.*, 1979), with the result that the WUE of C_4's is about two to three times higher than that of C_3's under the same environmental conditions (Gifford, 1974; Rawson *et al.*, 1977). However, as C_4 species are normally found in higher-temperature environments than C_3's, their higher WUE may not be apparent in their natural conditions. There is some evidence that C_4 plants adapted to growing in cooler climates maintain their low internal resistance so that their WUE should be lower than that of C_3's in the same environment (Long and Woolhouse, 1978).

Discussions of WUE often imply that if water is a potential limiting factor for production during a growing season, then those plants with the maximal WUE will be at an advantage. However, a high WUE may not be an advantage if some of the available water remains unused at the end of the growing season. Under these circumstances it is more desirable to achieve maximum production for the amount of available water. The way in which stomata operate in order to accomplish this has been elucidated in a model of optimal stomatal behaviour in a changing climate developed by Cowan (1977) and Cowan and Farqhar (1977). The principle of optimization is simply one of balancing the risk of water use against the benefits of carbon fixation in a changing environment, and the model predicts that optimal behaviour of the stomata is achieved when $\partial E/\partial P_n = \lambda$ is a constant. In other words, in economic terms the stomata respond to environmental change during the day so as to keep the marginal cost of water loss for carbon gain a constant. The ratio λ is a physiological parameter, dependent on the amount of water available to the plant, which becomes smaller as the availability of water decreases. The effect is to move towards an increasing tendency for midday stomatal closure, a phenomenon observed in many species, as the amount of available water decreases, and this will maximize WUE in drought conditions. The theory of optimization demands in some circumstances that stomata respond directly to changes in environmental conditions so that the actual rate of transpiration decreases

The hydrological cycle

when environmental changes would normally tend to increase the rate of transpiration (Cowan, 1982). This requires that the stomata control water loss by a feedforward process and respond directly to low atmospheric humidity, for example. This type of response is discussed further on p. 229.

6.2.6 Uptake of water by the plant

Compared with the water content of the plant canopy, the water loss through the stomata is extremely high. Consequently, the water lost in transpiration must be replaced quickly by water uptake from soil. Fortunately, the water storage capacity of most soils is quite high, although this varies with soil type. Neglecting extremes, the available water is about 0.20 ± 0.05 of soil volume (Rutter, 1975). Water conduction towards roots is generally considered to be very slow below field capacity, so the only water effectively available to the plant is that within the rooting zone. As a broad generalization, for a rooting depth of 1 m this is approximately 150–250 mm of water and at evapotranspiration rates of 3–5 mm day^{-1} this reserve would be used in 1–2 months (Rutter, 1975). The amount of soil water available to the plant is always less than the total storage capacity of the soil, and depends on the ability of the plant to maintain a gradient in water potential between the root and soil as the soil water potential declines. The lower limit of soil water potential, below which plants can no longer extract water, is termed the 'permanent wilting point' or the 'wilting percentage'. However, the permanent wilting point cannot be regarded as a constant, because wilting depends on the loss of turgor from leaves and this is determined by the osmotic properties of plants, which can vary with time and between species (Slatyer, 1957). Nevertheless, most crop plants have osmotic potentials in the range from -1.5 to -2.0 MPa, so for practical purposes the water content of the soil at -1.5 MPa is near to the point at which soil moisture usually becomes severely limiting (Kramer, 1983).

As water is extracted from the soil by the roots, appreciable gradients in water potential develop. These gradients are determined by a combination of climatic conditions, the distribution of the roots and the soil characteristics, so it is impossible to make general statements about their magnitude. An example of the gradients that can develop in a dry summer in a temperate climate below a sward of perennial ryegrass is shown in Fig. 6.3. The depth to which roots of grass species can penetrate to extract water is variable, but Garwood and Sinclair (1979) found that the depth of penetration corresponded closely with the dry matters yields obtained from six unirrigated grasses cut at intervals of 6 weeks (Table 6.1). The distribution of root biomass with depth in all cases showed a large concentration of roots in the 0–10 cm horizon and a substantial reduction in root density with depth. However, the roots of tall fescue (*Festuca arundinaceae*), which penetrated the soil most deeply, also had a more even distribution down the profile (Garwood and Sinclair, 1979).

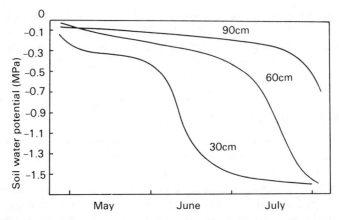

Figure 6.3 The decline in soil water potential at three depths below a sward of perennial ryegrass (*Lolium perenne*) covered by rain shelters for three summer months. (From Russell, 1977, and based on data of Garwood and Williams, 1967.)

TABLE 6.1 *The effective rooting depth and above-ground yields of unirrigated grass swards cut at intervals of 6 weeks at Hurley, England (from Garwood and Sinclair, 1979)*

Species	Effective rooting depth (cm)	Yield (t ha^{-1})
Poa trivialis	40	0.1
Phleum pratense	70	1.1
Dactylis glomerata	70	2.0
Lolium perenne	80	2.3
Festuca arundinacea	>100	3.3

Reference to equation (6.1) shows that the rate of water movement through the plant is proportional to the difference in water potential between the soil and leaf. In its simplified form this equation shows that if the water capacitance of the plant is small and soil water is freely available, then this rate of movement is equal to the rate of transpiration and absorption. To maintain a constant rate of absorption, the gradient in water potential must remain constant as long as the frictional resistances, which occur both in the plant and soil, do not change. There has been much discussion in recent years over where the largest resistance to water movement from soil to shoot is located. It now appears that it is most likely to be located in the roots of the plant (Blizzard and Boyer, 1980) or at the soil–root interface (Faiz and Weatherley, 1978). The reason for the high resistance at the soil–root interface appears to be the incomplete contact between the root and soil particles, but the properties of this interface are, at present, poorly understood (Passioura, 1981). As far as the radial flow of water across the roots is concerned, little is known about the actual route or where the major resistances lie, although they are probably relatively large (Newman, 1976; Passioura, 1981). Axial flow of water occurs almost entirely in the xylem, and in the grasses where xylem vessels are relatively less well developed than in dicots, it has been suggested that *this* is the location of the major resistance between soil and shoots (Greacen et al., 1976; Passioura, 1981). Certainly, when the development of nodal roots in blue grama (*Bouteloua gracilis*) seedlings was prevented by drought the large hydraulic resistance of the small xylem vessels of the seminal roots prevented leaves of the seedling from getting adequate supplies of water (Wilson et al., 1976). Small changes in the radius of xylem vessels can have a very large effect on their permeability, because water flow in the xylem can be described approximately by Poiseuille's equation, in which the volume of flow is dependent on the fourth power of the radius (Greacen et al., 1976).

There is also much recent evidence that the plant resistance is not constant and that resistances to flow are a function of the rate of water loss by the plant (Ng et al., 1975; Davies, 1985). These functions are complex, but they do not detract from the more general observation that if rates of uptake are to be maintained as water in the soil is depleted and Ψ_s falls, then some adjustment must take place in leaf water potential. In essence, leaf water potential falls as soil water potential falls and the water potential of the leaf is a function of the supply potential (Ψ_s), the transpirational flux of water and the liquid-phase resistance to flow of water from soil to the evaporating site within the leaf ($r_{s,l}$), so that

$$\Psi_l = \Psi_s - \text{flux}\,(r_{s,l}) \tag{6.5}$$

The decline in leaf water potential results in water deficits or water stress developing in the leaf cells. The effects of these deficits are complex, and we will now look at how the decline in leaf water potential during evaporative loss of water from the plant can affect the plant, particularly in relation to its growth and productivity.

6.3 DEVELOPMENT AND EFFECTS OF WATER STRESS

We have seen that transpiration is dependent on the specific humidity gradient between the intercellular spaces of the leaf and the bulk air, and that this loss of water can be maintained by the difference in water potential between the soil and leaf if the hydraulic resistances of the soil and plant are sufficiently low and the soil water potential is high. The rate of transpiration will vary throughout the day not only due to changes in the specific humidity gradient, which is dependent on water content of the air and its temperature as well as that of the leaf, but is also due to changes in leaf (primarily stomatal) conductance (equation (6.3)). In non-stressed plants the stomata will open in response to increasing irradiance and in some cases close in response to a fall in atmospheric humidity. This will result in a daily cycle of plant water potential even when soil water potentials are high (equation (6.5)). In the absence of added soil water, both plant and soil water potentials will decrease if the rate of transpiration is maintained. Eventually the plant will reach the point where leaf cell turgor falls to zero, and the leaves will usually wilt. At this point the leaf water potential (Ψ_l) will be equal to the leaf solute potential. This is because the water potential at any point in the soil/plant system (Ψ_w) can be expressed as

$$\Psi_w = \Psi_\pi + \Psi_t + \Psi_m \tag{6.6}$$

where Ψ_π is the solute or osmotic potential, Ψ_t is the turgor or pressure potential and Ψ_m is the matric potential due to the capillary and molecular forces associated with cell walls or soil particles. The soil water potential at which this wilting occurs was referred to as the permanent wilting point on p. 215, and as it is clearly a function of the solute potential of the plant it is not necessarily the same value for all plants, as has often been assumed (Slatyer, 1957). It is generally agreed that the point of zero turgor is about the lowest value the plant water potential can reach, at least in non-scleorophylls. However, small values of negative turgor have been recorded on some occasions (e.g. Turner, 1974). The changes in soil and plant water potential which occur as the soil dries out are shown schematically in Fig. 6.4. It shows the diurnal fluctuations in leaf water potentials with a recovery only to the soil

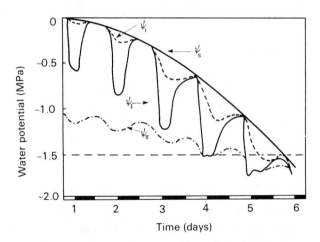

Figure 6.4 Diagram showing the changes in leaf water potential (ψ_l), root water potential (ψ_r) and leaf osmotic potential (ψ_π) which are likely to occur in a transpiring plant rooted in a soil which is allowed to dry from a soil water potential (ψ_s) near zero to a water potential at which wilting occurs (-1.5 MPa). Evaporative conditions are assumed to be similar on each day, and transpiration occurs only during the hours of light (light bars, daylight; dark bars, darkness). (Adapted from Slatyer, 1967; Turner and Begg, 1978.)

Development and effects of water stress

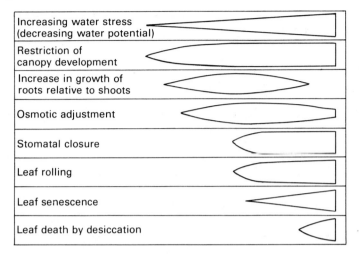

Figure 6.5 The sensitivity of anatomical and physiological changes in crop plants to the gradual development of water stress in the field. The starting position of each band indicates the threshold level of water stress for eliciting the change, and the width of the band at any time represents the relative magnitude of the change. (Adapted from Bradford and Hsiao, 1981.)

water potential at night. The inclusion of the leaf osmotic potential shows when leaf turgor reaches zero and wilting occurs. It is convenient to use leaf water potential (Ψ_l) as an indicator of the degree of water stress that develops when soil water availability declines or the rate of transpiration is high. However, as can be seen from Fig. 6.4a, low level of water stress may develop even when soil water is freely available and the air is moist, because of the resistance to flow of water that exists at some point in the plant. The development of a Ψ_l at which wilting occurs will obviously result in harmful effects on the plant, and ultimate death of the leaves if it continues, but it is also the case that potentials which fall to somewhere between zero and the permanent wilting point affect a number of other plant processes. In many cases these effects can be viewed as adaptive changes to the development of water stress, which enable the plant to survive, and in some cases continue to grow, when water supply is restricted. A summary of some of these changes is illustrated in Fig. 6.5, which emphasizes the difference in the sensitivity of the physiological and morphological processes which are affected by water stress. The effect of water stress on some of these processes will now be discussed in more detail.

6.3.1 Water stress and cell growth and division

The plant process most sensitive to water stress appears to be cell growth, defined as the irreversible enlargement or expansion of cells (Hsiao and

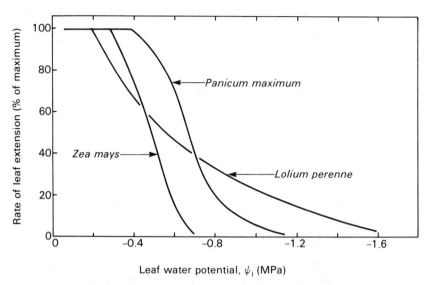

Figure 6.6 The relationship between the rate of leaf extension and leaf water potential (ψ_l) for three grasses. (Adapted from Acevedo *et al.*, 1971; Lawlor, 1972; Ludlow and Ng, 1976.)

Acevedo, 1974). In the vegetative grass plant the main areas of cell enlargement are just behind the meristems at the base of growing leaves and at the tips of extending roots. It is at these points that the effect of water stress on cell growth should be investigated but, because of the problems associated with working on such small amounts of plant tissue, the more general effects on whole leaf and root extension have usually been studied. The level of stress at which leaf extension is affected is variable, but in some cases a fall of Ψ_l to -0.7 MPa results in a complete inhibition of leaf elongation as shown in Fig. 6.6 (Acevedo *et al.*, 1971). The reason for this extreme sensitivity seems to be that cell expansion is largely a physical process driven by the hydrostatic pressure within the cells – that is, the turgor pressure. In a simple form an equation for growth of cells can be written as

$$dV/dt = Eg(\Psi_t - \Psi_{t,thr}) \tag{6.7}$$

where dV/dt is the change in cell volume (V) with respect to time (t), Eg is a coefficient termed gross extensibility, Ψ_t is the turgor pressure and $\Psi_{t,thr}$ is the threshold Ψ_t below which growth will not occur (Green, 1968). To date it has not been possible to measure turgor pressure (Ψ_t) directly in the growing cells of leaves, which in monocots are located at the base of the leaf, ensheathed by the older leaves. However, Ψ_t can be estimated from the difference between water potential (Ψ_w) and osmotic potential (Ψ_π), although these measurements are still very difficult to make. In one experiment, where careful measurements were made on the 50 mm basal segment (growing zone) of maize

Development and effects of water stress

leaves during rapid development of water stress, leaf elongation was found to be linearly related to Ψ_t above a threshold ($\Psi_{t,thr}$) of approximately 0.4 MPa, as predicted by equation (6.7) (Fig. 6.7) (Hsiao *et al.*, 1985). These results also show that the dependence of elongation on Ψ_t above $\Psi_{t,thr}$ is quite steep, as a reduction of 0.25 MPa in Ψ_t was sufficient to reduce elongation from its maximum to zero. Unfortunately, the simplicity of equation (6.7) may be misleading, as other studies of monocot leaf growth have shown that as water stress developed, leaf elongation slowed and then stopped but Ψ_t of the growing zone remained apparently unchanged (Matsuda and Riazi, 1981; Michelena and Boyer, 1982). The question of what reduces leaf growth in these cases remains unresolved, but there is evidence that, particularly when stress develops relatively slowly, both Eg and $\Psi_{t,thr}$ can shift as Ψ_w declines (Van Volkenburgh and Cleland, 1984). A further complication is that when Ψ_w declines, then it is

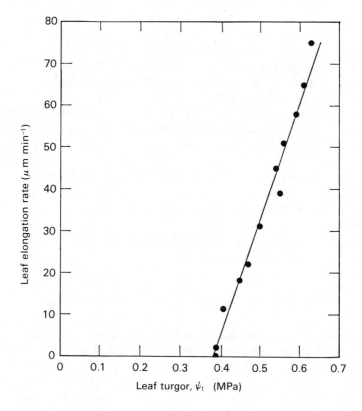

Figure 6.7 The relationship between elongation rate of the fifth leaf of maize (about 30 cm long) and the estimated turgor potential (ψ_t) in the growing zone. Elongation rate was reduced from the maximum by fast development of water stress. (From Hsiao *et al.*, 1985.)

possible to maintain turgor by a corresponding 'osmotic adjustment' or lowering of the cell solute potential (Ψ_π) (see equation (6.6)).

The lowering of solute potential can arise both from the concentration of solutes as water is withdrawn from the vacuoles and cell volume decreases, and the active accumulation of solutes in the cell. Turner and Jones (1980) suggest that the term 'osmotic adjustment' should be used only to describe the lowering of solute potential due to net accumulation of solutes in response to water stress. However, if a lowering of Ψ_π is accompanied by a decrease in cell size, as is often the case, then a net accumulation of solute per cell may simply be due to the smaller relative size of the vacuole (Simmelsgaard, 1976). A lowering of solute potential in response to water stress (decreasing Ψ_l) has been demonstrated in a number of tropical pasture grasses (Wilson et al., 1980) and in the temperate grasses *Dactylis glomerata* (Gavande and Taylor, 1967) and *L. perenne* (Jones et al., 1980b). However, in all cases the measurements of water status were made in the exposed, non-growing leaf blades rather than the basal, elongating part of the leaf. More recently it has been demonstrated that differences in water status exist along grass leaves and that the water status of the elongating region cannot be precisely inferred from measurements in the exposed blade (Matsuda and Riazi, 1981; Michelena and Boyer, 1982). Also, as already mentioned, the maintenance of turgor brought about by the decrease in Ψ_π is not always associated with the continued expansion of cells, which is usually measured indirectly by observations of leaf extension. Michelena and Boyer (1982) have shown that complete turgor maintenance in growing cells at the base of maize leaves does not result in the maintenance of growth rate, and Wilson and Ludlow (1983a) have also reported that leaf growth in *Cenchrus ciliaris* ceases under water stress despite an osmotic adjustment of 0.65 MPa.

The solutes which accumulate during osmotic adjustment have generally been assumed to be mainly soluble carbohydrates (Wilson and Ng, 1975; Munns et al., 1979; Kigel and Dotan, 1982), which are produced in photosynthesis but not utilized in leaf expansion as water stress develops, although inorganic ions are also accumulated. Proline is important in osmotic adjustment in halophytes (Stewart and Lee, 1974), but its accumulation appears to be insufficient to account for lowering of Ψ_π in grasses (Chu et al., 1976). However, Ford and Wilson (1981) have shown that inorganic ions, and potassium in particular, are the most important solutes involved in osmotic adjustment in stressed leaves of three tropical grasses, while sugars are apparently not important. They point out that the minor role of soluble carbohydrates in osmotic adjustment in tropical grasses is not unexpected because, unlike temperate species, they do not accumulate sugars under other circumstances (Wilson and Ford, 1973).

It is therefore clear that the plant's response to water stress in terms of cell growth is a complex interaction of a decrease in threshold turgor, a degree of osmotic adjustment to maintain turgor and possibly an increase in gross extensibility of the cells. All of these changes can maintain growth under water stress, but it now appears that osmotic adjustment is the major strategy adopted by plants to counteract the fall in water potential as stress develops (Turner and Jones, 1980). The ability to adjust to water stress does, however, depend very strongly on the rate at which stress develops, and this is the reason why experiments on the effects of water stress on growth in controlled environments and under field conditions have often yielded conflicting information (Sheehy *et al.*, 1975; Turner and Begg, 1978; Jones *et al.*, 1980b). The difference is probably due to the fact that plants grown in controlled environments usually have a restricted volume of soil and a dense root system. As a consequence the amount of stored water in the soil is limited and when water stress is induced by withholding water, then stress develops much more rapidly than in the field. The overnight recovery of Ψ_l which occurs during the early stages of drying in the field (see Fig. 6.4), but not in pot-grown plants in controlled environments, allows time for osmotic adjustment to occur. The importance of the reserves of soil water under stress development in the field has been demonstrated by growing grass swards in boxes containing soil with a low water-holding capacity. Under these circumstances the response of *L. perenne* to developing water stress is very similar to observations made with the same species in controlled environments (Sheehy *et al.*, 1975; Jones *et al.*, 1980b). This would suggest that levels of irradiance, the other major difference between controlled environments and the field, do not have significant effects on the pattern of osmotic adjustment. An extreme example of the effect of rate of drying on osmotic adjustment is shown in Fig. 6.8 for sorghum. In plants which were water stressed over several weeks the turgor was maintained, but if they were dried over several hours there was no osmotic adjustment. The same type of response has been shown for several grasses (Wilson *et al.*, 1980) and in Table 6.2 this is illustrated by observing the differences in slope of the relationship between turgor potential and water potential ($\Delta\Psi_t/\Delta\Psi_l$) for rapidly and slowly dried plants. Evidence of osmotic adjustment is shown by the shallow slopes determined for slowly dried plants.

Although the above discussion has emphasized that osmotic adjustment is more likely to occur if stress develops slowly, allowing time for the accumulation of solutes, there is some evidence that short-term osmotic adjustment can also occur on a daily cycle. This enables leaves to continue growth during the middle of the day when Ψ_l declines due to high transpiration rates (Hsiao *et al.*, 1976). The importance of this diurnal osmotic adjustment has not been fully investigated under field conditions, and there is still conflicting information on whether leaf growth is restricted when evaporative demand is high during the middle of the day but soil water potential has not declined significantly (Boyer, 1968; Jones *et al.*, 1980a).

Research on the effect of water stress on cell growth has to a large extent emphasized its influence on *leaf* expansion and extension and its effect on canopy development, but there have been some demonstrations of water stress leading to similar osmotic adjustment in growing roots (Greacen and Oh, 1972; Sharp and Davies, 1979). For example, Greacen and Oh (1972) found that the roots of wheat seedlings were able to maintain full turgor by osmotic adjustment as the soil water potential (Ψ_s) fell from −0.3 to −0.8 MPa. As a conse-

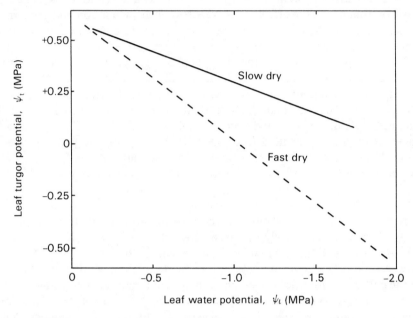

Figure 6.8 The relationship between leaf water potential (ψ_l) and turgor potential (ψ_t) for sorghum plants dried slowly over several weeks to allow osmotic adjustment, or dried quickly over several hours to minimize osmotic adjustment. (From Turner and Jones, 1980.)

TABLE 6.2 *The effect of the rate of drought development on the slope of the relationship between turgor potential and water potential ($\Delta\psi_t/\Delta\psi_l$) for grass leaves. A difference in the values of $\Delta\psi_t/\Delta\psi_l$ between rapidly dried plants and plants allowed to dry slowly is used as evidence of osmotic adjustments. (From Turner and Jones, 1980.)*

Species	$\Delta\psi_t/\Delta\psi_l$	
	Rapid dry	Slow dry
Sorghum bicolor cv. RS610	0.65	0.36
Sorghum bicolor cv. Shallu	0.59	0.29
Cenchrus ciliaris	0.71	0.46
Panicum maximum var. *trichoglume*	0.75	0.47
Heteropogon contortus	0.70	0.47

Development and effects of water stress

quence of this turgor maintenance the roots continued to extend at the same rate throughout the drying cycle. Clearly, osmotic adjustment allows root growth to continue as stress develops, and this enables a greater volume of soil to be explored or leads to a greater density of roots in a fixed volume of soil. Both of these responses will enable the plant to maintain rates of transpiration and increase the amount of water which can be extracted from the soil. Furthermore, there is some evidence that water stress can result in an absolute increase in root growth or an increase in depth of penetration of the roots into the soil (Sharp and Davies, 1979; Molyneux and Davies, 1983). This suggests that turgor maintenance in roots is possibly more effective in allowing continued cell expansion than the similar mechanism in leaves. The importance of this strategy for monocultures of crop plants has been questioned (Passioura, 1981), as it might only be useful when one species competes with another for available water. A larger root system may simply increase the rate at which stress develops (Troughton, 1974). Also, a large root system will inevitably divert assimilates below ground at the expense of the shoot, and will have the effect of reducing WUE. In mixed grass swards the differential sensitivity of species to water stress appears to be largely determined by the depth to which roots can penetrate and the extent of root branching. Garwood and Sinclair (1979) found that after a period of water stress the roots of tall fescue (*F. arundinacea*) were more numerous at depth than those of timothy (*P. pratense*), cocksfoot (*D. glomerata*) or perennial ryegrass (*L. perenne*), and at the same time it was more tolerant of the dry conditions and yielded more (Table 6.1). Early morning values of Ψ_l also showed that leaves of tall fescue (-0.86 MPa) were less stressed than cocksfoot (-0.97 MPa), perennial ryegrass (-1.52 MPa) or timothy (-1.24 MPa), clearly showing that this is a strategy of stress avoidance rather than stress tolerance.

It has been known for many years that plants can lower their osmotic potential in response to water stress, but the significance of these observations was not appreciated initially (Turner and Jones, 1980). However, the recent demonstrations that osmotic adjustment is widespread in cultivated plants, including grasses, have to some extent revised our ideas on the sensitivity of plants to water stress. They have shown that osmotic adjustment is an important mechanism in relation to drought tolerance which appears to be a more general phenomenon than was once thought. Nevertheless, there is no doubt that cell expansion is still relatively sensitive to water shortage, and a reduction in leaf growth in particular is the first major response when water becomes limiting (Hsiao *et al.*, 1976). Indeed, Wilson and Ludlow (1983a, b) have shown that although water stress induced an osmotic adjustment of up to 1.0 MPa in the three tropical grasses *Panicum maximum, Cenchrus ciliaris* and *Heteropogon contortus,* this only delayed the loss of turgor in water-stressed leaves by about 4 days.

Closely associated with the process of cell expansion is, of course, cell division, and there is much evidence to suggest that this process is less sensitive to water stress than cell expansion is (Slatyer, 1957; Hsiao, 1973). However, if stress is prolonged this may not be the case (Hsiao, 1973) and McCree and Davies (1974) have shown that in sorghum cell division and enlargement are equally sensitive to water stress. Nevertheless, if cells do not expand during water stress and if in some circumstances small cells also accumulate near the meristems, there is no doubt that rapid cell expansion occurs when stress is relieved. This may, in part, explain the many observations that rapid growth of grass can follow on rewatering after a period of water stress (Horst and Nelson, 1979; Chu *et al.*, 1979; Norris and Thomas, 1982; Kigel and Dotan, 1982). In some cases the maximum rate of leaf extension during recovery was up to 20% higher than the non-stressed control plants of *Bromus catharticus* (Chu *et al.*, 1979). Work with water stressed *Panicum maximum*, a tropical C_4 grass, however, showed no evidence of stimulation of growth following relief of stress (Ng *et al.*, 1975).

The sensitivity of cell division to water stress will also influence the initiation of new tillers on the grass plant. A reduction in tiller numbers in water stressed grass swards is commonly observed (Luxmoore and Millington, 1971; Jones *et al.*, 1980a; Norris and Thomas, 1982), and this reduction is due mainly to cessation of tillering rather than to a faster rate of tiller death. A marked increase in tiller emergence has also been recorded when water stress is relieved (Chu *et al.*, 1979). This probably occurs because although under water stress tiller initiation continues, these tillers do not emerge until cell expansion starts when stress is relieved.

6.3.2 Water stress and stomatal conductance

It has been pointed out already that the stomata are the major variable resistance controlling gaseous exchange between the cells of the leaf and the air (p. 207), and consequently they are very important regulators of water loss from the plant. Stomata have been described by Mansfield and Davies (1981) as 'miniature sense organs, constantly perceiving the environment outside the leaf and adjusting their apertures to achieve the best compromise between the plant's requirement for CO_2 and its need to conserve water'. When the stomata close they increase the resistance of (decrease the conductance to) gaseous diffusion, and this is an important method by which plants can reduce loss of water when stress develops. Of course, stomatal closure while reducing water loss will at the same time reduce CO_2 uptake in photosynthesis, and this will lead to lower carbon gains and weight increase by the plant. Stomata are very sensitive to many environmental and internal factors including irradiance and CO_2 concentration (Raschke, 1975), but in relation to the effects of water stress their response to plant water potential and the dryness of the air (saturation deficit) seem to be the most important. A sensitivity to water potential might be expected as the movement of the stomata is brought about by the difference in turgor between the guard cells and the other epidermal cells (Willmer, 1983).

Development and effects of water stress

However, the turgor of these cells and their water potential is very difficult to measure, and consequently the response of stomata to the water potential of the whole or bulk leaf is usually studied, although this may bear no direct relationship to the water potential of the guard cells. The justification for this approach is that it is possible to speculate on a link between bulk leaf water potential and stomatal behaviour (Davies *et al.*, 1981), although the exact nature of this link is not clear.

(a) Response to leaf water potential

As leaf water potential declines during the development of water stress there is usually a range over which there is no effect on the stomata, then below a threshold value the conductance declines in a linear or curvilinear fashion until the stomata close (Hsiao, 1973; Turner and Begg, 1978; Ludlow, 1980) (Fig. 6.9). At the point of stomatal closure the leaf conductance is equal to cuticular conductance. The view that stomata have a clear threshold Ψ_l above which they are unaffected by water status and below which they close rapidly has been questioned by Jones (1983). He suggests that many of the responses recorded are artefacts of the instruments used for measuring conductance and the use of resistance rather than conductance measurements. Nevertheless, the response of stomata to leaf water potential can be viewed as a feedback mechanism because when Ψ_l falls below a critical value the stomata begin to close and the reduction in transpiration allows Ψ_l to recover. As a consequence

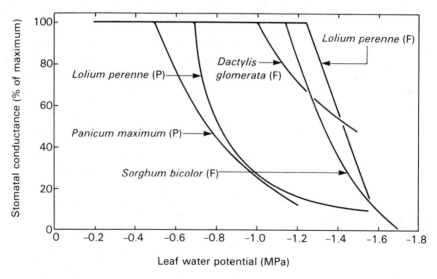

Figure 6.9 The relationship between stomatal conductance and leaf water potential for a number of grasses grown either in pots or containers with limited soil reserves of water (P) or under field conditions (F). (Adapted from Jackson, 1974; Jones *et al.*, 1980a; Ludlow and Ng, 1976; Ludlow, 1980; Sheehy *et al.*, 1975.)

stomata may re-open and then close in a cyclic pattern throughout the day (Willmer, 1983).

The sensitivity of stomata to declining Ψ_l often depends on the conditions of growth, and as with leaf extension, plants grown in the field are usually less sensitive than those grown in controlled environments (Table 6.3). Again, the reason for this difference in sensitivity seems to reflect the difference in the *rate* at which water stress develops. Support for this comes from the experiments of Ludlow (1980) using the C_4 grass *Heteropogon contortus* where different relationships between leaf conductance and Ψ_l were obtained over a 5-week drying cycle (Fig. 6.10). The responses were obtained by cutting stressed tillers at different stages of the drying cycle, rehydrating them, and allowing them to dry to various levels of Ψ_l before measurement of stomatal conductance. The degree of stomatal adjustment varies considerably between species, and Ludlow (1980) suggests that it is least marked in temperate species and highly selected agricultural crops. Consequently, the amount of stomatal adjustment, measured as the difference in Ψ_l when the stomata closed on plants stressed slowly or rapidly, ranged from 1.2 to 3.6 MPa for C_4 pasture grasses to only 0.3 or 0.4 MPa for temperate agricultural crops (Ludlow, 1980). This observation is supported by experiments on perennial ryegrass, which shows only a small stomatal adjustment of about 0.5 MPa between water-stressed plants grown in the field and controlled environment (Sheehy *et al.*, 1975; Jones *et al.*, 1980b) (Table 6.3).

The mechanism underlying this stomatal adjustment when it does occur, is probably associated with osmotic adjustment, particularly in the guard and other epidermal cells. This allows turgor differences between the guard cells and epidermal cells to be maintained and stomata to stay open as Ψ_l declines. However, as mentioned above, direct measurements of turgor in guard cells are difficult to make, so the main evidence for osmotic adjustment comes from a correlation between bulk leaf tissue determinations of osmotic adjustment

TABLE 6.3 *Stomatal response to leaf water potential (ψ_l) for grass species grown in the field* (F) *and controlled environments* (C)

Species	Threshold ψ_l (MPa)	Closed ψ_l (MPa)	References
Panicum maximum var. *trichoglume* (C)	−0.6	−1.2	Ng *et al.* (1975)
Lolium perenne (C)	−0.8	−1.5	Sheehy *et al.* (1975)
Lolium perenne (F)	−1.0	−2.0	Jones *et al.* (1980b)
Lolium perenne (F)	−1.0	−2.0	Jackson (1974)
Dactylis glomerata (F)	−1.0	−1.5	Jackson (1974)
Astrebla lappacea (C)	−1.5	>−3.5	Doley and Trivett (1974)

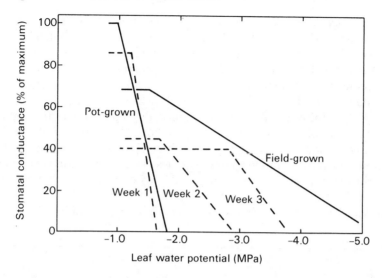

Figure 6.10 Differences in the relationship between stomatal conductance and leaf water potential in the grass *Heteropogon contortus* grown under different water stress conditions in pots and in the field. In the pots water stress developed over several days, whereas in the field it took 5 weeks. The broken lines show the changes in the relationship brought about by cutting tillers after 1, 3 and 5 weeks of drying in the field, re-hydrating then dehydrating to determine the relationship between stomatal conductance and leaf water potential. (From Ludlow, 1980.)

and stomatal adjustment. Ludlow (1980) has reported a strong correlation between Ψ_l at the point of stomatal closure and the osmotic potential of leaves of three C_4 pasture grasses at zero turgor. However, it has also been suggested that the stress hormone abscisic acid (ABA) is involved in stomatal adjustment, as it is produced in large amounts under water stress, and stomata are very sensitive to increasing levels of ABA (Davies *et al.*, 1981). Davies (1978) has suggested that stomatal adjustment results from a decline in the threshold water potential for the initiation of ABA synthesis which will lead to stomatal closure. There is at present little experimental support for this proposal, but the observation that field-grown plants show diurnal variations in ABA would suggest that they may influence stomatal activity, at least to some extent.

(b) Response to humidity

There have been many demonstrations over the last 15 years that stomata respond directly to changes in ambient humidity (Lange *et al.*, 1971; Lösch and Tenhunen, 1981). In experiments where the water content of the ambient air varied in otherwise constant environmental conditions the stomata closed as the water vapour pressure deficit between leaf and air increased. Mansfield and Davies (1981) have referred to this response of stomata to the water vapour pressure deficit of the atmosphere as the plants 'first line of defence' in maintaining a high Ψ_l as water stress develops. The 'second line of defence' is the closure of the stomata in direct response to declining Ψ_l. Unlike the feedback response to bulk leaf water potential, the stomata respond directly to humidity in a 'feedforward' manner (Cowan, 1982), so that stomatal closure occurs before deficits develop in the rest of the leaf (p. 214). The response tends to be very rapid and the effect is to prevent damage through excessive transpiration, particularly to plants growing in areas where there are periods of high evaporative demand. Consequently, a direct humidity response might be expected to be most common amongst species that grow and survive in arid and semi-arid areas (Ludlow and Ibaraki, 1979; Lösch and Tenhunen, 1981). However, from a survey of a large number of species, Sheriff (1977) found that the presence or absence of a humidity response is not dependent on the normal

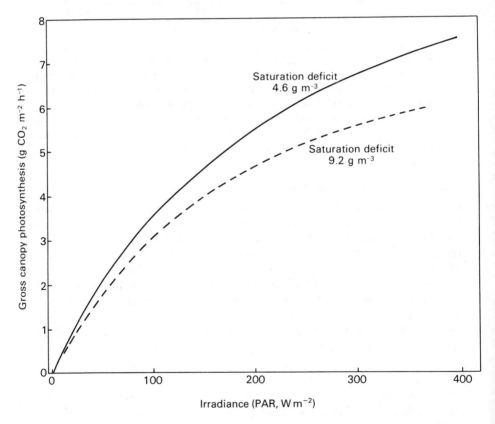

Figure 6.11 The effect of increasing the water vapour saturation deficit of the air on the photosynthesis of a sward of perennial ryegrass (*Lolium perenne*). Stomatal conductance was not measured, but it is assumed that the response is due to partial stomatal closure at the higher saturation deficit. (Parsons and Woledge, unpublished.)

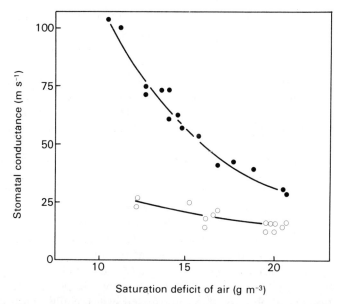

Figure 6.12 The relationship between water vapour saturation deficit of the air and stomatal conductance in *Panicum maximum* in the field at leaf water potentials of −0.5 MPa (●) and −1.2 MPa (○). (From Ludlow, 1980.)

habitat of the plant. There is some evidence that *temperate* pasture grasses do respond to humidity, as is illustrated by Fig. 6.11, which shows the effect of reducing saturation deficit of the air on photosynthesis in *L. perenne*. In this case it is assumed that the reduction in photosynthesis is due to partial stomatal closure.

As with stomatal response to Ψ_l, the sensitivity to humidity can change depending on the plant's immediate prehistory. In general the stomata become less sensitive to leaf–air vapour pressure deficit as water stress develops; as can be seen in Fig. 6.12 for the C_4 grass *Panicum maximum* (Ludlow, 1980). Ludlow does not consider this to be an indication of adjustment to stress but simply a lack of sensitivity in already closed stomata. Stomata also become more sensitive to CO_2 as water stress develops (Raschke, 1975), but the significance of this in the natural environment is unclear.

6.3.3 Water stress and photosynthesis

The rate of leaf photosynthesis is directly proportional to the leaf-to-air CO_2 concentration differences times the boundary layer, stomatal and internal conductances to CO_2 diffusion (see equation (6.4)). Clearly, levels of water stress which reduce stomatal opening will also reduce the rate of photosynthesis. In addition, canopy photosynthesis is dependent, at least up to full light interception, on the area of leaf intercepting light for photosynthesis (see Chapter 2). Consequently, water stress which reduces leaf expansion will also reduce canopy photosynthesis. The 'indirect' effects of water stress on photosynthesis are important in the early stages of development of stress because, as we have seen, both stomatal conductance and, in particular, leaf expansion are sensitive to a relatively small decline in leaf water potential. The consequence is that the decline which is observed in leaf photosynthesis as water potential falls follows very closely the decline in stomatal conductance.

In addition to these 'indirect' effects of water stress on photosynthesis, there are more direct effects which are not normally observed until the leaf water potential falls considerably below that required to induce stomatal closure (Hsiao and Acevedo, 1974). These direct effects are upon the activity of the chloroplasts (Boyer, 1976), and can be viewed as reducing the internal conductance to CO_2 diffusion. There are few reports of these direct or non-stomatal effects of water stress on photosynthesis in grasses. Ludlow and Ng (1976) using *Panicum maximum* showed that the stomata began to close at -0.6 MPa, but internal conductance was not affected until Ψ_l fell to -1.0 MPa. While Jones *et al.*, (1980a) found that in field-grown *L. perenne* most of the 40% reduction in leaf photosynthesis in stressed swards could be attributed to stomatal closure and not a decrease in internal conductance. Also, extreme levels of water stress down to -9.0 MPa in *Panicum maximum*, although inhibiting photosynthesis, did not result in permanent damage because photosynthesis recovered after rewatering (Ludlow and Ng, 1976). Once again the sensitivity of photosynthesis to water stress, like all processes described so far, can vary depending largely on the rate at which stress develops and this is presumably related to allowing sufficient time to elapse for the plants to develop some level of tolerance. This is illustrated by the differences observed between simulated swards of *L. perenne* stressed rapidly over a period of 10 days in a controlled environmental when canopy photosynthesis declined by more than 80% (Sheehy *et al.*, 1975), and more-slowly stressed field swards where canopy photosynthesis only declined by about 40% (Jones *et al.*, 1980a).

6.3.4 Water stress and other processes

So far we have looked at some of the more extensively investigated plant processes which are affected by water stress and, to a greater or lesser extent, allow the plant to adapt to these conditions. There are, of course, a number of other processes both morphological and physiological which are influenced by water stress and can ultimately lead to a change in growth and productivity of the grass sward (see Fig. 6.5).

Amongst the morphological changes which occur under water stress the reduced expansion of leaves and the increased growth of roots relative to shoots has already been discussed. An additional morphological change in grasses is leaf rolling, which is a common response to water stress (Begg, 1980).

Rolling results in a marked reduction in effective leaf area, which in turn reduces the heat load on the plant and lengthens the effective pathway for water vapour diffusion (Johns, 1978; O'Toole and Cruz, 1980). The result is a reduction in transpiration by as much as 46–63% (Parker, 1968) when evaporative demand is high but a rapid recovery as stress is relieved. Leaf rolling in grasses is due to volume changes in the bulliform cells distributed longitudinally in rows along the adaxial sides of leaves. It appears to start at about the same Ψ_l as stomatal closure, but the fact that Ψ_l, stomatal conductance and leaf rolling change together causes difficulty in evaluating the role of leaf rolling in maintenance of Ψ_l (O'Toole and Cruz, 1980). Johns (1978) suggests, however, that the superior herbage yield of *F. arundinacea* compared with other temperate grasses under water stress is due partly to its ability to roll its leaves into tight spirals.

As water stress increases it also reduces leaf area by accelerating both the rate of senescence of the physiologically older leaves and their shedding (Begg, 1980). The level of water stress which triggers leaf senescence is probably dependent on nitrogen nutrition and the stage of development (Bradford and Hsiao, 1982), but experimental evidence is lacking. In the tropical C_4 grass *Panicum maximum* it has been demonstrated that although increasing water stress causes progressive leaf death, it also suspends the ageing of the youngest leaves (Ludlow, 1975). These leaves can survive levels of Ψ_l down to -9.0 MPa and after rewatering achieve photosynthetic rates similar to control leaves of the same physiological age. During recovery, stomatal conductance exerted greater control than internal or residual conductance over net photosynthesis although Ludlow *et al.*, (1980) were unable to explain what controls the level of stomatal conductance.

Additional physiological responses which may be of importance in grass production are translocation, nutrient uptake and the more general term, herbage quality. The rate of translocation of assimilates in the phloem appears to be relatively insensitive to water stress (Wardlaw, 1969) as the velocity of assimilate movement in *L. temulentum* was unaffected by leaf water potentials down to -3.0 MPa. However, the allocation of current assimilates and the mobilization of stored assimilates to metabolic sinks in the plant can be markedly affected (Fischer and Turner, 1978). Clearly, the inhibition of leaf expansion by mild water stress, which we have already looked at in some detail, will mean that the activity of a major sink for assimilates will be reduced. In addition water stress appears to cause an increased allocation of assimilates to roots, increasing the root/shoot ratio. Consequently, the main effect of water stress on the process of translocation is directed towards changes in source and sink activity rather than a direct effect on the translocation system.

It has been demonstrated frequently that the uptake of nitrogen and phosphorus is reduced when plants are water stressed (D'Aoust and Tayler, 1968; Colman and Lazenby, 1975; Begg and Turner, 1976). When water is non-limiting for grass growth there is often a highly significant linear relationship between nitrogen recovered in the plants and that applied. This relationship becomes curvilinear even when water stress is relatively mild (Colman and Lazenby, 1975). Consequently, water stress can act indirectly to reduce growth of the plant by reducing nutrient uptake. This is particularly important if most of the nutrients are in the upper horizons of the soil which dry out first during stress development. In these circumstances the penetration of the roots to lower levels may enable water uptake to continue, but these levels may be deficient in nutrients. The importance of this effect has been demonstrated by Garwood and Williams (1967) who obtained a 100% increase in yield of *L. perenne* by injecting nitrogen to a depth of 75 cm when the upper layer of the soil was dry.

The influence of water stress on herbage quality has been investigated in the C_4 grass, *Panicum maximum* (Wilson and Ng, 1975). In general these authors found that water stress caused a substantial decrease in dry matter digestibility, but this was not associated with changes in content of cell-wall material or nitrogen. Clearly, these observations are important in relation to utilization of grasses, but there appears to be no other similar work on other grass species.

6.4 IMPLICATIONS FOR YIELD

From the foregoing discussion it can be seen that the effects of water stress on grass growth are difficult to quantify because of the many fundamental processes that are influenced. It is important also to appreciate the difference between two broad strategies exhibited by grasses in their adaptation to water stress. These are strategies of: (a) avoidance, where responses like leaf shedding, leaf rolling and stomatal closure to reduce water loss predominate, and (b) tolerance, where the main response is the development of low osmotic potentials. To a large extent the response to water deficits must be dependent on severity, duration and time during the life-cycle of occurrence of stress. Consequently, stress tolerance may predominate at certain times while at others stress avoidance is the overriding strategy. In addition, we have looked largely at the effects of stress on the individual plant rather than the community or canopy.

One approach to a better understanding of the ultimate effects on yield is through the use of a simple model which considers the influences of water deficits as a series of factors which interact with each other. This has been done by Hsiao and Acevedo (1974), and Fig. 6.13 illustrates the possible effects of mild water deficits on leaf area development, CO_2 assimilation and root growth in any crop. An important conclusion from this type of analysis is that mild water stress can have a much more damaging effect when LAI is low because, as we have seen, leaf expansion is more sensitive to water stress than photosynthesis. When LAI is higher, canopy assimilation is more closely related to the potential for leaf photosynthesis and mild water stress at this time will have

Implications for yield

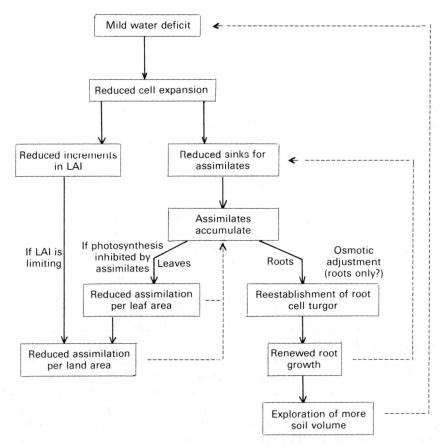

Figure 6.13 The possible effects of *mild* water deficits on leaf area development, CO_2 assimilation and root growth. Broken lines indicate negative feedback effects. The period of water deficit is assumed to be short and to be mild enough so as not to affect stomata and photosynthesis directly. (Adapted from Hsiao and Acevedo, 1974.)

less effect on canopy growth, other things being equal. This simple and systematic approach to the effects of water stress illustrates how the influence of stress on one plant process can rapidly interact with several other processes to produce a whole complex of responses. Some of these responses feed back to actually reduce the expected effect, as is illustrated by the osmotic adjustment which occurs in the roots. This eventually makes more water available to the plant, but has the effect of increasing the rate at which soil water is depleted and stress develops. Clearly the integration of these effects of water stress on grass growth could best be dealt with using a physiological model of the grass crop as described in Chapter 7, but this has not yet been attempted.

6.5 CONCLUSION

This chapter has attempted to explain why continued availability of water is important for grass growth, and some of the effects of a lack of water on the physiology of the grass plant. We now have a relatively thorough understanding of what the effect of water stress is on the grass plant, and one of the most interesting aspects of the work in recent years has been the many demonstrations of how grass plants adapt to water stress. The indeterminate nature of growth in grass crops makes this adaptation particularly important because it allows for continued response to changing conditions of water availability. From the point of view of maximizing productivity this adaptation may have its limitations because adaptation may be for continued survival rather than growth. Also, the yield of grass over the season will in addition be influenced by the rate and degree of recovery after water stress is relieved (Ludlow et al., 1980) and the competitive interaction of grasses with contrasting characteristics of water stress avoidance or tolerance (Harris and Lazenby, 1974; Sala et al., 1982; Sheriff and Ludlow, 1984). However, there is some evidence that it may be possible to select for cultivars with better adaptability which could continue to grow through periods of water stress. Also, the demonstration of the extreme sensitivity of leaf extension to water stress suggests that selection for continued leaf growth at increased water stress would be worthwhile.

Much of the work discussed here has involved experiments on single plants and direct extrapolation to the field situation must be treated with caution. The initial confusion caused by experiments on stressed plants grown in small containers illustrates this point. The interaction of individuals under field conditions and the effects of stress in mixed swards are two examples where more work is needed at the physiological level.

REFERENCES

Acevedo, E., Hsiao, T. C. and Henderson, D. W. (1971) Immediate and subsequent growth responses of maize leaves to changes in water status. *Pl. Physiol.*, **48**, 631–6.

Al-Nakshabandi, G. A. (1983) The potential evapotranspiration of short grass (*Cynodon dactylon*) as related to the estimated potential evaporation and evapotranspiration from meterorological data in the coastal region of Kuwait. *J. Arid Envir.*, **6**, 33–8.

Begg, J. E. (1980) Morphological adaptations of leaves to water stress, in *Adaptation of Plants to Water and High Temperature Stress* (eds N. C. Turner and P. J. Kramer), Wiley, New York, pp. 33–42.

Begg, J. E. and Turner, N. C. (1976) Crop water deficits. *Adv. Agron.*, **28**, 161–216.

Blizzard, W. E. and Boyer, J. S. (1980) Comparative resistance of the soil and the plant to water transport. *Pl. Physiol.*, **66**, 809–14.

Boyer, J. S. (1968) Relationships of water potential to growth of leaves. *Pl. Physiol.*, **42**, 213–7.

References

Boyer, J. S. (1974) Water transport in plants: mechanism of apparent changes in resistance during absorption. *Planta*, **117**, 187–207.

Boyer, J. S. (1976) Photosynthesis in low water potentials. *Phil. Trans. R. Soc. Lond.*, *B* **273**, 501–512.

Bradford, K. J. and Hsiao, T. C. (1982) Physiological responses to moderate water stress, in *Encyclopedia of Plant Physiology, New Series*, Vol. 12B (eds O. L. Lange, P. S. Nobel, C. B. Osmond and H. Ziegler), Springer-Verlag, Berlin, 263–324.

Chu, T. M., Aspinall, D. and Paleg, L. G. (1976) Stress metabolism. VII. Salinity and proline accumulation in barley. *Austral. J. Pl. Physiol.*, **3**, 219–28.

Chu, A. C. P., McPherson, H. G. and Halligan, G. (1979) Recovery growth following water deficits of different duration in Prairie grass. *Austral. J. Pl. Physiol.*, **6**, 255–63.

Colman, R. L. and Lazenby, A. (1975) Effect of moisture on growth and nitrogen response by *Lolium perenne*. *Pl. Soil*, **42**, 1–13.

Cowan, I. R. (1977) Stomatal behaviour and environment, in *Advances in Botanical Research*, Vol. 4 (eds R. D. Preston and H. W. Woolhouse), Academic Press, London, pp. 117–228.

Cowan, I. R. (1982) Regulation of water use in relation to carbon gain in higher plants, in *Encyclopedia of Plant Physiology, New Series* Vol. 12B (eds O. L. Lange, P. S. Nobel, C. B. Osmond and H. Ziegler), Springer-Verlag, Berlin, pp. 589–613.

Cowan, I. R. and Farquhar, G. D. (1977) Stomatal function in relation to leaf metabolism and environment. *Symp. Soc. Exp. Biol.*, **31**, 471–505.

D'Aoust, M. J. and Tayler, R. S. (1968) The interaction between nitrogen and water in the growth of grass swards. I. Methods and dry matter results. *J. Agric. Sci., Camb.*, **70**, 11–7.

Davies, W. J. (1978) Some effects of abscisic acid and water stress on stomata of *Vicia faba* L. *J. Exp. Bot.*, **29**, 175–82.

Davies, W. J. (1985) Transpiration and the water balance of plants, in *Water and Solutes in Plants* (eds F. C. Steward, James F. Sutcliffe and John E. Dale), Plant Physiology; A Treatise. Vol. 9, Academic Press, London, pp. 49–154.

Davies, W. J., Wilson, J. A., Sharp, R. E. and Osonubi, O. (1981) Control of stomatal behaviour in water-stressed plants, in *Stomatal Physiology* (eds P. G. Jarvis and T. A. Mansfield), SEB Seminar Series No. 8, Cambridge University Press, London, pp. 162–85.

Day, W., Legg, B. J., French, B. K., Johnston, A. E., Lawlor, D. W. and Jeffers, W. de C. (1978) A drought experiment using mobile shelters: the effect of drought on barley yield, water use and nutrient uptake. *J. Agric. Sci., Camb.*, **91**, 599–623.

Doley, D. and Trivett, N. B. A. (1974) Effects of low water potentials on transpiration and photosynthesis in Mitchell grass (*Astrebla lappacea*). *Austral. J. Pl. Physiol.*, **1**, 539–50.

Doyle, C. J. (1981) Economics of irrigating grassland in the United Kingdom. *Grass, Forage Sci.*, **36**, 297–306.

Faiz, S. M. A. and Weatherley, P. E. (1978) Further investigations into the location and magnitude of the hydraulic resistances in the soil:plant system. *New Phytol.*, **81**, 19–28.

Fischer, R. A. and Turner, N. C. (1978) Plant productivity in the arid and semi-arid zones. *A. Rev. Pl. Physiol.*, **29**, 277–317.

Ford, C. W. and Wilson, J. R. (1981) Changes in levels of solutes during osmotic adjustment to water stress in leaves of four tropical pasture species. *Austral. J. Pl. Physiol.*, **8**, 77–91.

Gardner, C. M. K. and Field, M. (1983) An evaluation of the success of MORECS; a meteorological model, in estimating soil moisture deficits. *Agric. Met.*, **29**, 269–84.

Garwood, E. A. and Sinclair, J. (1979) Use of water by six grass species, 2. Root distribution and use of soil water. *J. Agric. Sci., Camb.*, **93**, 25–35.

Garwood, E. A. and Williams, T. E. (1967) Growth, water use and nutrient uptake from the subsoil by grass swards. *J. Agric. Sci. Camb.*, **69**, 125–310.

Gavande, S. A. and Taylor, S. A. (1967) Influence of soil water potential and atmospheric evaporative demand on transpiration and the energy status of water in plants. *Agron. J.*, **59**, 4–7.

Gifford, R. M. (1974) A comparison of potential photosynthesis, productivity and yield of plant species with differing photosynthetic metabolism. *Austral. J. Pl. Physiol.*, **1**, 107–17.

Grace, J. (1983) *Plant–Atmosphere Relationships*, Outline Studies in Ecology, Chapman and Hall, London.

Greacen, E. L. and Oh, J. S. (1972) Physics of root growth. *Nature, New Biol.*, **235**, 24–5.

Greacen, E. L., Ponsama, P. and Barley, K. P. (1976) Resistance to water flow in the roots of cereals, in *Water and Plant Life, Ecological Studies No. 19* (eds O. L. Lange, L. Kappen and E.-D. Schulze), Springer, Berlin, pp. 86–100.

Green, P. B. (1968) Growth physics in *Nitella*: a method for continuous *in vivo* analysis of extensibility based on a micro-manometer technique for turgor pressure. *Pl. Physiol.*, **43**, 1169–84.

Harris, W. and Lazenby, A. (1974) Competitive interaction of grasses with contrasting temperature responses and water stress tolerances. *Austral. J. Agric. Res.*, **25**, 227–46.

Honert, T. H. van den (1948) Water transport in plants as a catenary process. *Discussion Faraday Soc.*, **3**, 146–53.

Horst, G. L. and Nelson, C. J. (1979) Compensatory growth of tall fescue following drought. *Agron. J.*, **71**, 559–63.

Hsiao, T. C. (1973) Plant responses to water stress. *A. Rev. Pl. Physiol.*, **24**, 519–70.

Hsiao, T. C. and Acevedo, E. (1974) Plant responses to water deficits, water-use efficiency, and drought resistance. *Agric. Met.*, **14**, 59–84.

Hsiao, T. C., Acevedo, E., Fereres, E. and Henderson, D. W. (1976) Stress metabolism; water stress, growth and osmotic adjustment. *Phil. Trans. R. Soc. Lond.*, **B 273**, 479–500.

Hsiao, T. C., Silk, W. K. and Jing, J. (1985) Leaf growth and water deficits: Biophysical effects, in *Control of Leaf Growth* (eds N. R. Baker, W. J. Davies and C. K. Ong), S.E.B. Seminar Series No. 27, Cambridge University Press, London, pp. 239–266.

Jackson, D. K. (1974) The course and magnitude of water stress in *Lolium perenne* and *Dactylis glomerata.*, *J. Agric. Sci. Camb.*, **82**, 19–27.

Jarvis, P. G. (1981) Stomatal conductance, gaseous exchange and transpiration, in *Plants and their Atmospheric Environment* (eds J. Grace, E. D. Ford and P. G. Jarvis), B.E.S. Symposium No. 21, Blackwell Scientific, Oxford, pp. 175–203.

References

Johns, G. G. (1978) Transpirational, leaf area, stomatal and photosynthetic responses to gradually induced water stress in four temperate herbage species. *Austral. J. Pl. Physiol.*, **5**, 113–25.

Jones, H. G. (1983) *Plants and Microclimates – a Quantitative Approach to Environmental Plant Physiology*, Cambridge University Press, Cambridge.

Jones, M. B., Leafe, E. L. and Stiles, W. (1980a) Water stress in field-grown perennial ryegrass. I. Its effect on growth, canopy photosynthesis and transpiration. *Ann. Appl. Biol.*, **96**, 87–101.

Jones, M. B., Leafe, E. L. and Stiles, W. (1980b) Water stress in field-grown perennial ryegrass. II. Its effect on leaf water status, stomatal resistance and leaf morphology. *Ann. Appl. Biol.*, **96**, 103–10.

Kerr, J. P. and Beardsell, M. F. (1975) Effect of dew on leaf water potentials and crop resistance in a *Paspalum* pasture. *Agron. J.*, **67**, 596–9.

Keulen, H. van (1975) *Simulation of Water Use and Herbage Growth in Arid Regions*, Centre for Agricultural Publishing and Documentation, Wageningen.

Kigel, J. and Dotan, A. (1982) Effect of different durations of water withholding on regrowth potential and non-structural carbohydrate content in Rhodes grass (*Chloris gayana* Kunth). *Austral. J. Pl. Physiol.*, **9**, 113–20.

Körner, Ch., Scheel, J. A. and Bauer, H. (1979) Maximum leaf diffusive conductance in vascular plants. *Photosynthetica*, **13**, 45–82.

Kozlowski, T. T. (ed.) (1984) *Flooding and Plant Growth.*, Academic Press, London.

Kramer, P. J. (1980) Drought, stress and the origin of adaptation, in *Adaptation of Plants to Water and High Temperature Stress* (eds N. C. Turner and P. J. Kramer), Wiley, New York, pp. 7–20.

Kramer, P. J. (1983) *Water Relations of Plants*, Academic Press, London.

Lange, O. L., Lösch, R., Schulze, E.-D. and Kappen, L. (1971) Responses of stomata to changes in humidity. *Planta*, **100**, 76–86.

Lawlor, D. W. (1972) Growth and water use of *Lolium perenne* II. Plant growth. *J. Appl. Ecol.*, **9**, 99–105.

Long, S. P. and Woolhouse, H. W. (1978) The response of net photosynthesis to light and temperature in *Spartina townsendii* (*sensu lato*), a C_4 species from a cool temperate climate. *J. Exp. Bot.*, **29**, 803–14.

Lösch, R. and Tenhunen, J. D. (1981) Stomatal responses to humidity – phenomenon and mechanism, in *Stomatal Physiology* (eds P. G. Jarvis and T. A. Mansfield), S.E.B. Seminar Series No. 8, Cambridge University Press, London, pp. 137–61.

Ludlow, M. M. (1975) Effect of water stress on the decline of leaf net photosynthesis with age, in *Environmental and Biological Control of Photosynthesis* (ed. R. Marcelle), Dr W. Junk, The Hague, pp. 123–34.

Ludlow, M. M. (1980) Adaptive significance of stomatal response to water stress, in *Adaptation of Plants to Water and High Temperature Stress* (eds N. C. Turner and P. J. Kramer), Wiley, New York, pp. 123–38.

Ludlow, M. M. and Ibaraki, K. (1979) Stomatal control of water loss in siratro (*Macroptilium atropurpureum* (DC.) Urb.) a tropical pasture legume. *Ann. Bot.*, **43**, 639–42.

Ludlow, M. M. and Ng, T. T. (1976) Effect of water deficit on carbon dioxide exchange and leaf elongation rate of *Panicum maximum* var. *trichoglume*. *Austral. J. Pl. Physiol.*, **3**, 401–13.

Ludlow, M. M., Ng, T. T. and Ford, C. W. (1980) Recovery after water stress of leaf gas exchange in *Panicum maximum* var. *trichoglume*. *Austral. J. Pl. Physiol.*, **7**, 299–313.

Luxmoore, R. J. and Millington, R. J. (1971) Growth of perennial ryegrass (*Lolium perenne*) in relation to water, nitrogen and light intensity. *Pl. Soil*, **34**, 269–81.

McCree, K. J. and Davies, S. D. (1974) Effect of water stress and temperature on leaf size and on size and number of epidermal cells in grain sorghum. *Crop Sci.*, **14**, 751–5.

McMillan, W. D. and Burgy, R. H. (1960) Interception loss from grass. *J. Geophys. Res.*, **65**, 2389–94.

Mansfield, T. A. and Davies, W. J. (1981) Stomata and stomatal mechanisms, in *Physiology and Biochemistry of Drought Resistance in Plants* (eds L. G. Paleg and D. Aspinall), Academic Press, Australia, pp. 315–346.

Matsuda, K. and Riazi, A. (1981) Stress-induced osmotic adjustment in growing regions of barley leaves. *Pl. Physiol.*, **68**, 571–6.

Merriam, R. A. (1961) Surface water storage on annual ryegrass. *J. Geophys. Res.*, **66**, 1833–8.

Michelena, V. A. and Boyer, J. S. (1982) Complete turgor maintenance at low water potentials in the elongating region of maize leaves. *Pl. Physiol.*, **69**, 1145–9.

Molyneux, D. E. and Davies, W. J. (1983) Rooting pattern and water relations of three pasture grasses growing in drying soil. *Oecologia*, **58**, 220–4.

Monteith, J. L. (1965) Evaporation and environment, in *The State and Movement of Water in Living Organisms* (ed. G. E. Fogg), S.E.B. Symposium, Vol. 19. Cambridge University Press, London, pp. 205–34.

Monteith, J. L. (1973) *Principles of Environmental Physics,* Edward Arnold, London.

Morrison, J. and Idle, A. A. (1972) A pilot survey of grassland in south east England. *Grassland Res. Inst. Tech. Rep. 10.*

Munns, R., Brady, C. J. and Barlow, E. W. R. (1979) Solute accumulation in the apex and leaves of wheat during water stress. *Austral. J. Pl. Physiol.*, **6**, 379–89.

Ndawula-Senyemba, M. S., Brink, V. C. and McClean, A. (1971) Moisture interception as a factor in the competitive ability of bluebunch wheatgrass. *J. Range Mgmt.*, **24**, 198–200.

Newman, E. I. (1976) Water movement through root systems. *Phil. Trans. R. Soc. Lond. Ser. B*, **273**, 463–78.

Ng, T. T., Wilson, J. R. and Ludlow, M. M. (1975) Influence of water stress on water relations and growth of a tropical (C_4) grass, *Panicum maximum* var. *trichoglume*. *Austral. J. Pl. Physiol.*, **2**, 581–95.

Norris, I. B. and Thomas, H. (1982) Recovery of ryegrass species from drought. *J. Agric. Sci. Camb.*, **98**, 623–8.

O'Toole, J. C. and Cruz, R. T. (1980) Response of leaf water potential, stomatal resistance and leaf rolling to water stress. *Pl. Physiol.*, **65**, 428–32.

Parker, J. (1968) Drought-resistance mechanisms, in *Water Deficits and Plant Growth,* Vol. 1 (ed. T. T. Kozlowski), Academic Press, New York, pp. 195–234.

Passioura, J. B. (1981) Water collection by roots, in *The Physiology and Biochemistry of Drought Resistance in Plants* (eds L. G. Paleg and D. Aspinall), Academic Press, Australia, pp. 39–53.

Penman, H. L. (1948) Natural evaporation from open water, bare soil and grass. *Proc. R. Soc., Ser. A*, **194**, 120–46.

References

Philip, J. R. (1966) Plant water relations: some physical aspects. *A. Rev. Pl. Physiol.*, **17**, 245–68.

Powell, D. B. B. and Thorpe, M. R. (1977) Dynamic aspects of plant water relations, in *Environmental Effects on Crop Physiology* (eds J. J. Landsberg and C. V. Cutting), Academic Press, London, pp. 259–79.

Raschke, K. (1975) Stomatal action. *A. Rev. Pl. Physiol.*, **26**, 309–40.

Rawson, H. M., Begg, J. E. and Woodward, R. G. (1977) The effect of atmospheric humidity on photosynthesis, transpiration and water use efficiency of leaves of several plant species. *Planta*, **134**, 5–10.

Russell, R. S. (1977) *Plant Root Systems: Their Function and Interaction with the Soil*, McGraw-Hill, London.

Rutter, A. J. (1975) The hydrological cycle in vegetation, in *Vegetation and the Atmosphere*, Vol. 1 *Principles* (ed. J. L. Monteith), Academic Press, London, pp. 111–54.

Sala, O. E., Lauenroth, W. K. and Reid, C. P. P. (1982) Water relations: a new dimension for niche separation between *Bouteloua gracilis* and *Agropyron smithii* in north American semi-arid grasslands. *J. Appl. Ecol.*, **19**, 647–57.

Sharp, R. E. and Davies, W. J. (1979) Solute regulation and growth by roots and shoots of water-stressed maize plants. *Planta*, **147**, 43–9.

Sheehy, J. E., Green, R. M. and Robson, M. J. (1975) The influence of water stress on the photosynthesis of a simulated sward of perennial ryegrass. *Ann. Bot.*, **39**, 387–401.

Sheriff, D. W. (1977) The effect of humidity on water uptake by, and viscous flow resistance of, excised leaves of a number of species: physiological and anatomical observations. *J. Exp. Bot.*, **28**, 1399–407.

Sheriff, D. W. and Ludlow, M. M. (1984) Physiological reactions to an imposed drought by *Macroptilium atropurpureum* and *Cenchrus ciliaris* in a mixed sward. *Austral. J. Pl. Physiol.*, **11**, 23–34.

Silcock, R. G. and Wilson, D. (1981) Effect of watering regime on yield, water use and leaf conductance of seven *Festuca* species with contrasting leaf ridging. *New Phytol.*, **89**, 569–80.

Simmelsgaard, S. E. (1976) Adaptation to water stress in wheat. *Physiol. Pl.*, **37**, 167–74.

Slatyer, R. O. (1957) Significance of the permanent wilting percentage in studies of plant and soil relations. *Bot. Rev.*, **23**, 585–636.

Squire, G. R. and Black, C. R. (1981) Stomatal behaviour in the field, in *Stomatal Physiology* (eds P. G. Jarvis and T. A. Mansfield), S.E.B. Seminar Series No. 8, Cambridge University Press, London, pp. 223–46.

Stewart, G. R. and Lee, J. A. (1974) The role of proline accumulation in halophytes. *Planta*, **120**, 279–89.

Szeicz, G., Van Bavel, C. H. M. and Takami, S. (1973) Stomatal factor in the water use and dry matter production of sorghum. *Agric. Met.*, **12**, 361–89.

Thom, A. and Oliver, H. R. (1977) On Penman's equation for estimating regional evaporation. *Q. J. R. Met. Soc.*, **103**, 345–57.

Thompson, N. (1981) MORECS, in *The MORECS Discussion Meeting (April 1981)* (ed. C. M. K. Gardner), Institute of Hydrology Report Series No. 78, Wallingford, pp. 1–10.

Troughton, A. (1974) The development of leaf water deficits in plants of *Lolium perenne* in relation to the sizes of the root and shoot system. *Pl. Soil,* **40,** 153–60.

Turner, N. C. (1974) Stomatal response to light and water uptake under field conditions, in *Mechanisms of Regulation of Plant Growth* (eds R. L. Bieleski, A. R. Ferguson and M. M. Cresswell), Bulletin 12, The Royal Society of New Zealand, Wellington, pp. 423–32.

Turner, N. C. and Begg, J. E. (1978) Responses of pasture plants to water deficits, in *Plant Relations in Pastures* (ed. J. R. Wilson), CSIRO, Melbourne, pp. 50–66.

Turner, N. C. and Jones, M. M. (1980) Turgor maintenance by osmotic adjustment: a review and evaluation, in *Adaptation of Plants to Water and High Temperature Stress* (eds N. C. Turner and P. J. Kramer), Wiley, New York, pp. 87–103.

Volkenburgh, E. Van and Cleland, R. E. (1984) Control of leaf growth by changes in cell wall properties. *What's New Pl. Physiol.,* **15,** 25–8.

Wardlaw, I. F. (1969) The effect of water stress on translocation in relation to photosynthesis and growth. *Austral. J. Biol. Sci.,* **22,** 1–16.

Watt, T. A. and Haggar, R. J. (1980) The effect of height of water table on the growth of *Holcus lanatus* with reference to *Lolium perenne. J. Appl. Ecol.,* **17,** 423–30.

Willmer, C. M. (1983) *Stomata,* Longman, London.

Wilson, A. M., Hyder, D. N. and Briske, D. D. (1976) Drought resistance characteristics of Blue Gamma seedlings. *Agron. J.,* **68,** 479–84.

Wilson, D. (1971) Selection responses of stomatal length and frequency, epidermal ridging and other leaf characteristics of *Lolium perenne* L. 'Grassland Ruanui'. *N. Z. J. Agric. Res.,* **14,** 761–71.

Wilson, D. (1975a) Leaf growth, stomatal diffusion resistances and photosynthesis during droughting of *Lolium perenne* populations selected for contrasting stomatal length and frequency. *Ann. Appl. Biol.,* **79,** 67–82.

Wilson, D. (1975b) Stomatal diffusion resistances and leaf growth during droughting of *Lolium perenne* plants selected for contrasting epidermal ridging. *Ann. Appl. Biol.,* **79,** 83–94.

Wilson, J. R. and Ford, C. W. (1973) Temperature influences on the *in vitro* digestibility and soluble carbohydrate accumulation of tropical and temperate grasses. *Austral. J. Agric. Res.,* **24,** 187–98.

Wilson, J. R. and Ludlow, M. M. (1983a) Time trends for change in osmotic adjustment and water relations of leaves of *Cenchrus ciliarus* during and after water stress. *Austral. J. Pl. Physiol.,* **10,** 15–24.

Wilson, J. R. and Ludlow, M. M. (1983b) Time trends of solute accumulation and the influence of potassium fertilizer on osmotic adjustment of water-stressed leaves of three tropical grasses. *Austral. J. Pl. Physiol.,* **10,** 523–37.

Wilson, J. R. and Ng, T. T. (1975) Influence of water stress on parameters associated with herbage quality of *Panicum maximum* var. *trichoglume. Austral. J. Agric. Res.,* **26,** 127–36.

Wilson, J. R., Ludlow, M. M., Fisher, M. J. and Schulze, E.-D. (1980) Adaptation to water stress of leaf water relations of four tropical forage species. *Austral. J. Pl. Physiol.,* **7,** 207–20.

CHAPTER 7
Physiological models of grass growth

J. E. Sheehy and I. R. Johnson

7.1 INTRODUCTION

A mathematical model can be thought of as a concise mechanism for providing a numerical description of a process or an object. In agriculture, models are usually concerned with processes, which may be associated with problems as different as economic issues and the biochemistry of nitrate reduction. In this chapter we shall be concerned with modelling the growth of the grass crop using physiological data.

Some readers might be sceptical about the value of modelling. Clearly, for improvements to be made in the efficiency of grassland agriculture a good understanding of the grass crop is essential. One benefit derived from modelling is the exposure of gaps in knowledge at the sub-model level, i.e. the individual processes, such as photosynthesis and nutrient uptake, which contribute to the crop model. As well as ultimately being used as predictive tools, models of crop growth provide a means by which knowledge of individual plant and crop processes may be systematically brought together to increase our understanding of the integrated system. The whole grass plant integrates environmental and management factors in producing a harvestable yield, and in this chapter some of the surprisingly large gaps in current knowledge concerning the physiology of grass plants will be highlighted.

Another useful aspect of modelling is that many data relevant to a complex process can be described concisely in terms of a few model parameters, thus simplifying discussions of the process. For example, data describing the relationship between two sets of measurements can often be described by a linear equation established using regression techniques. The data and their relationship can then be described in terms of the regression coefficient and constant. Thus, in addition to their usefulness in revealing gaps in knowledge and condensing experimental data, models also have the capacity for prediction. It is this combination of attributes which makes them such powerful and valuable tools.

The development of models will play a significant role in future improvements in grassland production and utilization. As computers become more common in the agricultural industry, the use of crop models as aids to resource management will increase. The construction of models to meet this demand presents an immediate challenge to physiologists and mathematicians.

7.2 MODEL TERMINOLOGY

Thornley (1976) divided mathematical models into two categories: mechanistic and empirical. Mechanistic models describe the behaviour of the plant or crop in terms of how their various parts actually work. An example of a mechanistic model is the Münch hypothesis, which describes the movement of assimilates in the phloem. The model is based on the assumption that differences in the concentrations of solutes in the plant result in pressure gradients which force the phloem contents to move. Models based on the Münch hypothesis are used to describe the flow of assimilates between the regions of assimilation and utilization.

Often, empirical models are formulated by applying statistical techniques to experimental data to establish the most satisfactory equation for describing, and therefore summarizing, the data. The empirical models, or equations, of the various processes in a system may then be combined to provide a mechanistic description of the whole system. For example, as will be seen later, a mechanistic model of canopy photosynthesis is constructed from empirical equations which describe light interception and attenuation through the canopy, and the single-leaf photosynthesis response to irradiance.

A good example of an empirical model is the linear regression equation describing the relationship between crop growth rate and percentage light interception. Monteith (1981b) observed that some of the worst examples of empirical models involved the prediction of crop yields from weather data. As a consequence of the total lack of mechanistic structures, such models are unable to give good predictions for extreme and abnormal weather conditions, the very conditions in which good predictions would be of most value. Thus, to make good predictions of forage crop growth, in the complex weather patterns they normally experience, it is desirable that crop models be mechanistic. However, the distinction between mechanistic and empirical models is not always clear (Thornley, 1976), and in physiological models of crop production there are usually several sub-models, some of which may be mechanistic and some empirical.

7.3 REQUIREMENTS FOR CROP PHYSIOLOGICAL MODELS

The energy costs of growing or constructing a crop are large, approximately 2000 MJ of solar energy (400–2400 nm) are absorbed during the production of 1 kg of dry matter. Clearly, the factors affecting the energy budget of the crop are of primary importance during the modelling of growth. The average

Crop physiological models

efficiency of conversion of photosynthetically active radiation (PAR, 400–700 nm) into harvestable biomass is approximately 2%; the other 98% is dissipated as heat in the environment. The largest component of the dissipated heat is that associated with transpiration and the latent heat of vaporization ($2450\,\text{J}\,\text{g}^{-1}$) ensures that the crop remains cool. Sensible heat losses and re-radiation of energy account for the solar energy not consumed in transpiration.

A crop obeys the First Law of Thermodynamics, which states that energy is conserved. Thus, the radiant energy absorbed by the crop drives photosynthesis, transpiration and convective heat losses. The equation describing the energy budget is

$$R_a = C + \lambda E + P$$

where R_a is the total radiation (shortwave plus longwave) absorbed by the crop, C is the sensible heat loss (convection), λ is the latent heat of vaporization and E is the rate of transpiration; P, the energy consumed in photosynthesis, is assumed to be negligible. With the exception of unusual crops such as leafless peas, it is the leaves which intercept and utilize PAR in photosynthesis. The amount of shortwave radiation absorbed by the crop depends on the shortwave irradiance, the leaf area per unit ground area and the architecture of the crop. The absorption or loss of longwave radiation depends on the thermal properties of the soil, air and crop and their temperatures. The sensible heat loss is a function of the temperature gradient between the crop and the air, the windspeed and the aerodynamics roughness of the crop; the last two of these terms can be combined to yield a crop aerodynamic resistance. The gradient in water vapour density between the air inside the leaves and the bulk atmosphere, an average crop stomatal resistance and crop aerodynamic resistance govern the rate of loss of latent heat from the crop.

Meterorological data and the energy budget equation are the basis of water use, irrigation and transpiration calculations. Sub-models describing the relationship between transpiration, soil water status and stomatal resistance are used to estimate plant water potential. The influence of water stress on the growth and physiology of the crop can be estimated using the values of water potential (see Chapter 6). Examples of such models can be found in the work of Lambert and Penning de Vries (1971), Penning de Vries (1972), Goudriaan and Waggoner (1972) and Zur and Jones (1981).

The carbohydrates constructed during photosynthesis provide the energy and the carbon skeletons – the building materials for the synthesis of roots, stems, leaves and seeds. Temperature influences the rate of most biochemical processes, and it is not surprising that it affects both growth and senescence of plant organs. The water relations of a crop, which are largely determined by the availability of water and the transpiration burden, have a direct effect on the growth of leaves and other organs, as well as having an indirect effect on growth through the stomatal regulation of photosynthesis. Thus, it can be seen that the rate of photosynthesis, the division of the photosynthate between the various plant organs, mineral nutrition and the rate of growth of the various plant organs are all influenced strongly by environmental conditions.

In the field the land area occupied by the crop is divided between a large number of plants. The aerial portions of the plants overlap, and the amount of mutual shading depends on the size of the plants; the roots share the same nutrient and water supply. The consequence of this sharing of, or competition for, resources is to produce plants which have totally different morphologies from unshaded plants growing with abundant resources. Crop plants are often unable to express their individual potential for growth because they are severely substrate limited. Thus, a successful crop model must have a number of physiological sub-models, the parameters of which are a function of the macroclimate, the microclimate, crop morphology and time. The sceptic at this point might exclaim, with some justification, that the exercise is bound to be unrealistic as it embraces problems across the whole of plant science research. However, it should be noted that it is only research at the level of whole organs, plants and crops that is usually relevant to whole crop models.

Our aim, therefore, is to try and highlight the various processes which must be described by sub-models in order to construct whole crop models. It is likely that much of the detail presented in this chapter will be changed and modified as our understanding of crop physiology increases and the modelling techniques improve; however, the general approach and strategy will still be followed.

7.4 LIGHT INTERCEPTION AND PHOTOSYNTHESIS

Light interception and photosynthesis has been the subject of much research in recent years. Reviews of models in this area have been presented by Acock *et al.* (1971), Thornley (1976), Charles-Edwards (1981) and Johnson and Thornley (1984) have recently described a model of instantaneous and daily canopy photosynthesis in some detail. In this section we will outline a simple approach for modelling photosynthesis in forage crops. Throughout the discussion, the term irradiance will be taken to mean photosynthetically active radiation (PAR).

In the irradiance distribution as shown in Fig. 7.1, the cumulative leaf area index, L (leaf area per ground area), is measured from the upper surface. I_0 and $I(L)$ are the irradiances at the surface of the canopy and at depth represented by L, with units $W\,m^{-2}$ PAR. Monsi and Saeki (1963) demonstrated that the light attenuation through the canopy could be described by the Beer–Lambert law:

$$I(L) = I_0 e^{-kL} \tag{7.1}$$

k, the extinction coefficient, is a dimensionless parameter and depends on such factors as leaf angle and leaf transmission as well as non-randomness of leaf

Light interception and photosynthesis

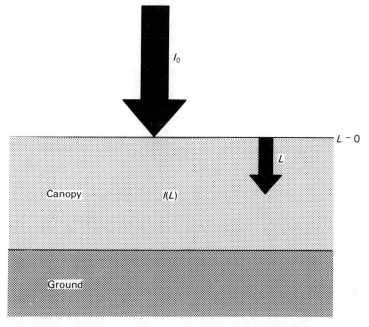

Figure 7.1 Light attenuation through the canopy (see text for details).

distribution, and characterizes the light attenuation properties of the crop in question. Although empirical, equation (7.1) can be derived from

$$\frac{1}{I}\frac{dI}{dL} = -k \tag{7.2}$$

which states that the specific rate of irradiance attenuation is constant so that, for a small increase in L, a fixed proportion of the irradiance is attenuated. Monteith (1965) developed a similar model for light attenuation in the canopy by considering the fraction, S, of light which was not intercepted by leaves in a unit L layer. The remaining fraction, $1-S$, is intercepted and a further fraction of this, m, is transmitted. Under these assumptions, the analogous equation to equation (7.1) is

$$I(L) = I_0[S + (1-S)m]^L \tag{7.3}$$

and it follows, comparing equations (7.1) and (7.3), that

$$e^{-k} = S + (1-S)m \tag{7.4}$$

or, alternatively,

$$S = \frac{e^{-k} - m}{1 - m} \tag{7.5}$$

Thus, the two models are similar and, henceforth, we will restrict attention to equation (7.1), as it is analytically more convenient.

Equation (7.1) gives the irradiance per unit horizontal area in the canopy at a given value of L. In order to calculate the photosynthetic rate of the leaves at that level it is necessary to derive an expression for the light flux density incident per unit area of leaf. The following method is due to Saeki (1963).

As L changes by an increment ΔL, let the corresponding change in I be ΔI, (if L is positive, ΔL will be negative and vice versa). Let I_1 be the mean irradiance incident on the surface of the leaves in ΔL. Neglecting scattered light, this can be partitioned into absorbed and transmitted fractions

$$(1 - m) I_1, m I_1,$$

respectively, where m is the fraction of light transmitted (approximately 10% of light is transmitted, Sheehy and Cook, 1977). The light which is lost from I is the absorbed fraction (that which is used for photosynthesis), so that

$$\Delta I = - \Delta L (1 - m) I_1$$

The negative sign ensures that ΔI is negative when ΔL is positive. Taking the usual limit in the calculus gives

$$I_1 = - \frac{1}{(1-m)} \frac{dI}{dL} \tag{7.6}$$

(Note that this is perfectly valid when $m = 0$). Thus, combining eqns (7.1) and (7.6):

$$I_1 = \frac{k}{1-m} I_0 e^{-kL} \tag{7.7}$$

There are two important points which are worth emphasizing at this stage:

(i) k is not independent of m so that, for example, the effect of varying m may not be analysed by simply changing the value of m in equation (7.7)
(ii) Near the top of the canopy $e^{-kL} = 1$ and so, to a first approximation

$$I_1 = \frac{k}{1-m} I_0,$$

and hence, if $k/(1-m)$ is greater than unity the model breaks down. In such cases an alternative expression to the Beer–Lambert law for light attenuation is required. In practice this problem is not encountered when considering grass crops. For example, Sheehy and Peacock (1975) observed values of k in the region of 0.5 for grass so that, with $m = 0.1$, this gives $k/(1-m) = 0.56$.

Equation (7.7) relates the irradiance per unit horizontal area at the surface of the canopy and the light properties of the canopy to the irradiance incident on the leaves in the canopy. In order to calculate canopy photosynthetic rates it is now necessary to relate the single-leaf photosynthetic rate to PAR. Respiration

Light interception and photosynthesis

is not considered here, as it is discussed later, but the term gross photosynthesis is the sum of net photosynthesis and dark respiration.

The most simple and widely used empirical equation for describing the rate of gross photosynthesis in single leaves is the rectangular hyperbola

$$P_g = \frac{\alpha I_1 P_m}{\alpha I_1 + P_m} \tag{7.8a}$$

P_g is the rate of gross photosynthesis, kg CO_2 m^{-2} (leaf) s^{-1}, and α the photochemical efficiency, kg CO_2 J^{-1}, P_m is the rate of photosynthesis at saturating PAR (I).

The initial slope of the curve is

$$\left. \frac{dP_g}{dI_1} \right|_{I_1 = 0} = \alpha \tag{7.9a}$$

and the asymptote is

$$P_g(I_1 \to \infty) = P_m \tag{7.9b}$$

P_m depends upon the ambient CO_2 concentration, but although this is important in greenhouses where the CO_2 concentration may be controlled, it is less relevant in field crops where CO_2 concentrations vary little.

Equation (7.8a) is widely used due to its simplicity. However, it is generally accepted that this curve does not always provide a good fit to experimental data (Marshall and Biscoe, 1980; Monteith, 1981a). The most versatile semi-empirical curve for describing single-leaf photosynthesis is the non-rectangular hyperbola which was first developed by Rabinovitch (1951), and where P_g is given by the lower root of the quadratic

$$\theta P_g^2 - (\alpha I_1 + P_m)P_g + \alpha I_1 P_m = 0 \tag{7.8b}$$

i.e.

$$P_g = \frac{1}{2\theta} \{\alpha I_1 + P_m - [(\alpha I_1 + P_m)^2 - 4\theta \alpha I_1 P_m]^{1/2}\} \tag{7.8c}$$

This equation has one more parameter than the rectangular hyperbola, and this parameter is constrained by $0 \leq \theta \leq 1$. However, the initial slope and asymptate are independent of θ and are still given by equations (7.9 a,b). For $\theta = 0$ equation (7.8b) reduces to equation (7.8a), and with $\theta = 1$ one obtains the Blackman (1905) limiting response:

$$P_g = \begin{cases} \alpha I_1 & I_1 < P_m/\alpha \\ P_m & I_1 > P_m/\alpha \end{cases}$$

Intermediate values of θ generate response curves lying between these two extremes, as illustrated in Fig. 7.2 for $\theta = (0, 0.5, 0.9, 1)$. As might be anticipated, the analysis for calculating canopy photosynthesis is greatly simplified

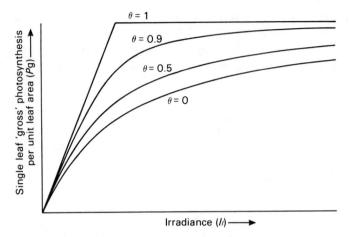

Figure 7.2 The relationship between leaf 'gross photosynthesis' and irradiance in arbitrary units. The parameter θ in equation (7.8b) is indicated in the figure.

by taking $\theta = 0$: that is, using the rectangular hyperbola. This is the only case we shall consider here, and the interested reader should consult Johnson and Thornley (1984) for a more complete discussion of the problem.

The canopy gross photosynthetic rate P_c, kg CO_2 m^{-2} (ground) s^{-1}, is given by

$$P_c = \int_0^L P_g \, dL \tag{7.10}$$

Combining equations (7.7), (7.8) and (7.10) and integrating yields

$$P_c = \frac{P_m}{k} \ln \left(\frac{\alpha k I_0 + P_m(1-m)}{\alpha k I_0 e^{-kL} + P_m(1-m)} \right) \tag{7.11}$$

This simple equation gives the canopy photosynthetic rate in terms of the canopy architecture, the photosynthetic properties of the single leaves, the light flux density from the environment and the leaf area index. The initial slope is

$$\left. \frac{dP_0}{dI_0} \right|_{I_0 = 0} = \frac{\alpha}{1-m} (1 - e^{-kL}) \tag{7.12a}$$

and the asymptote is

$$P_c(I_0 \to \infty) = P_m L \tag{12.7b}$$

representing the situation where all the leaves are fully light intercepting (although seldom realized in practice, such expressions give good checks for mathematical consistency). Equation (7.11) is illustrated in Fig. 7.3.

Respiration

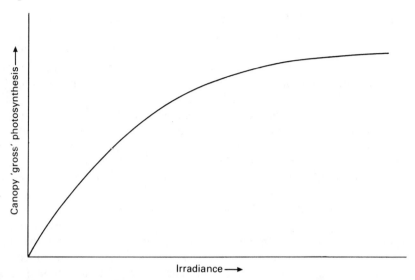

Figure 7.3 The relationship between canopy 'gross' photosynthesis and irradiance in arbitrary units.

The total daily photosynthesis is given by

$$P_d = \int_{daylength} P_c \, dt$$

and Johnson and Thornley (1984) give details for evaluating P_d for both variable and constant conditions of within-day irradiance and temperature.

7.5 RESPIRATION

The contemporary view of respiration as part of the essential process of building and maintaining plant biomass owes much to the work of McCree (1970) and Penning de Vries (1972). McCree observed that the rate of plant respiration, measured as CO_2 efflux, was affected by plant size and that it increased with increasing plant photosynthesis. He described the respiratory loss of CO_2, R, using a simple two-component equation which was quantified by Thornley (1970) and may be expressed as

$$R = (1 - Y_G) P_g + MW \tag{7.13}$$

where P_g is the daily rate of gross photosynthesis and W the weight of the crop, Y_G is the efficiency with which photoassimilate is converted to plant material and is dimensionless (typically $Y_G \approx 0.75$), and M is a maintenance respiration rate constant (units day^{-1}).

Penning de Vries (1972) charted the biochemical pathways leading to plant products, starting with glucose as the primary assimilate. He calculated the costs of biosynthesis for higher plants by estimating the amount of glucose required as skeletons in amino acids and other compounds, and included the additional glucose required as an energy source. Thus, total budgets for O_2 consumption, CO_2 production and heat production were estimated. From this work it was apparent that the total respiration associated with growth depended on the biochemical nature of the products synthesized, and was independent of temperature. However, the rate of synthesis (growth) is usually temperature dependent. Penning de Vries estimated that approximately 25% of the photoassimilate was consumed supplying energy for biosynthesis; this respiration is often referred to as synthetic respiration. The energy, in the form of glucose, expended in maintaining the integrity of the plant is largely associated with protein turnover (Penning de Vries, 1972). The corresponding efflux of CO_2 associated with this maintenance of the organization of the plant is known as maintenance respiration. The amount and rate of protein turnover is temperature dependent, so that maintenance respiration is temperature dependent. There are a number of important and obvious conclusions on crop production we can derive from the work of Penning de Vries and McCree:

(1) during the early stages of crop growth the plants are growing rapidly but are small, so that synthetic respiration is high and maintenance respiration is low;
(2) when the crop is fully grown the maintenance component can be greater than the synthetic component of respiration;
(3) to improve the quality of the crop, in terms of perhaps protein content, an increased respiratory cost must be paid.

There are two principal limitations to the simple model outlined by McCree (1970). The first is that not all structural biomass is likely to require maintenance. The second is that there is no provision for reserve materials or the recycling of those materials. Furthermore, maintenance respiration is concerned largely with protein turnover, and as the biomass ages the protein content relative to the total biomass falls and the simple McCree approach can overestimate maintenance respiration.

Thornley (1977a) presented a mechanistic model which incorporated the concept of two different forms of substrate utilization. The model represented schematically in Fig. 7.4 has three plant compartments. The first compartment is the entry point for photosynthate, and contains stored substrate. Substrate moves from this compartment into two others. One contains non-degradable biomass (lignin and other structural tissue such as cell wall tissue) and the other degradable biomass (e.g. enzymes and amino acids). Thus, degradable materials can be recycled via the storage compartments into the growth process.

The model has three components: the storage dry weight, W_S, the degrad-

Respiration

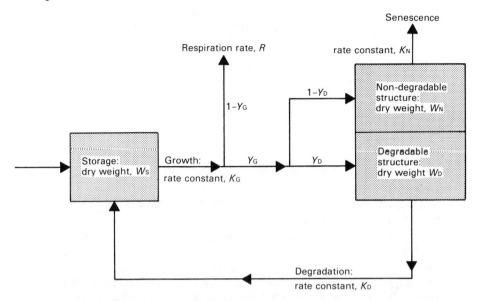

Figure 7.4 Schematic representation of Thornley's recycling respiration model (see text for details).

able component of the structural dry weight, W_D, and the non-degradable structural dry weight, W_N. The structural dry weight, W_G, is given by

$$W_G = W_D + W_N \tag{7.14}$$

and the total dry weight

$$W = W_S + W_G = W_S + W_D + W_N \tag{7.15}$$

The storage pool, W_S, is supplied with assimilate at a rate P_g the daily gross photosynthetic rate. Storage material is utilized for growth with a rate constant k_G. The conversion efficiency from storage to structure is Y_G, so for each unit of substrate utilized Y_G units of structure are produced. Each unit of structure is further partitioned between degradable, Y_D and non-degradable structure, $1 - Y_D$; it is assumed that the degradable component has a degradation constant k_D and that the non-degradable component senesces and is removed from the system with a rate contant k_N. The system may be described by the three first-order differential equations

$$\frac{dW_S}{dt} = P_g - k_G W_S + k_D W_D \tag{7.16a}$$

$$\frac{dW_D}{dt} = Y_D Y_G k_G W_S - k_D W_D \tag{7.16b}$$

$$\frac{dW_N}{dt} = (1 - Y_D) Y_G k_G W_S - k_N W_N \qquad (7.16c)$$

where t is time (days). Adding these equations gives

$$\frac{dW}{dt} = P_g - (1 - Y_G) k_G W_S - k_N W_N \qquad (7.17)$$

which implies

$$R = (1 - Y_G) k_G W_S \qquad (7.18)$$

This model allows the crop growth to respond to the environment and for the respiration to respond, in turn, in a physiologically meaningful way. It has been applied successfully to a model of tree growth (McMurtrie, 1981), to the data of Ryle *et al.* (1976) for uniculm barley and maize, by Barnes and Hole (1978), and a simplified version of this model with two compartments has been applied to grass (Johnson *et al.*, 1983).

The compartmental model does not differentiate explicitly between respiration associated with the synthesis of new structure (growth) and resynthesis of degraded structure (maintenance). Barnes and Hole (1978) went some way in reconciling equation (7.13) (McCree's equation) with Thornley's compartmental model by relating the maintenance respiration to the protein content of the plant or crop, and this analysis was further clarified and extended by Thornley (1982).

7.6 PARTITION OF ASSIMILATES

Assimilates produced in photosynthesis in the leaves are transported or partitioned to various parts of the plant where they are used in growth and maintenance. For growth to take place the assimilates must move to an area of meristematic activity. Despite there being a close and obvious coupling between the partitioning of assimilate to a meristem and the rate of assimilate utilization at that meristem, it is worthwhile analysing the processes separately. Thus, it must be clearly understood that in this section we are considering the partitioning of assimilates between the various meristematic regions and not the utilization of assimilate by those meristems.

A plant contains the genetic information which enables its resources to be organized into the three-dimensional form characteristic of its species. Such autonomic factors impose a hierarchy of priorities for partitioning to the various meristematic regions. The exact nature of the factors governing the partitioning and utilization of assimilates in plants are not well understood. Consequently, precise, universally applicable mathematical descriptions of partitioning and utilization of assimilates by plants cannot at present be made.

Environmental control of assimilate partitioning is most easily recognized in

Partition of assimilates

species that have a requirement for vernalization or a critical photoperiod before they become reproductive. Furthermore, the plant rapidly readjusts its partitioning strategy following a change in environmental conditions, establishing a new pattern within 3–4 days of the change (Ryle and Powell, 1976). Management factors are also important; the morphology and weight of individual spaced plants grown with no limitations of water or nutrients are markedly different from those of plants grown at greater densities. Mutual shading results in the potential size and shape of an individual plant not being realized. Indeed, modern forage crop production is concerned with the productivity per unit ground area, so a balance is struck between productivity per unit ground area and the weight of an individual plant. It is important to realize that individual plants in crops are usually light-limited due to competition, and therefore their meristems are unable to express their full potential for growth because they are photoassimilate limited. The distinction between crop plants and spaced plants is important when modelling crop growth.

That most plants can switch assimilates between the construction of leaves, stems, storage organs, reproductive parts and roots depending on environmental conditions led White (1937) and Brouwer (1963) to postulate the concept of a functional balance. If the growth of the plant reflects a functional balance, then, for example, an increased supply of nitrogen without a corresponding increase in carbohydrate supply would lead to an increase in leaf carbohydrate utilization. This would result in a decrease in the amount of carbohydrate available for further root growth and shoots would develop more than the roots. It is interesting to note that following defoliation a grass plant has far more root than is required for the supply of nutrients and water to the shoots. Thus, it might be expected that there would be little or no partitioning of photoassimilate to the root following defoliation. Indeed, carbon substrate may be mobilized in the roots and partitioned to the shoots.

As leaves age, their pattern of photoassimilate distribution changes. Young leaves export a greater fraction of their photoassimilates to the terminal meristem, whereas older leaves export a greater fraction to the roots. Because young leaves are large and photosynthetically efficient they contribute far more, in an absolute sense, than older leaves to the total carbon economy of the plant. It is unlikely that such differences can easily be incorporated in a crop model.

In grasses, it is important to take account of the assimilate partitioned to tillers (see Chapter 3). Perenniality is achieved in the Gramineae by tillering, and approximately 30% of the plant's photoassimilate is consumed in this growth process (Ryle and Powell, 1976). Another distinctive feature of the perennial grass crop is the constant turnover of ephemeral leaves; the death of such leaves can lead to a considerable reduction in the harvested dry matter (see Chapter 4). Both tillering and senescence are important processes which the modeller interested in the grass crop should consider.

256   Physiological models of grass growth

The leaves, stems, reproductive parts and roots of the grass crop experience different micro-environments. Furthermore, the various organs have an age structure which affects the physiology. Such complexity is usually ignored in mathematical models of assimilate partitioning. To reduce the problem to manageable proportions it is often assumed that there are well-defined organs

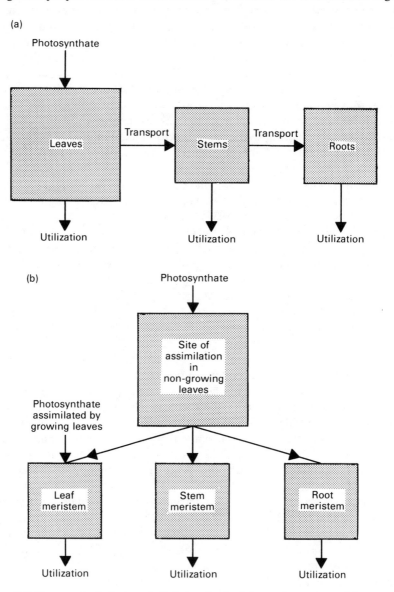

Figure 7.5 Transport of photoassimilate in plants: (a) a series system, (b) a parallel system.

Partition of assimilates

containing pools of photosynthate separated by well-defined pathways and that the organs all exist in a uniform environment.

Thornley (1977b) has given a review of some of the mathematical techniques used when considering models of root–shoot interactions, and he considered

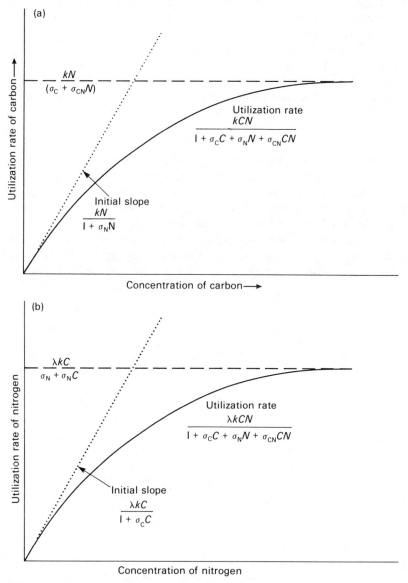

Figure 7.6 Relationship between rate of utilization of carbon and nitrogen substrate and the concentrations of carbon and nitrogen substrate. (Redrawn from Thornley, 1976.)

the problem in terms of assimilate utilization and assimilate transport. He described utilization and active transport using a Michaelis–Menten equation and passive transport using the concept of a concentration gradient and a resistance. Thornley (1972a) used a simple series arrangement (Fig. 7.5(a)) to investigate the growth of leaves, stems and roots limited only by photosynthate, although a parallel system may be more realistic and would provide more flexibility (Fig. 7.5(b)).

The influence of carbon and nitrogen as factors controlling growth and the utilization of the substrates were considered by Thornley (1972b) using Michaelis–Menten type equations which are shown in Fig. 7.6(a) and (b). In the case of grass crop plants which are usually photosynthate-limited it is possible to simplify the Michaelis–Menten equation and assume a linear relationship between utilisation rate and photosynthate concentration (Fig. 7.7).

Sheehy *et al.* (1980) described the partitioning of photosynthate in a grass crop, unrestricted by water or nutrients, using simple linear equations. They assumed that crop plants were 'starved' of photosynthate and that partitioning was a function of photosynthesis. Light and temperature then affected partitioning via their effect on photosynthesis.

The model was extremely simple and effective for the grass crop, and it is worth examining the arguments used in its construction. Thornley (1976) described the Münch hypothesis for sugar transport which in isothermal

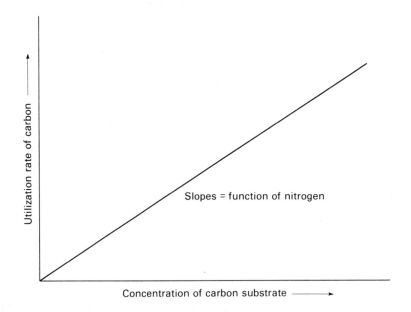

Figure 7.7 The relationship between the rate of utilization of carbon and the concentration of carbon substrate for plants growing in dense swards.

Partition of assimilates

conditions can be described using an equation of the form

$$F = \sigma X_1 (X_1 - X_2) \tag{7.19}$$

where F is the flux of assimilate, $X_1 > X_2$ are the densities of the substances in regions 1 and 2 of the plant and σ is a constant, in part describing resistance of the pathway separating the regions. If at region 2 the substance is utilized rapidly, converted to another form or transported across a membrane then $X_1 \gg X_2$ and

$$F = \sigma X_1^2 \tag{7.20}$$

If we assume that the rate of transport of photosynthate out of the leaves is proportional to the rate of photosynthesis and that for a crop model diurnal changes may be ignored and daily rates of photosynthesis may be taken to be a constant (Sheehy et al., 1979), we may write

$$F_T = \beta P_d^2 \tag{7.21}$$

where F_T is the total amount of photosynthate transported from leaves in a day, P_d is the total amount of photosynthate fixed in a day in crop photosynthesis and β is a constant. Rearranging equation (7.21) it can be seen that

$$\frac{F_T}{P_d} = \beta P_d \tag{7.22}$$

so that the fraction of photosynthate transported daily to a stem or a root may, in simple terms, be assumed to be proportional to the daily rate of photosynthesis. Assuming the leaves, tillers and roots to be arranged in a parallel system (Fig. 7.5(b)), the constant of proportionality for tillers would be β_T and for roots β_R. Following defoliation the concept of a fractional balance would lead to the leaves retaining the photosynthate for the regrowth of new leaves. Sheehy et al. (1980) used this concept to formulate a relationship for the fraction of daily photosynthesis allocated to the assimilate pool for leaf growth:

$$L_p = 1 - \gamma P_d \tag{7.23}$$

where L_p is the proportion of a day's photosynthate allocated to the assimilate pool for leaf growth and γ is a constant. Note that $\gamma = \beta_R + \beta_T$ and that each of these constants is a fraction of nutrients and water. The values used were $\beta_R = 0.5, \beta_T = 0.5$ and $\gamma = 1.0$, with dimensions $kg^{-1} CH_2O\ m^2$ day.

Ryle and Powell (1976) showed that it took 3 or 4 days for the partitioning of photosynthate to assume new steady-state values following an abrupt change in photosynthesis. In order to take account of the delay in the change of the pattern of photosynthate partitioning following a change in environmental conditions, Sheehy et al. (1980) used the average of daily crop photosynthesis for the most recent three days in the partitioning equations.

Clearly, the leaves are the most important source of carbon for growth. However, matters are not so straightforward when nitrogen is considered. The form of nitrogen can vary; soils contain nitrate, ammonium and other organic nitrogen compounds (see Chapter 5). The factors affecting the supply of nitrogen in British grassland agriculture have been examined using a model (Sofield, 1980). In intensive grassland systems the amount of nitrogen needed for high productivity is such that most of it must be in nitrate form. Nitrate can be reduced in different organs, the availability of nitrate to the roots is intimately associated with transpiration, and the relationship between the metabolic activity of roots and nitrate uptake is complex. There is evidence that the uptake of NO_3^- by roots requires photoassimilate (Clement *et al.*, 1978). Furthermore, NO_3^- has to be reduced before it is available as a substrate for growth so that, if it is assumed that much of the nitrogen in grasses is reduced using light energy in the leaves (Klepper *et al.*, 1971), then the leaves must be assumed to be the source of nitrogen for growth. However, the role and amount of nitrate reduction in the roots, using photoassimilate, remains obscure, and until this question is resolved, modelling efforts have to be based on assumptions.

Although mineral elements other than nitrogen are vital for plant growth, nitrogen is the most important in practical agriculture, so we shall confine our attention to this element in the form of nitrate (NO_3^-). Other mineral elements may be treated in a similar manner. The entry of nitrate into the roots is an active process requiring energy, and hence incurs a respiratory cost, in addition to the cost of reducing the nitrate to ammonia. If nitrate is reduced using light energy in the leaves, the source of nitrogen for plant growth is the shoots not the roots. Under such conditions it is perhaps easiest to consider the partitioning of nitrogen as being proportional to that of carbon, so that nitrogen is distributed via the phloem system. To ensure a functional balance, such as a preference for root growth at low intake levels of nitrate, a system of priorities (Thornley, 1977b) or threshold utilization responses may be considered. For example, the effect of limiting the supply of NO_3^- to the plant might invalidate the assumption $X_1 \gg X_2$ which was used to obtain equation (7.20). To investigate such ideas we must consider utilization.

7.7 RATE OF UTILIZATION

The partitioning of assimilates between the various plant meristems is intimately related to the utilization of the assimilates at those meristems. Consequently, the concept of a functional balance controlling the partitioning of assimilates may be described in terms of assumptions concerning assimilate utilization at the various meristems. To simplify the model it may be assumed that growth depends on the utilization of the two substrates carbon and nitrogen (Fig. 7.7). Their rate of utilization is strongly influenced by water stress and temperature, and the model has to take account of these environmental variables. The utilization rates of carbon and nitrogen can be expressed

Rate of utilization

mathematically as

$$U_C = f(C, N, S_\psi, T) \tag{7.24a}$$

and

$$U_N = \phi(N, C, S_\psi, T) \tag{7.24b}$$

where U_C and U_N are the utilization rates of carbon and nitrogen respectively, f and ϕ represent mathematical functions and C, N, S_ψ and T represent carbon concentration, nitrogen concentration, a water stress factor and temperature. To make progress it is assumed that nitrate reduction in the roots is negligible. It is important to note that if much nitrate reduction occurred in the roots, then the structure of the sub-models described in this section would have to be reconsidered.

In environmental conditions suitable for growth it is unlikely that the energy costs of nitrate uptake processes in the roots is a major factor limiting growth, although at low rates of photosynthesis nitrate uptake may well be regulated. In the absence of data and to further simplify this section the energy cost of the uptake process is assumed to be small compared with the costs of reduction and is ignored in this chapter (Fig. 7.8). Furthermore, it is assumed that nitrate reduction is never light limited and that it does not compete with photosynthesis.

When the supply of nitrogen falls below some value optimal for growth, we assume that a priority for assimilate utilization is accorded to root growth (Robson and Parsons, 1978), and hence also for nitrogen utilization. The ratio of carbon to nitrogen is fixed for each type of biomolecule, but the ratio of one type of biomolecule to another may change. Consequently, changes in

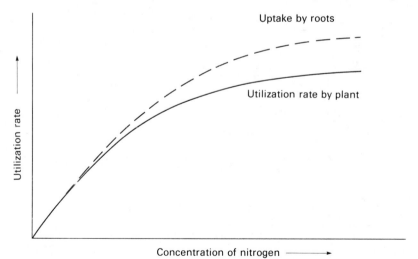

Figure 7.8 The rate of uptake and utilization rate of nitrogen. At low concentrations the rate of utilization and uptake are limited by photoassimilate.

molecular composition can lead to variation in the nitrogen content of whole organs (Robson and Deacon, 1978). Nitrogen stress can result in premature remobilization of nitrogen from leaves, although the contribution of remobilized nitrogen to the nitrogen economy of grass plants is not well understood and cannot easily be incorporated in simple models. It is likely to be of most consequence when nitrogen markedly limits growth; a circumstance of limited practical importance in intensively managed grasslands. At present there is little detailed information concerning roots and their rate of uptake of nitrogen in field crops, although some data are available for plants grown in flowing culture solution. This is no doubt due to the complexities of making such measurements in the field, and the complex processes taking place in the soil such as leaching, denitrification, bulk transport of nitrogen and diffusion. Consequently, it is difficult to give quantitative estimates for nitrate uptake in field grown crops.

Thornley (1972a,b, 1977b) presented equations describing the utilization of two substrates and a method of according different priorities to the various meristems (Fig. 7.9). Crop plants are limited by competition with their neighbours and environmental constraints. In this section we will again assume that the mature crop plant is light limited and therefore unable to photosynthesize at a rate comparable with its genetic potential. This light limitation ensures that each meristem is substrate limited and allows the use of a much simplified version of the more general Thornley analysis; it permits the use of linear

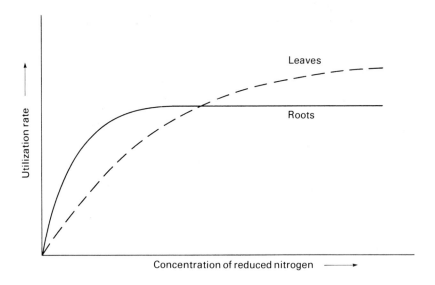

Figure 7.9 The utilization of nitrogen in roots and leaves showing the roots to have a higher priority at low concentrations of nitrogen substrate.

Rate of utilization

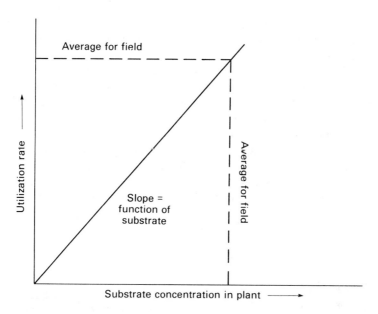

Figure 7.10 The utilization rate of carbon and nitrogen substrate in grass plants growing in the field.

equations (Fig. 7.10). The carbon to nitrogen ratio of the roots are assumed to be constant, and at low availabilities of soil nitrate the utilization rate of nitrogen will be assumed to be greater in roots than in leaves or sheaths (Fig. 7.9).

Using the simple linear equations described above, the utilization rates of carbon in the leaves, stems and roots can be written as

$$U_{CL} = f_L(N)\, C_L \tag{7.25a}$$

$$U_{CS} = f_S(N)\, C_S \tag{7.25b}$$

$$U_{CR} = f_R(N)\, C_R \tag{7.25c}$$

where U_{CL}, U_{CS} and U_{CR} are the utilization rates of carbon in the leaves, sheaths and roots, and $f_L(N)$, $f_S(N)$ and $f_R(N)$ are functions of nitrogen relating utilization rate and carbon concentration at the different meristems, and C is the concentration of carbon at a meristem. Similarly, the utilization rates of nitrogen in the leaves, sheaths and roots can be written as

$$U_{NL} = \phi_L(C)\, N_L \tag{7.26a}$$

$$U_{NS} = \phi_S(C)\, N_S \tag{7.26b}$$

$$U_{NR} = \phi_R(C)\, N_R \tag{7.26c}$$

where ϕ_L, ϕ_S and ϕ_R are functions of carbon and N is the concentration of nitrogen at a meristem. The exact form of equations (7.25a)–(7.26c) has to be determined using experimental data.

The influence of water stress on plant growth is complex, and to simulate its effects a model must have a section dealing with the water budget (Lambert and Penning de Vries, 1971). The water availability in the soil and the crop microclimate interact to determine the water budget and the gradient in water potential between the roots and the shoots. When transpiration occurs, shoot water potentials are more negative than root potentials, and under conditions of high transpiration this may induce a shift in favour of root growth (Brouwer, 1966). To incorporate the effects of water stress, the utilization rates described in equations (25a)–(26c) can be multiplied by a water stress factor (Fig. 7.11)

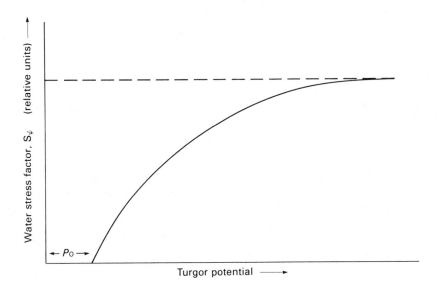

Figure 7.11 The influence of water stress on the utilization of assimilates. When the turgor potential is high, water stress is negligble and the utilization of assimilates is independent of water stress. If the turgor potential falls below some threshold value P_0, the rate of utilization is zero.

described in terms of water potential or turgor potential. Care has to be taken in selecting data to construct a water stress factor, as Leafe *et al.* (1978) observed the effects of water stress on the expansion of grass leaves and indicated that there were differences between the water relations of stress 'hardened' plants in the field and those in controlled environments (see Chapter 6).

Rate of utilization

One possible form for the stress factor, S_ψ, is

$$S_\psi = \begin{cases} \dfrac{(P - P_0)}{K_1 + K_2(P - P_0)} \\ 0 \quad P \leq P_0 \end{cases} \tag{7.27}$$

where P_0 is some threshold value of turgor potential, P is average turgor potential for the day, K_1 is a constant with the dimensions of turgor potential and K_2 is a dimensionless constant. Thus, equation (7.25a) could be modified to take account of water stress:

$$U_{CL} = \begin{cases} f_L(N) \, C_L \, S_\psi \\ \dfrac{f_L(N) \, C_L (P - P_0)}{K_1 + K_2(P - P_0)} \end{cases} \tag{7.28}$$

In a vegetative grass crop the production of a new leaf is generally balanced by the death of an old leaf, and there are approximately three live leaves per tiller (Bean, 1964; Davies, 1971). The rate of leaf appearance is largely dependent on temperature, and Sheehy et al. (1980) described the data of Peacock (1975a,b) using an empirical equation

$$DT = 150 \tag{7.29}$$

where D is the time in days for a leaf to expand and T is average daily temperature in degrees Celsius. This can also be taken to read that $(1/D)$th of a leaf appears each day, and Sheehy et al. (1980) assumed that for leaves the daily rate of photoassimilate use could be written as

$$\frac{\text{Pool size}}{D}$$

substituting from equation (7.29) for D this expression can be rewritten as

$$(\text{Pool size}) \frac{T}{150}$$

Thus, equations (7.25a)–(7.26c) may be written in the form of equation (7.28) with the temperature function included:

$$U_{CL} = \frac{f_L(N) \, C_L (P - P_0) \, T}{150 [K_1 + K_2(P - P_0)]} \tag{7.30}$$

This assumes that all the meristems are affected by temperature in the same manner (Fig. 7.12).

In this section we have ignored the influence of hormones and other such regulatory substances (Wareing, 1977). The change from vegetative to reproductive growth requires a change in the utilization of assimilates with the importance of stems increasing and roots and tiller decreasing, such changes may well be the result of an alteration in the balance of growth substances.

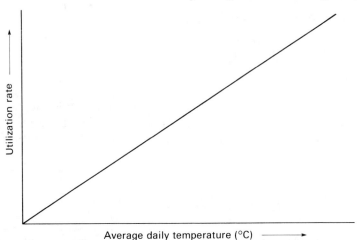

Figure 7.12 The influence of temperature on the rate of utilization of assimilates at a meristem. The temperature in degrees Celsius is the average of a daily maximum and minimum temperature.

However, they can be best modelled by imposing a change on the model parameters as suggested by Sheehy *et al.* (1979).

7.8 TRANSFORMATION OF ASSIMILATE INTO PLANT TISSUE

Carbohydrates and mineral elements are the substrates which are converted into plant organs in the synthetic processes. In this section the factors governing the conversion process will be described.

7.8.1 Leaf area

The light-harvesting power of leaves is crucial for growth, and it is somewhat surprising that models of leaf area development have largely been ignored relative to models of light distribution in leaf canopies. Lainson and Thornley (1982) used a mechanistic model to investigate some of the problems of modelling leaf expansion for cucumbers. The ratio of leaf area to leaf dry weight (specific leaf area) in the appropriate units was considered by Sheehy *et al.* (1979) to be an approximation of the leaf area formed per unit weight of assimilate utilized after synthetic respiration. Specific leaf area is strongly influenced by temperature and irradiance (Blackman, 1956; Robson and Jewiss, 1968; Robson, 1972; Eagles, 1973), but Robson and Parsons (1978) observed that specific leaf area was largely unaffected by nitrogen stress. Less is known about the influence of water stress on specific leaf area. Assuming the effects of water stress are separate from those of irradiance and temperature,

Transformation of assimilate

the relationship between specific leaf area, δ, and environmental conditions may be written as

$$\delta = \delta_\psi(P)\, \delta_{\mathrm{IT}}(I, T) \tag{7.31}$$

with units m^2 leaf kg^{-1} leaf dry wt, δ_{IT} is the specific leaf area irradiance temperature function, I is the irradiance received by the leaf, T is average temperature on the day of emergence, δ_ψ is the specific leaf area water stress function and P is the average turgor potential on the day of emergence. Sheehy et al. (1980) used a simple empirical equation to describe the effects of light and temperature in relatively unstressed grass crops on specific leaf area:

$$\delta_{\mathrm{IT}}(I, T) = (\delta_1 T + \delta_2) \exp(-\delta_3 I) \tag{7.32}$$

Since leaves in a grass sward develop in the shaded conditions of the interior of a grass sward, Beer's law, equation (7.1), was used to describe the irradiance received by a leaf on the day of emergence, so that equation (7.32) becomes

$$\delta_{\mathrm{IT}}(I, T) = (\delta_1 T + \delta_2) \exp[-\delta_3 I_0 \exp(-kL)] \tag{7.33}$$

To take account of water stress it is necessary to define a relationship between specific leaf area and water stress, represented by turgor potential. An equation compatible with that suggested for assimilate utilisation (Fig. 7.11) would have the form

$$\delta_\psi(P) = \frac{(P - P_0)}{C_1 + C_2(P - P_0)} \tag{7.34}$$

where P is average turgor potential for the day, P_0 is some threshold value of turgor potential, C_1 is a constant with the dimensions of turgor potential and C_2 is a dimensionless constant. Such an equation might be suitable, but would require experimental data to support it (Fig. 7.11). The use of envrionmental conditions on the day of emergence to determine the physical properties of the leaf was somewhat arbitrary and so, in the absence of experimental data, the simplest assumption was made.

It should be possible to model the growth of individual leaves, and Lainson and Thornley (1982) have recently suggested some new approaches which might lead to a less empirical and more mechanistic approach.

7.8.2 Stem

In this context stem is defined as true stem, sheath and developing tillers, in other words those portions of the shoot other than leaves. Sheehy et al. (1979; 1980) did not describe the precise manner in which the crop invested assimilate in 'stem' as it was felt that this portion of the crop would play a negligible role in photosynthesis. However, they did indicate that there was sufficient data in the literature to enable such a description to be made if thought necessary.

7.8.3 Root

There is a great deal more information on shoots than roots, especially for field grown crops. The factors controlling the surface area and density of roots, their longevity and their respiratory activity in relation to nutrient uptake are not well understood. Thus, the physical and physiological properties of a piece of root constructed from unit weight of assimilate cannot yet be adequately described, and this remains an obstacle to progress in the modelling of the grass crop.

7.9 REGROWTH

Most models necessarily start their predictions at some particular point in time and take no account of events before that time. Sheehy *et al.* (1979) assumed that 15% of the 'stem' weight remaining after defoliation was assimilate available for leaf growth after deductions for respiration (Davies, 1965; Davidson and Milthorpe, 1966; Peacock, 1975b). During the first 4 days of the simulated regrowth the reserve assimilate was utilized in an exponential fashion. To minimize the effects of the initial assumption, Sheehy *et al.* (1980) simulated crop growth and a defoliation and then studied the simulated regrowth in detail.

7.10 DEATH OF TISSUE

Plants pass through a series of phases during their life-cycle, and it is obvious that the orderly change requires a complex control system which integrates time and takes account of environmental factors; the role of growth substances in control have been discussed by Wareing (1977). The factors controlling senescence could be many, varied and complex and the best description a modeller can provide, at present, is a simple chronology of the process. The chronology must be sufficiently accurate to ensure that the simulated physiological activity of the various organs declines at a realistic rate depending on time and environmental conditions. One possible approach is to invent a bio-clock which runs fast or slow depending on environmental conditions. Experimentally observed phases of organ activity could then be related to simulated bio-time and the relationship used to regulate simulated senescence.

Another method would be to ascribe to a small portion of the carbohydrate utilized for organ growth some property enabling it to be used as a bio-clock. Some portion of the assimilate could be used to construct a degradable form of substrate which would experience a respiratory loss of weight and, as time passed, the weight of this substrate (bio-clock) would decrease until some threshold value was reached and death occurred. In the absence of well-defined methods, the method used to describe senescence depends on the availability of experimental data and the inventiveness of the modeller.

One marked characteristic of the grass crop is the ephemeral nature of its leaves. In a vegetative grass crop the production of a new leaf is generally balanced by the death of an old leaf and there are, on average, approximately three live fully expanded leaves per tiller (Bean, 1964; Davies, 1971). Johnson and Thornley (1983) described this process by dividing the crop into leaf categories corresponding to growing leaves, first fully expanded leaves, second fully expanded leaves and senescing leaves. First-order kinetics were used for the flux of material between the different compartments. The model provides a good description of vegetative grass growth and has been used to analyse grass growth and utilization under grazing (Johnson and Parsons, 1985) with some illuminating results, as discussed by Parsons in Chapter 4.

7.11 CONCLUDING REMARKS

We have outlined some modelling approaches to several different aspects of plant and crop growth as applied to grass. These models provide a framework for looking at grass growth and utilization in response to the environment and different managements. In recent years there has been an increasing interest in modelling crop growth, and the subject is advancing at a considerable rate. Nevertheless, while there are changes in the specific detail of these models, the general underlying structure described here should provide a sound basis for further study.

ACKNOWLEDGEMENTS

The authors would like to thank Miss Julia Prickett for drawing the diagrams and Mrs Harriet Bristow for typing the manuscript.

REFERENCES

Acock, B., Thornley, J. H. M. and Warren-Wilson, J. (1971) in *Potential Crop Production* (eds P. F. Wareing and J. P. Cooper), Heinemann, London, pp. 43–75.

Barnes, A. and Hole, C. C. (1978) A theoretical basis of growth and maintenance respiration. *Ann. Bot.*, **42**, 1217–21.

Bean, E. W. (1964) The influence of light intensity on the growth of an S37 cocksfoot sward. *Ann. Bot.*, **28**, 427–43.

Blackman, F. F. (1905) Optima and limiting factors. *Ann. Bot.*, (old series) **19**, 281–95.

Blackman, G. E. (1956) Influence of light and temperature on leaf growth, in *The Growth of Leaves* (ed. F. L. Milthorpe), Proc. 3rd Easter School Agric. Soc., Univ. Nottingham, 151–67.

Brouwer, R. (1963) Some aspects of the equilibrium between overground and underground plant parts. *Jaarb. Inst. Biol. Scheik. Onder*, **2**, Landbewass., 31–9.

Brouwer, R. (1966) Root growth of cereals and grasses, in *The Growth of Cereals and Grasses* (eds F. L. Milthorpe and J. P. Iving), Proc. 12th Easter School Agric. Sci., Univ. Nottingham, pp. 153–66.

Charles-Edwards, D. A. (1981) *The Mathematics of Photosynthesis and Production*, Academic Press, London.

Clement, C. R., Hopper, M. J., Jones, L. H. P. and Leafe, E. L. (1978) The uptake of nitrate by *Lolium perenne* from flowing nutrient solution. II. Effect of light, defoliation and relationship to CO_2 flux. *J. Exp. Bot.*, **29**, 1173–83.

Davidson, J. L. and Milthorpe, E. L. (1966) Leaf growth in *Dactylis glomerata* following defoliation. *Ann. Bot.*, **30**, 73–184.

Davies, A. (1965) Carbohydrate levels and regrowth in a perennial ryegrass. *J. Agric. Sci., Camb.*, **65**, 213–21.

Davies, A. (1971) Changes in growth rate and morphology of perennial ryegrass swards at high and low nitrogen levels. *J. Agric. Sci., Camb.*, **77**, 123–34.

Eagles, C. F. (1973) Effect of light intensity on the growth of natural populations of *Dactylis glomerata* L. *Ann. Bot.*, **37**, 253–62.

Goudriaan, J. and Waggoner, P. E. (1972) Simulating both aerial microclimate and soil temperature from observation above the foliar canopy. *Neth. J. Agric. Sci.*, **20**, 104–24.

Johnson, I. R. and Parsons, A. J. (1985) A theoretical analysis of grass growth under grazing. *J. Theor. Biol.*, **112**, 345–67.

Johnson, I. R. and Thornley, J. H. M. (1983) Vegetative crop growth model incorporating leaf area expansion and senescence, and applied to grass. *Plant, Cell, Envir.*, **6**, 721–9.

Johnson, I. R. and Thornley, J. H. M. (1984) A model of instantaneous and daily canopy photosynthesis. *J. Theor. Biol.*, **107**, 531–45.

Johnson, I. R., Amiziane, T. E. and Thornley, J. H. M. (1983) A model of grass growth. *Ann. Bot.*, **51**, pp. 599–609.

Klepper, L., Flesher, D. and Hageman, R. H. (1971) Generation of reduced nicotinamide adenine dinucleotide for nitrate reduction in green leaves. *Pl. Physiol.*, **48**, 580–90.

Lainson, R. A. and Thornley, J. H. M. (1982) A model for leaf expansion in cucumber. *Ann. Bot.*, **50**, 407–25.

Lambert, J. R. and Penning de Vries, F. W. T. (1971) Dynamics of water in the soil–plant–atmosphere system: a model named Troika. *Rep. no. 3*, Dept. Theor. Prod. Ecol., Univ. Wageningen.

Leafe, E. L., Jones, M. B. and Stiles, W. (1978) The physiological effects of water stress on perennial ryegrass in the field. *Proc. 13th Int. Grassland Congr., Leipzig, 1977*, Section 1–2, pp. 165–84.

Marshall, B. and Biscoe, P. V. (1980) A model for C_3 leaves describing the dependence of net photosynthesis on irradiance. I. Derivation. *J. Exp. Bot.*, **31**, 41–8.

McCree, K. J. (1970) An equation for the rate of respiration of white clover plants grown under controlled conditions, in *Prediction and Measurement of Photosynthetic Productivity* (ed. I. Setlik), Pudoc, Wageningen, pp. 221–9.

McMurtrie, R. (1981) Suppression and dominance of trees with overlapping crowns. *J. Theor. Biol.*, **89**, 151–174.

Monsi, M. and Saeki, T. (1953) Über den Lichtfaktor in den Pflanzengesell schaften und seine Bedentung für die Stoffproduktion. *Jap. J. Bot.*, **14**, 22–52.

Monteith, J. L. (1965) Light distribution and photosynthesis in field crops. *Ann. Bot.*, **29**, 17–27.

References

Monteith, J. L. (1981a) Does light limit crop production? in *Physiological Processes Limiting Plant Productivity* (ed. C. B. Johnson), Butterworths, London, pp. 23–38.

Monteith, J. L. (1981b) Climatic variation and the growth of crops. *Q. J. R. Met. Soc.,* **107,** 749–74.

Peacock, J. M. (1975a) Temperature and leaf growth in *Lolium perenne.* I. The thermal microclimate: its measurement and relation to crop growth. *J. Appl. Ecol.,* **12,** 99–114.

Peacock, J. M. (1975b) Temperature and leaf growth in *Lolium perenne.* II. The site of temperature perception. *J. Appl. Ecol.,* **12,** 115–23.

Penning de Vries, F. W. T. (1972) Respiration and Growth, in *Crop Processes in Controlled Environments* (eds A. R. Rees, K. E. Cockshull, D. W. Hand and R. G. Hurd), Academic Press, London, pp. 327–46.

Rabinowitch, E. I. (1951) *Photosynthesis and Related Processes,* Vol. 2, Part 1. Interscience, New York.

Robson, M. J. (1972) The effect of temperature on the growth of S170 tall fescue (*Festuca arundinacea*). 1. Constant temperature. *J. Appl. Ecol.,* **9,** 643–53.

Robson, M. J. and Deacon, M. J. (1978) Nitrogen deficiency in small closed communities of S24 ryegrass. II. Changes in the weight and chemical composition of single leaves during their growth and death. *Ann. Bot.,* **42,** 1199–213.

Robson, M. J. and Jewiss, O. R. (1968) A comparison of British and North African varieties of tall fescue (*Festuca arundinacea*). I. Leaf growth during winter and the effects on it of temperature and daylength. *J. Appl. Ecol.,* **4,** 475–84.

Robson, M. J. and Parsons, A. J. (1978) Nitrogen deficiency in small closed communities of S24 ryegrass. I. Photosynthesis, respiration, dry matter production and partition. *Ann. Bot.,* **42,** 1185–97.

Ryle, G. J. A. and Powell, C. E. (1976) Effect of rate of photosynthesis on the pattern of assimilate distribution in the graminaceous plant. *J. Exp. Bot.,* **27,** 189–99.

Ryle G. J. A., Cobby, J. M. and Powell, C. E. (1976) Synthetic and maintenance respiratory losses of $^{14}CO_2$ in uniculm barley and maize. *Ann. Bot.,* **40,** 571–86.

Saeki, T. (1963) Light relations in plant communities, in *Environmental Control of Plant Growth* (ed. L. T. Evans), Academic Press, New York, pp. 79–94.

Sheehy, J. E. and Cook, D. (1977) Irradiance distribution and CO_2 flux in forage grass canopies. *Ann. Bot.,* **41,** 1017–29.

Sheehy, J. E. and Peacock, J. M. (1975) Canopy photosynthesis and crop growth rate of eight temperate forage grasses. *J. Exp. Bot.,* **26,** 679–91.

Sheehy, J. E., Cobby, J. M., and Ryle, G. J. A. (1979) The growth of perennial ryegrass: a model. *Ann. Bot.,* **43,** 335–54.

Sheehy, J. E., Cobby, J. M. and Ryle, G. J. A. (1980) The use of a model to investigate the influence of some environmental factors on the growth of perennial ryegrass. *Ann. Bot.,* **46,** 343–65.

Sofield, I. (1980) A computer simulation model of the soil-plant-and-nitrogen relationships in a cut grass sward. Ph.D. Thesis, Reading University.

Thornley, J. H. M. (1970) Respiration, growth and maintenance in plants. *Nature,* **227,** 304–5.

Thornley, J. H. M. (1972a) A model to describe the partitioning of photosynthate during vegetative plant growth. *Ann. Bot.,* **36,** 419–30.

Thornley, J. H. M. (1972b) A balanced quantitative model for root:shoot ratios in vegetative plants. *Ann. Bot.*, **36**, 431–41.
Thornley, J. H. M. (1976) *Mathematical Models in Plant Physiology*, Academic Press, London.
Thornley, J. H. M. (1977a) Growth maintenance and respiration: a reinterpretation. *Ann. Bot.*, **41**, 1191–203.
Thornley, J. H. M. (1977b) Root: shoot interactions, in *Integration of Activity in the Higher Plant* (ed. P. H. Jennings), *Symp. Soc. Exp. Biol.*, **31**, 367–89.
Thornley, J. H. M. (1982) Interpretation of respiration coefficients. *Ann. Bot.*, **49**, 257–9.
Wareing, P. F. (1977) Growth substances and integration in the whole plant, in *Integration of Activity in the Higher Plant* (ed. P. H. Jennings), *Symp. Soc. Exp. Biol.*, **31**, 337–65.
White, H. L. (1937) The interaction of factors in the growth of *Lemna* XII. The interaction of nitrogen and light intensity in relation to root length. *Ann. Bot.*, **1**, 649–54.
Zur, B. and Jones, J. W. (1981) A model for the water relations, photosynthesis, and expansive growth of crops. *Wat. Resour. Res.*, **17**, (2), 311–20.

DEFINITION OF SYMBOLS

Requirements for crop physiological models

C	sensible heat loss from crop (convection)	$W\,m^{-2}$
E	rate of transpiration	$kg\,H_2O\,m^{-2}\,s^{-1}$
P	energy consumed in photosynthesis	$kg\,CO_2\,m^{-2}\,s^{-1}$
R_a	total radiation	$W\,m^{-2}$
λ	latent heat of fusion	$J\,kg^{-1}$

Light interception and photosynthesis

I_0	instantaneous light flux density on the surface of the canopy	$W\,m^{-2}$
$I(L)$	instantaneous light flux density in the canopy at leaf area index L	$W\,m^{-2}$
I_1	instantaneous light flux density on the leaves at leaf area index L	$W\,m^{-2}$ leaf
k	extinction coefficient of canopy	
L	leaf area index	m^2 leaf m^{-2} ground
m	leaf transmission coefficient	
P_g	single leaf gross photosynthetic rate	$kg\,CO_2\,m^{-2}$ leaf s^{-1}
P_m	light saturated gross photosynthetic rate	$kg\,CO_2\,m^{-2}$ leaf s^{-1}
P_c	canopy gross photosynthetic rate	$kg\,CO_2\,m^{-2}$ ground s^{-1}

Definition of symbols

S	fraction of light passing through unit leaf area index layer which is unintercepted	
α	leaf photochemical efficiency	kg CO_2 J^{-1}
θ	single-leaf photosynthesis parameter	

Respiration

k_D	degradation rate constant	day^{-1}
k_G	growth rate constant	day^{-1}
k_N	senescence rate constant	day^{-1}
M	maintenance respiration rate constant	day^{-1}
P_g	rate of gross photosynthesis	kg CO_2 m^{-2} day^{-1}
R	rate of respiration	kg CO_2 m^{-2} day^{-1}
W	crop dry weight	kg dry weight m^{-2}
W_G	crop structural dry weight	kg dry weight m^{-2}
W_S	crop storage dry weight	kg dry weight m^{-2}
W_D	degradable component of crop structural dry weight	kg dry weight m^{-2}
W_N	non-degradable component of crop structural dry weight	kg dry weight m^{-2}
Y_G	relative yield of structural dry matter as a result of growth	
Y_D	relative yield of degradable structural dry matter	

Partition of assimilates

F	flux of assimilate	kg CH_2O m^{-2} s^{-1}
F_T	total daily flux of assimilate from leaves	kg CH_2O m^{-2} day^{-1}
L_p	proportion of days photoassimilate allocated to leaf growth	
P_d	total daily gross photosynthetic rate	kg CH_2O m^{-2} day^{-1}
X_1, X_2	substrate densities	kg CH_2O m^{-3}
β	flux constant	kg^{-1} CH_2O m^2 day
β_T, β_R	flux constants for tillers and roots	kg^{-1} CH_2O m^2 day
γ	flux constant	kg^{-1} CH_2O m^2 day
σ	flux constant	kg^{-1} CH_2O m^4 s^{-1}

Rate of utilization

C	the concentration of carbohydrate at some meristem	kg CH_2O m^{-3}
C_L	concentration of carbohydrate in the leaves	kg CH_2O m^{-3}

C_S	concentration of carbohydrate in the stems	kg CH$_2$O m^{-3}
C_R	concentration of carbohydrate in the roots	kg CH$_2$O m^{-3}
D	time it takes a leaf to expand fully	days
$f(N)$	a rate which is some function of the nitrogen status of the plant	m day^{-1}
K_1	a constant	J kg^{-1}
K_2	a constant	
N	the concentration of nitrogen at some meristem	kg nitrogen m^{-3}
P	turgor potential	J kg^{-1}
P_0	a threshold value of turgor potential	J kg^{-1}
S_ψ	water stress factor	
T	average daily temperature	°C
U_C	rate of utilization of carbohydrate by the crop	kg CH$_2$O m^{-2} day^{-1}
U_N	rate of utilization of nitrogen by the crop	kg nitrogen m^{-2} day^{-1}
U_{CL}	the rate of utilization of carbohydrate by the leaves	kg CH$_2$O m^{-2} day^{-1}
U_{CS}	the rate of utilization of carbohydrate by the stems	kg CH$_2$O m^{-2} day^{-1}
U_{CR}	the rate of utilization of carbohydrate by the roots	kg CH$_2$O m^{-2} day^{-1}
U_{NL}	the rate of utilization of nitrogen by the leaves	kg nitrogen m^{-2} day^{-1}
U_{NS}	the rate of utilization of nitrogen by the stems	kg nitrogen m^{-2} day^{-1}
U_{NR}	the rate of utilization of nitrogen by the roots	kg nitrogen m^{-2} day^{-1}
$\phi(C)$	a rate which is some function of the carbohydrate status of the crop	m day^{-1}

Transformation of assimilate into plant tissue

C_1	a constant	J kg^{-1}
C_2	a constant	
I_0	instantaneous light flux density at the surface of the canopy	W m^{-2}
K	canopy extinction coefficient	
L	leaf area index	m^2 leaf m^{-2} ground
P	turgor potential	J kg^{-1}
P_0	threshold turgor potential	J kg^{-1}

Definition of symbols

δ	specific leaf area	$m^2\,kg^{-1}$
δ_ψ	specific leaf area function defining dependence on the water potential	
δ_{IT}	specific leaf area function defining dependence on irradiance and temperature	
δ_1	specific leaf area constant	$m^2\,kg^{-1}$
δ_2	specific leaf area constant	$m^2\,kg^{-1}$
δ_3	specific leaf area constant	$m^2\,kg^{-1}$
ψ	water potential	$J\,kg^{-1}$

CHAPTER 8
The effects of pests and diseases on grasses

R. T. Plumb

8.1 INTRODUCTION

In this chapter pests and pathogens of temperate grass crops are considered, together with their effects on their hosts as individual plants and as crops, and how these effects are caused. The control of pests and pathogens will not be considered, although inevitably many of the data available on crop losses have been obtained from experiments in which this was attempted. While such information is valuable in demonstrating what losses can be caused, it is often of little practical use in grazed or conserved crops because the quantities of chemicals used would leave unacceptably toxic residues and often be prohibitively expensive. This is less true for seed crops. The most obvious solution to pest and pathogen problems is to breed resistant cultivars, but grasses, because they are outbreeding, are usually not as extensively damaged as genotypically uniform crops such as cereals. Indeed, in grass crops, damage is only rarely visible, and is then usually caused by the sporadic occurrence of a pest or pathogen that normally causes little damage and does not justify effort from plant breeders in trying to incorporate resistance to it.

Many pests and pathogens of grass are known (Sampson and Western, 1954; Smith, 1965; Heard, 1972; O'Rourke, 1976; Williams, 1984), but there is generally little information on their distribution and the damage they cause. Of the 125 species of rusts on nearly 400 species of grasses in the USA, few are thought to be of economic importance (Fischer, 1953a); 140 species of smuts infect 300 species of grass in the USA (Fischer, 1953b) and more than 100 pathogens are known to cause disease in turf grasses (Lefebvre, *et al.*, 1953). As many of these pathogens and pests occur simultaneously there is the difficulty, not only of determining whether they decrease sward productivity, but also of determining which component(s) of the pathogen spectrum is the principal cause of any loss in yield.

There are many reasons for this lack of knowledge, not least of which is the comparative lack of interest that has been shown in pests and pathogens of grass crops. This stems from the absence of any consistently apparent effect of pests and diseases, and the difficulty of measuring any effects, either overt or covert, that they might have. Pathogens and pests affect grass in many ways: by decreasing the total quantity of seed or dry matter produced, by changing its nutritional value and by changing its acceptability to stock. These are direct effects caused on single plants or single-species swards, but in mixed swards additional effects are seen. Depredations by pest and pathogen can lead to changes in sward composition, either by changing the balance of sown species or by allowing weed species to invade. In either case yield, as measured by dry matter, may not change but the quality of the produce will. When only a few isolated plants are affected, compensation by adjacent unaffected plants may overcome any adverse effects.

Pests and pathogens also affect establishment: on average only about 20% of seeds sown eventually produce a plant (Michail and Carr, 1966; Henderson and Clements, 1980), whilst the longevity of grasses with relatively short persistence may be increased if pathogens or pests are controlled (Clements and Henderson, 1979). However, the factor which probably has most influence on pest and pathogen incidence and the damage they cause is management. The choice of cultivar(s) to be sown, the time of sowing and method of establishment, the amount of fertilizer used, the height of cutting, the intensity of grazing and whether the grass is grown in rotation or is permanent all influence the likelihood of pest and pathogen occurrence and whether they affect yield.

There are therefore few rules that can be followed when trying to anticipate problems from pests and pathogens. Work that has been done has demonstrated, for very few pests and diseases, that complex interactions exist and it is difficult to clarify their individual effects. The complexity of the interactions is doubtless one of the reasons why so few investigations have been made into the physiological basis of the effects of pests and pathogens. What follows may demonstrate how little is known about how damage is caused, and illustrate those affects that are known.

8.2 INSECT PESTS

Pests damage grasses in many ways and at many stages from seed to established sward. Heard (1972) listed the most important pests of grasses in Britain, and it is an indication of changes in knowledge that he put aphids as the most important pests followed by leatherjackets (*Tipula* spp.) and the larvae of seed midges, whereas Clements (1980) considered frit-fly (*Oscinella frit*) the principal cause of yield loss in lowland ryegrass-dominated swards in Britain.

Coulson and Butterfield (1978) calculated that on limestone and alluvial grassland the biomass of soil fauna, not all components of which are damaging, was 2.5–5 times the biomass of the grazing sheep. Even the definition of a pest

Insect pests

is difficult, as earthworms – normally considered beneficial – can be classified as pests because their casts decrease the playing qualities and appearance of sports turf (Anon, 1977).

Occasionally individual pests can be identified as the damaging agent. For example, Raw (1952) reported extensive damage to grassland by the larvae of the garden chafer *Phyllopertha horticola* which feed on roots, especially of fine-leaved grasses. However, such occurrences are rare and most pest damage is less-readily attributable.

8.2.1 Pests affecting establishment

One root-feeding pest has already been mentioned, but others are more frequently present and more often damaging especially where new sowings are made on ploughed-up established grassland (Clements *et al.*, 1982b). Among the most numerous in Britain are stem-boring dipterous larvae, larvae of swift moths (*Hepialus lupulinus*), larvae (wireworms) of click beetles (*Agriotes* spp.) and larvae (leatherjackets) of craneflies (*Tipula* spp.).

After the eggs hatch, the larvae of *H. lupulinus* feed on the roots and invade the growing point of plants which often die. Larvae of *Agriotes* spp. are especially common under grass and, as well as biting off roots, often eat seeds thereby decreasing establishment. They feed most actively in the spring and are less active in the summer. Threshold numbers causing visible damage have been estimated at 375 000 larvae ha^{-1} (Anon., 1948); adults are less damaging than the larvae, but do eat leaves.

Larvae of *Tipula* spp. are found throughout Britain and northern Europe, but are especially common and damaging in wet regions. The larvae eat roots and underground parts of the stem as well as germinating seedlings, which can result in bare patches in newly sown swards.

Larvae of *Oscinella* spp., *Opomyza* spp., *Meromyza* spp., *Geomyza* spp. and *Chlorops pumilionis* are common in many regions and damage newly sown swards (Webley, 1960). When treated with phorate at 10 kg a.i. ha^{-1} 24–35% more seedlings of Italian ryegrass (*Lolium multiflorum*) survived than where no insecticide was used, and this increase was attributed to the control of *O. frit*, the larve of which bore into grass stems; however, only 12–29% of sown seeds managed to produce plants (Henderson and Clements, 1980). The commonly sown grasses are not equally susceptible to frit fly: timothy and cocksfoot are almost immune, but the ryegrasses, and Italian ryegrass in particular, seem especially favourable hosts (Henderson and Clements, 1979; Clements, 1980).

Slugs (*Arion hortensis*, *Deroceras reticulatum*) are ubiquitous but especially damaging to grass in cool wet regions and on heavy soils when an established sward is ploughed and re-seeded. Young plants are destroyed, and in many areas such slug damage makes a large contribution to the losses at establishment.

8.2.2 Damage to the sward

Damage by pests on established swards is rarely obvious but has been detected by the use of pesticide treatments (Henderson and Clements, 1974, 1977; Clements, 1980) (Fig. 8.1); the most consistent benefits of pesticides were seen on newly sown swards (Henderson and Clements, 1974). When these investigations were extended to many sites throughout Britain the yield was increased at all but one site by up to 32% (Henderson and Clements, 1977), even though at none of these sites was there an obvious pest problem. Some of the insects that were controlled are also vectors of viruses, and there was some evidence that decreased infection by viruses, especially the eriophyid mite-transmitted ryegrass mosaic virus (RMV), was correlated with yield response.

The largest yield increase was obtained at the only site where Italian ryegrass was grown, and yield increases were generally associated with an increase in the growth of the dominant grass, usually ryegrass, and a decrease in the invasion of unsown grasses and broad-leaved weeds. When Italian ryegrass and hybrid ryegrass (*L. multiflorum* × *L. perenne*) were untreated they almost died out within 2 years, but where pesticides had been applied they survived with little thinning of the sward into a fourth year (Clements and Henderson, 1979). It was concluded that the stem-boring or leaf-mining larvae were most likely to be the cause of the degeneration. Larvae of stem borers, mainly *Oscinella* spp. and *Geomyza tripunctata*, were more numerous where large amounts of nitrogen had been applied, and there were differences in the predominant species depending on the rate of nitrogen used (Moore and Clements, 1984).

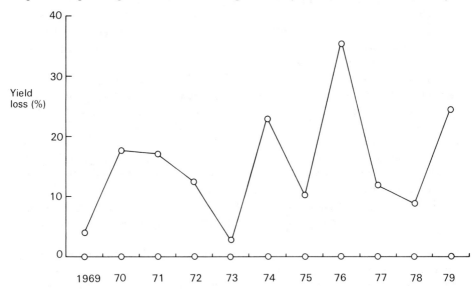

Figure 8.1 Grass yield losses due to insect pests, 1969–79. (Redrawn from Clements, 1980.)

Apart from losses due to death of plants, invasion by stem borers also decreases photosynthetic activity of plants that survive, but these effects together are insufficient to explain the crop losses observed (Clements et al., 1980). Mowat (1974) obtained yield increases of up to 59% when stem-boring larvae were controlled on a sward dominated by *Poa trivialis*.

There are fewer potential pests of upland grassland in Britain than of lowland ryegrass-dominated swards, and insecticides increased annual yields at only three of 13 sites and decreased yield at two (Clements et al., 1982a). The lack of response was attributed to the scarcity of frit fly and its favoured host, ryegrass.

These results clearly show that the yields of lowland grass swards, especially those of high productivity where ryegrass dominates, are considerably diminished by pests. Other preferred agricultural species such as timothy (*Phleum* spp.) and cocksfoot (*Dactylis glomerata*) are damaged less than the ryegrasses.

Leatherjackets cause much damage to established swards, although serious damage is sporadic (Heard, 1972). French (1969) estimated that in Britain, in bad years, 150 000 ha may be damaged, and Newbold (1981) calculated that a population of 2×10^6 leatherjackets ha^{-1} caused sufficient damage to justify the use of a pesticide to control them, although smaller populations have also been associated with damage. Based on Newbold's criteria, Blackshaw (1983), in Northern Ireland, calculated that almost 20 000 ha would benefit from leatherjacket control and that 10^5 ha were at risk from damage each year. Damage is usually most prevalent on thin crops and newly sown leys.

Wireworms, as well as damaging newly sown grass, also damage established swards, and Heath (1960) attributed an increase in yield of 21% to a decrease in *Agriotes* spp. larvae from 10^6 ha^{-1} to 2×10^5 ha^{-1}.

Aphids have a dual role as directly damaging pests and as vectors of viruses, and it is often difficult to separate these effects. In Britain cereal aphids have wide host ranges, and Gair (1953) reported damaging populations of the aphid *Rhopalosiphum padi* on *Phleum pratense* and *Festuca rubra*, and of *Metopolophium festucae* on perennial ryegrass, although *L. multiflorum*, *Alopecurus pratensis*, *F. pratensis* and *Poa annua* were also colonized. Dean (1973) found that *R. padi* showed a feeding preference for *L. perenne* and *F. pratensis*, *M. dirhodum* for *D. glomerata*, *Poa pratensis* and *F. pratensis*, and *Sitobion avenae* for *D. glomerata* and *L. perenne*. The least favoured grasses were *F. rubra* and *Holcus mollis*. Several cases of damage by *M. festucae* in the west of Scotland have been reported by Hill (1971), who found that this species multiplied more rapidly on *L. multiflorum* than on *P. pratensis* or *D. glomerata*. In North America the ecology of the greenbug *Toxoptera* (= *Schizaphis*) *graminum* which damages *Poa pratensis* and *D. glomerata* largely by the introduction of a toxin, possibly an enzyme, which breaks down chlorophyll, was reviewed by Wadley (1931). Aphids on grasses are probably a permanent component of the grassland fauna, but are best considered as sporadic pests, outbreaks often being associated with unusually favourable weather or poor management which allows grass to grow rank and aphids to multiply undisturbed. Averaged over years the losses caused by viruses transmitted by aphids are probably greater than the damage caused by direct aphid feeding.

Another virus vector, the eriophyid mite *Abacarus hystrix*, can reach large numbers (1000 + tiller^{-1}) in undisturbed swards and damage plants by causing rolling or drying of leaves, but where the crop is cut or grazed frequently their numbers are insufficient to cause damage (Gibson, 1976).

In North America several pests have been implicated as causes of damage to swards. When the army worm *Pseudaletia unipuncta* was controlled foliage dry matter of *Poa pratensis* was more than doubled (Starks and Thurston, 1962). Use of insecticides on *Cynodon dactylon* increased yield by 40%; most damage occurred when there was much young growth and the yield increase seemed to be associated with control of leafhoppers (Homoptera: Cicadellidae) and *Spodoptera frugiperda*, the fall army worm. Sod webworms (*Herpetogramma* spp. and *Crambus* spp.) damage *Poa pratensis* and *C. dactylon* (Reinert, 1976; Pass *et al.*, 1965). Turf grasses, especially *P. annua*, *P. pratensis* and *Agrostis* spp., are damaged by the grub *Ataenius spretulus*, which causes dead patches. Grasshoppers, *Melanoplus infantilis* and *Aulocara elliottii*, cause serious damage to rangeland grasses: at a density of 1 m^{-2} it was calculated that losses were 22 kg ha^{-1} (Hewitt, 1978), although much of this was cut but not eaten.

In Australasia damage by insects is also a problem. In northern New Zealand and large areas of South Australia and New South Wales *Teleogryllus commodus*, the black field cricket, at densities as low as 5 m^{-2} caused dry matter herbage losses of 1.8 kg ha^{-1} day^{-1}, equivalent to the herbage removed by 1.5 ewes. At densities as high as 60 m^{-2} the losses of 25.1 kg ha^{-1} day^{-1} were equivalent to the amount eaten by 21 ewes (Blank and Olsen, 1981). Nymphs of *T. commodus* eat the leaf margin and occasionally whole leaves, and adults sever ryegrass stems and eat freshly emerging tillers and seed heads (Blank and Olsen, 1981). In New Zealand the grass grub (*Costelytra* (= *Odontria*) *Zealandica*) was described by Dumbleton (1942) as the most destructive pest of pastures. The larvae feed on roots which they sever just below the surface. Third instar larvae feed preferentially on *A. tenuis*, *L. perenne*, and *H. lanatus* (Radcliffe, 1970). In damaged pastures the grass grub was often present in association with the subterranean grass caterpillar (*Oxycanus cerviciata*); the porina caterpillar (*Wiseana* spp.) damages hybrid ryegrass more than perennial and can cause losses of 30–50% dry matter.

Two further pests of New Zealand grassland are the black beetle (*Heteronychus arator* (= *sanctae-helenae*)) the larvae and adults of which cause damage, and *Hyperodes griseus* the Argentine stem weevil, the larvae of which tunnel into the base of tillers. Larvae of the black beetle damage the root system while adults, which normally feed on roots immediately below ground, can damage the crown of cocksfoot plants (Todd, 1959); however, this kind of damage is sporadic and usually follows warm springs (East *et al.*, 1981).

Nematodes

8.2.3 Damage to seed crops

While any damage by pests which removes foliage or damages shoots can potentially decrease seed yield, some pests directly damage seed. Cocksfoot, fescues, ryegrass and timothy seed crops can suffer severe attacks by the aphids *M. festucae* and *R. padi*. *Rhopalosiphum padi* appeared to be the most damaging in Britain but the losses may have been due to the transmission of barley yellow dwarf virus (Janson, 1959). Grass seed midges (*Contarinia* spp., *Dasineura* spp., *Stenodiplosis* spp.) are seed-borne, but damaging infestations are rare and usually confined to older crops. Damage is usually seen as empty florets, eaten by larvae which hatch from eggs laid in the seed head. Infestations of 50–80% have been recorded, and the accidental introduction of the foxtail midge (*Dasineura alopecuri*) had a disastrous effect on the production of meadow foxtail seed in New Zealand (Jones, 1945). Timothy flies (*Amaurosoma* spp.) are occasionally damaging (Coghill and Gair, 1954), but the fewer seeds produced in attacked ears are larger than those from undamaged crops.

8.3 NEMATODES

Many of the chemical treatments used to investigate damage by insect pests also control nematodes, and nematicides often control other pests. It is, therefore, extremely difficult to demonstrate unequivocally what damage nematodes cause to grasses. Knowledge of nematodes in grasses is summarized by Bezooijen (1979) and Cook and York (1980). Young seedlings are most vulnerable to attack, and the extent of damage is affected by time of sowing; Spaull *et al.* (1985) demonstrated that using aldicarb at sowing to control nematodes increased yields of *Lolium* spp. When *Helicotylenchus varicaudatus* and *Paratylenchus microdorus* were common, and Longidorid and Trichodorid nematodes were also present, aldicarb applied at 5 kg a.i. ha^{-1} increased dry matter yield at the first cut of a number of ryegrasses by an average of 71%; Italian ryegrass responded most and its yield was doubled at both first and second cuts when 10 kg a.i. ha^{-1} was used (Spaull and Clements, 1982). In the Netherlands the species associated with decreased growth and poor swards were *Tylenchorhynchus* spp., *Paratylenchus* spp. and *Helicotylenchus pseudorobustus* (Bezooijen, 1979). The cereal cyst nematode (*Heterodera avenae*) can affect ryegrass and cocksfoot when these crops are sown after oats, and *Heterodera* spp. have been implicated in damage to reseeded grassland (Cook and York, 1980).

Established grass seems to suffer little from nematode damage. However, the presence of *Meloidogyne naasi* and *Ditylenchus radicicolus* in galls on grass in a mixed white clover/grass sward weakened the roots of grasses which were then uprooted more easily during cutting or grazing; consequently the white clover component of the sward increased (Lewis and Webley, 1966). *Meloidogyne naasi* is widely distributed in Europe, North and South America and New Zealand, and has been implicated, with *Paratylenchus penetrans* and *Tylenchorhynchus agri,* in decreasing the yield of *Agrostis stolonifera.* When *Longidorus elongatus* was controlled in ryegrass, yield increased by 45% (Boag and Trudgill, 1977). Nematodes can damage turf grasses as well as agricultural species (Troll and Rohde, 1966; Eriksson, 1972).

An unusual feature of nematode parasitism of ryegrass in Australia is the occurrence of annual ryegrass toxicity. In this condition seeds are replaced by bottle-shaped galls containing nematode larvae. This restricts regeneration of swards from self-set seeds, but the principal problem is caused by the association of the galls with a bacterium which covers the inflorescence with a sticky, bright yellow slime. This makes the seed toxic to sheep and cattle many of which die after eating infested seed (Price *et al.,* 1979). The nematode is a species of *Anguina* (Price *et al.,* 1979; Stynes and Bird, 1980), and its function is as a vector of the bacterium *Corynebacterium rathayi* which produces the toxin in the walls of the nematode-induced galls (Stynes and Bird, 1980).

8.4 FUNGI

Fungal pathogens of grass crops are numerous and widespread, and probably no established grass crop is free of infection. Whether this results in damage depends on the host, its growth stage, the pathogen and the extent of infection, and how the crop is managed. Fungal diseases are rarely lethal, although by weakening their host they may bring about its elimination as a result of competition from healthier plants, or may predispose it to damage during severe weather. The effects of fungal diseases are often complex; fungi directly destroy leaf tissue, but may also change host physiology or produce toxins and thereby seriously affect herbage quality and palatability out of proportion to the leaf area damaged.

8.4.1 Fungi affecting establishment

The condition known as 'damping off' is associated with a number of pathogens any one or combination of which can cause seedling death. Among the most frequently isolated pathogens are *Fusarium, Drechslera, Pythium* and *Rhizoctonia* spp.; *Fusarium culmorum* and *F. nivale* caused more preemergence death of *Dactylis* and *Phleum* than of *Lolium* spp., and this has been attributed to the slow germination of cocksfoot seed and the small seed size of timothy (Holmes, 1979a). On perennial ryegrass the same two pathogens differed in their effect. *Fusarium culmorum* decreased emergence but had little effect on established seedlings, whereas *F. nivale,* which varies in pathogenicity, caused most damage to established seedlings (Holmes, 1976). *Fusarium culmorum* and *Cylindrocarpon radicicola* were associated with pre- and post-emergence damping off of grass seedlings, but *Pythium* spp. were

Fungi

isolated from most seedlings which failed to emerge (Michail and Carr, 1966). There is some host specialization, with *F. culmorum, C. radicicola* and a *Pythium* sp. being most frequently associated with ryegrass; *C. radicicola* with *Phleum;* and *P. ultimum* with *Festuca* and *Dactylis*. Isolates of the pathogens killed up to 70% of seedlings of the same species from which they had been isolated. Turf grasses also suffer from *Fusarium* damage. In California, swards of 100% *P. pratensis* suffered 30% loss of seedlings from *F. roseum* infection, but when 10–15% of the seed sown was *L. perenne* no symptoms were seen (Gibeault *et al.*, 1980). Also in the USA *Drechslera sorokiniana* and *Curvularia geniculata* both attack germinating seeds of *F. rubra* and their pathogenicity is apparently enhanced when large quantities of nitrogen are used. *Drechslera sorokiniana* decreased the rate of seedling emergence and increased seedling death, *C. geniculata* reduced seedling emergence and there was some evidence of competition between them (Madsen and Hodges, 1980a, b). In Britain some *Drechslera* spp. have been associated with post-emergence damping-off (Carr, 1971).

In New Zealand *Ligniera pilorum* was identified as the cause of root-hair proliferation and poor establishment of ryegrass (Latch, 1966b) and a *Rhizoctonia* sp. was associated with post-emergence damping off of turf grass seedlings (Christensen, 1979). When *L. perenne* and *L. multiflorum* seeds were treated with various fungicides, the numbers of emerging seedlings were increased by up to 150% (Falloon, 1980).

Thus, while it is clear that a small proportion, rarely 25%, of seeds sown produce mature plants, it is far from clear which pathogens have the most effect. It seems probable that a combination of pest and pathogen attack accounts for a large proportion of the losses and that adverse soil conditions exacerbate this damage as well as resulting directly in poor establishment.

8.4.2 Fungi affecting the established sward

In North America leaf area affected by fungal disease was reported to be 14.8% for *Poa pratensis,* of which more than half was due to one pathogen, 9.9% for *D. glomerata,* 11.4% for *Bromus inermis* and 3.2% for *Phleum pratense* (Elliott, 1962). In surveys in England, 47 of 154 fields of temporary ryegrass were judged to be damaged by disease, and in 45% of these fields 25% or more of their production was estimated to have been lost (Heard and Roberts, 1975).

Fungi causing diseases of agricultural and turf grasses have been documented by Sampson and Western (1954) and Smith (1965), and the *US Department of Agriculture Yearbook for 1953* contains several articles on diseases of grasses. Since the publication of these works, husbandry practice has changed and this has influenced disease occurrence. Nevertheless, while more has been learnt about disease distribution and new pathogens have been reported, the principal pathogens remain the same. The following account is limited to the economically important diseases of grasses.

(a) Rusts

Rusts, mainly *Puccinia* and *Uromyces* spp., are ubiquitous on grasses of agricultural importance. Pre-eminent are *Puccinia coronata* (crown rust), which occurs in many specialized forms (*formae speciales*) and attacks a wide range of important grasses including *Lolium* and *Festuca* spp. (Eshed and Dinoor, 1981), and *P. graminis* (black stem rust) which also occurs in many *formae speciales* and damages *Phleum, Dactylis, Lolium, Festuca, Poa* and many other grass hosts. *Puccinia striiformis* damages cocksfoot severely and infects many other cultivated grasses. *Uromyces dactylidis* mainly affects *Dactylis glomerata*, but *formae speciales* also infect *Festuca, Poa, Agrostis* and *Cynosurus* spp.

Crown rust (Puccinia coronata) occurs on grasses in all temperate regions of the world, and the characteristic bright orange pustules containing urediniospores are probably present throughout the year. However, most damage occurs when high humidity, often associated with dew, encourages spread i.e. in late spring and late summer and autumn. Damage in Britain and New Zealand is normally greatest at the latter times, when favourable conditions for disease spread and relatively slow growth rates of grasses can result in widespread uniform infection.

Puccinia coronata is considered the most important fungal disease of grasses in Britain (Heard, 1972) and of ryegrass in New Zealand (Latch, 1966a), and is one of the commonest diseases, often causing appreciable damage, of ryegrasses and *Festuca pratensis* in Ireland (O'Rourke, 1976). Infection first appears as scattered yellowish flecks on the leaves, which develop into pustules that erupt to expose bright orange urediniospores. These pustules are usually surrounded by a yellow halo, and at this stage the disease is most obvious. Badly affected leaves turn completely yellow and die back; whole tillers may be killed by severe infections.

Leaf area affected by crown rust can be large, but the useful yield from rusted ryegrass is decreased not only by loss in dry matter yield, but also by loss of quality and decreased palatability. In addition, in New Zealand rusted foliage is frequently colonized by the saprophytic fungus *Pithomyces chartarum*. This fungus produces a toxin, sporidesmin, which causes facial eczema when ingested by sheep. The problem seems unique to New Zealand, although a condition similar to facial eczema, not apparently associated with *P. chartarum*, occurs in Britain (Lacey, 1975).

Dry matter losses of up to 53% of single harvests of *L. perenne* have been reported from rust attack with a loss for the whole year of 36% (Fig. 8.2). These losses were associated with decreases of 30% in leaf area index and 20% in tiller number, and were greater where the grass was left long than where it was mown closely, and, in mixed swards of ryegrass and white clover, crown rust infection led to dominance by the white clover (Lancashire and Latch, 1966). Other management factors also appear to affect crown rust incidence, and although

Fungi

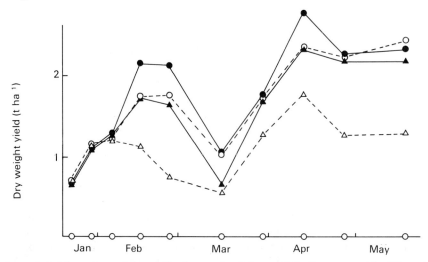

Figure 8.2 Mean dry weight yield of green leaf tissue of *Lolium* spp. with different amounts of crown rust (*Puccinia coronata* Corda.) (Redrawn from Lancashire and Latch, 1966.) ●———●, Grasslands Ariki (*L. perenne* × *multiflorum*), low rust; ▲———▲, Grasslands Ariki (*L. perenne* × *multiflorum*), high rust; ○---○, Grasslands Ruanui (*L. perenne*), low rust; △---△, Grasslands Ruanui (*L. perenne*), high rust.

the evidence is conflicting it is generally held that infection by rust is least where most nitrogen fertilizer has been used (Carr, 1971). In Ireland, O'Rourke (1975) found that crown rust infection of *L. perenne* decreased the dry matter yield of summer and autumn cuts by 9.5% but decreased tillering by 31.5%. Subsequent tests showed that not only was the rusted crop affected, but also that the dry matter of the regrowth after cutting, which was essentially free of infection, was decreased by 17.5% compared with the regrowth after cutting non-rusted foliage. In Britain, Carr (1979) reported a 23% decrease in regrowth after a rust attack which decreased tiller numbers by 21%. However, the estimated losses due to *P. coronata* in ryegrass were no more than 15–20%, and were confined to late conservation cuts.

Other studies have shown that the rejection of rusted foliage by sheep (Cruickshank, 1957) is a common phenomenon and diminishes the value of the grass that is produced. Cows also reject rusted grass, and if forced to eat it their milk production declines (Latch, 1966a).

Puccinia coronata infection of *L. multiflorum* decreases water-soluble carbohydrate and crude protein contents, and increases the non-digestible fibre fraction and silica content (Isawa *et al.*, 1974; O'Rourke, 1975). Similar results were found for *P. coronata* (var. *festucae*) infection of *F. pratensis* (Cagas, 1979). Such changes may help to explain the decreased digestibility of rusted foliage, and may also account for some of the interactions reported between rusts and other pathogens attacking grasses.

Black rust (Puccinia graminis) infects *Lolium* spp. but is generally most common as *P. graminis* f.sp. *phlei pratensis* on *Phleum pratense* and as *P. graminis* f.sp. *avenae* on cocksfoot. Infection is regularly present on ryegrass in France, Ireland, North America and New Zealand, where it can infect seed heads (Latch, 1966a). The uredinia usually develop on grass culms or leaf sheaths, rarely on leaf laminae, and produce reddish-brown spores. The effect of *P. graminis* on cocksfoot yield was similar to that of *P. coronata* on ryegrass, a severe infection causing a 37% decrease in tiller number, widespread death and a 36% decrease in fresh weight (Lancashire and Latch, 1969).

Stripe rust (Puccinia striiformis) is especially damaging to *D. glomerata* in the autumn, and decreases yield and palatability. Severe damage to seed crops of *P. pratensis* in the USA has also been reported (Hardison, 1963). *Puccinia striiformis* is distinguished from the other rusts by the elongate uredinia arranged in lines on the leaf blades. Susceptibility to stripe rust in *D. glomerata* is positively correlated with high carbohydrate content (Carr, 1971), but infection decreases water-soluble carbohydrate content and has other effects on host metabolism. When 30% of the leaf area of *D. glomerata* was affected by *P. striiformis* respiration was increased almost three-fold, photosynthetic activity and water-soluble carbohydrate content were almost halved, but soluble protein content was little affected. As might therefore be expected, enzymes such as acid phosphatase which are associated with respiration were increased by infection whereas those associated with photosynthesis, such as hexokinase, alanine-amino-tranferase and RuBP carboxylase/oxygenase were all decreased (Carr, 1975).

Infection of *D. glomerata* in New Zealand by stripe rust decreased shoot number by 37% compared with grass where fungicides had controlled infection. Fresh weight was decreased by a similar percentage, but no differences were found in infection where different amounts of nitrogen had been used. Such damage was proposed as one of the principal causes for the decline in the proportion of *D. glomerata* in mixed species swards (Lancashire and Latch, 1969).

Other Puccinia spp. P. poarum is widespread on *P. annua, P. pratensis* and *P. trivialis,* especially where these form substantial components of swards which are kept for hay or laxly grazed but no measurements have been made of losses due to this fungus.

Other rusts of grasses are described by Sampson and Western (1954), Cummins (1971) and Ullrich, (1977).

(b) Powdery mildew

Powdery mildew (Erysiphe graminis) in its many *formae speciales* is an ubiquitous pathogen of Gramineae, but there are few reports of damage attributed to mildew infection of grass swards although seed yields are affected and thousand grain weight decreased by infection (Cagas, 1981).

Mildew is a superficial parasite that produces haustoria and mycelium in the epidermal cells and a mass of conidial chains on the surface, giving severely affected leaves a white, powdery appearance. In later stages there is much chlorosis of leaves and black cleistothecia are produced. Infection occurs throughout the year, but is especially common after dry periods and where much nitrogen has been used. The disease also seems to develop quickly in overcast weather or shaded conditions (Gaskin and Britton, 1962).

In swards of *L. multiflorum,* where mildew was an important disease, fungicides increased hay yield and digestible dry matter by 25% but no consistent differences were noted in soluble carbohydrate content between diseased and fungicide-treated crops (Davies *et al.,* 1970). Control of mildew on *Poa pratensis* increased fresh weight yields by 14% over four cuts and seed yield by 35% when moderate infection was controlled (Frauenstein, 1977). Large effects of mildew infection on the yield of cocksfoot seed are also reported (Heard, 1972).

(c) Leaf spots

Drechslera spp. cause leafspots on a wide range of grasses in all cool, temperate regions, and are most common when the sward is rank. More than 30 species are known (Labruyère, 1977; Lam, 1982), many of them are seed-borne (Noble and Richardson, 1968) but only a few attack economically important grasses and are sufficiently prevalent to be a problem in grass crops.

Drechslera siccans usually causes oval, chocolate brown spots on leaf laminae, but in Scotland also causes a foot-rot (Sampson and Western, 1954). In severe attacks the mesophyll of the leaf lamina is completely destroyed and, in wet conditions, whole shoots may be killed (Wilkins, 1973a). Spores are produced from the centre of old lesions and on dead spikelets. Infection is most common on *Lolium* spp., but *Festuca* and *Dactylis* spp. are also attacked.

Slight infection of *L. multiflorum* by *D. siccans* decreased dry matter digestibility, water soluble carbohydrate and the total amino acid composition (Lam, 1985), and in Britain, when 15% of the leaf area of *L. multiflorum* was affected by *D. siccans,* dry matter yield loss was of the same order but infection of ryegrass in swards mixed with clover did not put the ryegrass at a competitive disadvantage (Cook, 1975). In New Zealand *D. siccans* is present throughout the year and is more damaging to individual plants than *D. dictyoides,* although the latter was more common and nationally more damaging (Latch, 1966a). In Sweden *Drechslera* spp. are common and a combined infection of *D. siccans* and *D. catenaria* caused a 15% yield loss in *L. perenne* (Stegmark, 1979).

Drechslera sorokiniana, the conidial stage of *Cochliobolus sativus,* is one of the most important diseases of *P. pratensis* in North America (Bean and Wilcoxson, 1964), and much interest has been shown in the effects of fertilizer nitrogen on free amino acid and sugar contents of grasses and the consequent effect on infection (Gibbs and Wilcoxson, 1972; Robinson and Hodges, 1977). The interactions are complex and no conclusions are yet possible (Robinson and Hodges, 1981).

Drechslera poae (= *Helminthosporium vagans*) is also prevalent in North America and Europe, where it is predominantly a pathogen of turf grasses. The dark purplish-red lesions which turn brown and finally white may spread to the base of the plants causing a severe foot- and root-rot.

Leafspot (*Cladosporium phlei*) is widespread in North America and Europe. Infection is confined to the genus *Phleum,* but several species are infected and yields of timothy have been decreased in North America (Roberts *et al.*, 1955). In Japan infection decreased the content of crude leaf protein by 26% (Sakuma and Narita, 1961).

Other *Drechslera* spp. also occur more or less frequently. *Drechslera festucae* infects *F. arundinacea* in the British Isles (O'Rourke, 1976); *D. phlei* infects *Phleum pratense* and *Dactylis glomerata* in Europe and North America; *D. catenaria* causes a net blotch on cocksfoot, and *D. ergothrospila* and *D. fugax* infect *Agrostis* spp. (Carr, 1971).

(d) Smuts

Smuts affect both leaves and ears of many grass species and many are seed-borne (Noble and Richardson, 1968). However, few of their grass hosts are agriculturally important.

Loose smut (*Ustilago avenae*) is common wherever *Arrhenatherum elatius* is common, and the seeds are replaced by black spore masses (Sampson and Western, 1954).

Stripe smut (*Ustilago striiformis*) is both soil and seed-borne and produces long streaks of black spore-masses along the length of the leaf. Infected plants make little vegetative growth and inflorescences are absent or stunted, although the pathogen only rarely infects the ovaries.

Other smuts include ear smut (*U. bullata*), which is common on *Bromus* spp.; stem or sheath smut (*U. hypodytes*), which infects some important grass species in North America and can render hay unfit for consumption (Fischer, 1938); and flag smut (*Urocystis agropyri*), which decreases inflorescence production and delays their development.

Bunts or seed smuts (*Tilletia* spp.) replace seeds by spores, retard growth and distort seed heads (Carr, 1971). Sampson and Western (1954) report four *Tilletia* spp. from Britain, and Fischer (1953b) 25 species from the USA. While these pathogens are potentially damaging, serious losses have not been reported.

Fungi

(e) Other fungal pathogens

Leaf blotch or *scald* (*Rhynchosporium* spp.) is present in most temperate regions and *R. secalis* and *R. orthosporum* are the most common species. They cause superficially indistinguishable symptoms of dark bluish-grey water-soaked spots which develop into lesions 25–30 mm in length, with pale centres and dark brown margins. Infection tends to be most severe in cool, wet regions and spores are dispersed by rain splash.

Rhynchosporium orthosporum has been recorded from *L. perenne*, *L. multiflorum* and *D. glomerata*, and was isolated in Wales twice as frequently as *R. secalis* (Wilkins, 1973b). *Rhynchosporium secalis* has been reported from *D. glomerata*, *P. pratense*, *Alopecurus pratensis*, *Bromus* spp. and *Agropyron repens* in Britain, and from *Lolium* spp., meadow-grasses and *Holcus mollis* in Europe (Carr, 1971). Host specialization within *Rhynchosporium* spp. appears to be widespread (Carr, 1971), and extensive crop damage has not been recorded, although heavily infected *L. multiflorum* yielded 25% more when protected by fungicides than did untreated grass and there was no yield response when there was little leaf blotch. Where *Rhynchosporium* spp. were controlled digestible dry matter was increased, but there was no obvious effect on water-soluble carbohydrate content (Davies *et al.*, 1970). However, Lam (1985) found that infection of *L. multiflorum* caused significant decreases in dry matter digestibility, total amino acid content, and water-soluble carbohydrate. In addition to direct effects on yield, infected foliage is rejected by cattle (Wilkins, 1973b) and sheep (Latch, 1966a). In North America infection by *R. secalis* had only a slight effect on the digestibility of *Bromus inermis* (Gross *et al.*, 1975).

Leaf fleck caused by *Mastigosporium* spp. is common in cool, moist conditions in temperate regions. The symptoms develop from small, water-soaked flecks on leaf laminae to purplish-brown flecks, 3–6 mm long, which may coalesce to cover much of the leaf surface.

Mastigosporium rubricosum is common on *D. glomerata* throughout the year, although infection is especially noticeable in early spring and autumn when it decreases yield and quality. In Ireland up to 50% of the leaf surface is often affected by the fungus (O'Rourke, 1976). When 10% of the leaf area was affected the water-soluble carbohydrate was decreased by 50% and respiration rate was much increased (Carr and Catherall, 1964). Application of potassium generally decreases infection whereas fertilizer nitrogen often increases disease severity. Other grasses, including perennial ryegrass, timothy, meadow foxtail, Yorkshire fog and *Agrostis* spp., are also susceptible but rarely suffer seriously. However, in a mixed sward infected *Agrostis* spp. were left ungrazed (O'Rourke, 1976).

Halo spot (*Selenophoma donacis*) infection is widespread on many grasses, but in Britain is generally most severe on timothy and cocksfoot (Carr, 1971). Early symptoms in timothy are small, round, dark purple spots on the leaf and culm. As infection develops the spots enlarge, may merge, and become bleached at their centres. When infection is severe the leaves turn purple. On cocksfoot symptoms are similar to those on timothy, but the purple colour is absent and lesions may develop on the panicles and damage the quality and yield of seed (Carr, 1971). *Selenophoma donacis* is seed-borne in *Agropyron, Dactylis* and *Festuca* spp. (Noble and Richardson, 1968). Nitrogenous fertilizers increase disease severity.

Snow mould is an all-embracing name for a disease of turf or pasture grasses which may be caused by several pathogens, and is often associated with prolonged snow cover. O'Rourke (1976) distinguishes two diseases, Fusarium patch and pink snow mould, caused by one of the commonest components of the pathogen complex causing snow mould, *Fusarium nivale* (*Micronectriella nivalis*). Agriculturally important grasses such as *Lolium* and *Phleum* spp. can be damaged by pink snow mould, named from the pale pink, aerial mycellium that is often produced. *Fusarium nivale* is nearly always present, but other fungi also occur depending on region. In Scandinavia *Typhula* spp., especially *T. incarnata* and *T. ishikariensis,* and *Sclerotinia borealis* are commonly found (Årsvoll, 1977; Makela, 1981), and *T. incarnata* has also been reported from Scotland. In northern Finland the principal grass affected is timothy; young crops were most damaged by *S. borealis,* while older pastures were affected most by *T. ishikariensis,* which was found on 55% of leys. *Fusarium* spp. damaged *F. pratensis* more than timothy (Makela, 1981). In Scotland *T. incarnata* suppressed *L. perenne* and *P. annua* and allowed *Trifolium repens* to dominate mixed swards. *Typhula incarnata* affected only leaves and did not suppress tillering on vigorous plants, but *F. nivale* can kill many tillers or entire plants. The area of grassland at risk from snow mould in Scotland was calculated as 150–200 000 ha, with up to an additional 2×10^6 ha of rough grazing also likely to be damaged (Gray and Copeman, 1975).

For details of other pathogens affecting swards but usually of little economic importance, the reader is referred to Latch (1966a), Carr (1971) and O'Rourke (1976).

8.4.3 Fungi affecting turf

Fungal diseases of turf grasses and the effects of environmental and cultural practices on their incidence has been comprehensively considered by Smith (1965). No other group of pathogens appear to be of any importance. Many of the pathogens already mentioned, especially *Drechslera* spp., also affect turf grasses, but only those particularly associated with turf are considered here.

(a) Fusarium patch

Fusarium patch (*Micronectriella nivalis* (stat. conid. *F. nivale*)) is probably the most widespread and damaging disease of turfgrasses, the fine-leaved grasses

Agrostis, Festuca and *Poa* spp. being particularly susceptible. Symptoms appear first as small water-soaked areas of turf a few centimetres in diameter which later turn yellow or brown and increase in size, often coalescing to form extensive brown areas. Although the fungus may invade the crowns of plants, it is usually as a result of winter damage that infected areas die.

The disease occurs in Britain on many types of turf, but its incidence is increased by the application of alkaline fertilizers or lime (Smith, 1965). Late season application of nitrogen, by encouraging lush growth, also favours the disease.

(b) Dollar spot

Dollar spot (Sclerotinia homoeocarpa) often occurs on *F. rubra* ssp. *rubra* in Britain and is reported from elsewhere (Smith, 1965). Early symptoms are small patches (1 cm in diameter) which may increase up to 5–6 cm with leaf blades becoming bleached. Infection is favoured by moist conditions and late application of nitrogen.

(c) Red thread

Red thread (*Laetisaria fuciformis* McAlp. Burdsall) (syn. *Corticium fuciforme* (Berk.) Wakefield) has recently been distinguished from other pathogens, especially *Limonomyces* spp. that cause pink diseases in turf grasses (Stalpers and Loerakker, 1982). Infection is present in most temperatre areas and many cultivated grasses can be damaged (Hims *et al.*, 1984); *Festuca rubra* seems especially susceptible whereas *F. ovina* is not affected (O'Rourke, 1976). The symptoms first appear as small areas, up to a few centimetres in diameter, of bleached leaves. In cool moist conditions, such as occur in autumn, these aras spread and assume a reddish tinge due to pigment in the fungus which is especially noticeable in the aggregations of mycelium (stroma) which bind leaves together, and to the development of colour in infected leaves. The disease is most prevalent where soil fertility is low, especially where calcium and nitrogen are deficient (Gould *et al.*, 1967). Applications of nitrogen can minimize the effects of the disease (Goss and Gould, 1971) but not invariably (Hims *et al.*, 1984).

(d) Ophiobolus patch

Ophiobolus patch is caused by *Gaeumannomyces graminis* var. *avenae*, the oat strain of *G. graminis*, the wheat strains of which are the cause of 'take-all' in cereals. Many grasses are susceptible but infection is especially common on *Lolium, Agrostis, Poa, Agropyron, Holcus* and *Festuca* spp. in sports and recreationai turf. The disease is favoured by cool moist conditions and where the soil has been treated to decrease soil acidity. This effect can be countered by using sulphate of ammonia as the source of nitrogen.

Symptoms first appear as small patches which may, in favourable conditions, enlarge to several metres in diameter when the plants in the centre are replaced by weed species. Infection damages the roots of plants around the marginal patches, which are easily pulled up and may be bronzed.

(e) Fairy rings

Fairy rings are produced by many fungi, mostly basidiomycetes, throughout the world (Smith, 1965); one of the commonest is *Marasmius oreades*. Fairy ring fungi grow radially, and their growth is associated with the breakdown of organic matter which is manifest as rings of stimulated grass growth. Immediately inside the outer area of stimulated growth there is often a bare zone. There is some doubt about the cause of this poor growth, but reasons suggested include parasitism of the grass by the fungus, dense mycelium in the soil preventing water penetration, and the production of toxic compounds (Smith, 1980). Rings also occur in high quality grassland (Hardwick and Heard, 1978), where they may decrease productivity by up to 50% by causing bare areas, and change the botanical composition by permitting the growth of weed species.

8.4.4 Fungi affecting seed crops

Many of the pathogens already described directly or indirectly affect seed yield, but some have their greatest efects on seed production and they are described here.

(a) Ergot

Ergot (*Claviceps purpurea*) is one of the longest-known and potentially most damaging diseases of grasses. It has a worldwide distribution and infects many species. The disease not only damages seed production by replacing seeds by the diagnostic, dark, horn-shaped resting bodies (sclerotia or ergots), but also causes severe disease in cattle that eat infected hay or in humans that eat bread made from flour extracted from infected grain (usually rye). These effects are caused by alkaloids in the ergots; their concentration varies, but in Ireland the average is 3% (O'Rourke, 1976). The minimum amounts of ergot reported to cause symptoms in animals range from 0.5 to 3% of total feed dry matter. There is doubt whether eating ergotized grain causes abortion in sheep and cattle, but the alkaloids are not present in the milk or meat of animals that have eaten infected grain (Sampson and Western, 1954).

The first signs of the disease in grasses is a sticky 'honeydew' on inflorescences which later develop the ergots; these may be up to 2 cm long on large-seeded grasses. The sticky exudate aids the dispersal of spores which stick to the bodies of visiting flies. Infection is most common on late inflorescences, and if left undisturbed the ergots fall from mature inflorescences and remain in the soil until spring when they germinate to produce ascospores which are wind-dispersed and infect grass inflorescences through stigmata. In New

Zealand up to 10% of seed can be replaced by ergots and infection is more common in *L. perenne* than in *L. multiflorum* or *Lolium* hybrids.

Despite its adverse effects, ergot does provide many alkaloids of great medicinal value, and the disease has been 'cultivated' for the purpose in many regions, usually by the artificial inoculation of rye.

(b) Blind seed

Blind seed caused by *Gloeotinia temulenta* is especially common in cool moist climates such as the northwestern USA, New Zealand, Great Britain and the Netherlands. The most commonly infected hosts are the ryegrasses, especially perennial ryegrass, but many grasses are susceptible.

Ascospores penetrate to the ovary from just below the stigma to cause the primary infection. Secondary infection is by macroconidia which form a pink slime on the ovary. Infection early in development destroys the embryo and no seed forms. Later infection causes 'blind seeds' which do not germinate and can act as perennating sources of the fungus. Even later infection fails to infect the embryo and seeds still germinate. Stock which eat hay containing diseased seed are unaffected (Wright, 1967). In New Zealand, where infection can affect 44% of seed (Latch, 1966b), nitrogen supplied as urea in spring or at ear emergence significantly decreased blind seed infection and increased the percentage of harvested seed germinating (Hampton and Scott, 1980).

(c) Choke

Choke (*Epichloë typhina*) is common in Europe but rare elsewhere. It infects its hosts systemically and survives in rhizomes or vegetative organs overwinter, only manifesting its presence by developing on the flowering shoots in the spring and summer; vegetative growth, if affected at all, is enhanced. Many grasses are susceptible, but the disease is of most importance on cocksfoot where disease incidence increases with the age of the stand (Large, 1954). The symptom which gives the disease its name is the appearance of the fungus fruiting body around the uppermost leaf sheath and stem in cocksfoot and around the branches of the inflorescence in fescues. This fruiting body, which is initially white but becomes orange when spores are produced, either completely prevents the emergence of the inflorescence or chokes off nutrient supplies to it. The consumption by animals of infected grass appears to have no adverse affects (Cunningham, 1958), although Bacon *et al.* (1977) associated toxicity with infection. Infected red fescue may yield much seed, but a large proportion (up to 99%) are infected and produce infected plants (Sampson and Western, 1954). Infection of cocksfoot caused an 8% decrease in fertile tiller number in a second-year crop and from 0 to 69% in third-year crops (Sampson and Western, 1954).

An unusual feature of *E. typhina* infection of *A. tenuis* is that while infection almost completely suppresses inflorescence production, it also causes a considerable increase in vegetative tiller density. This is an advantage in grazed swards, and in this case infection has been considered symbiotic (Bradshaw, 1959).

8.5 BACTERIA

Several bacteria infect grasses, but few seem to be of great economic importance. In parts of Europe, including Switzerland, France, Germany and Great Britain, and in New Zealand *Xanthomonas campestris* pv *graminis* causes bacterial wilt of ryegrass (Egli and Schmidt, 1982). Symptoms of infection are most characteristic on flowering tillers. Yellow marginal stripes develop on flag leaves before they wilt and become bleached. Some inflorescences die, producing scattered 'whiteheads' in affected crops. In vegetative tillers infected leaves show dark green, water-soaked spots. The bacteria are present in the xylem and may be spread by cutting. In Scotland infection was widespread on ryegrass, although its affect on crop yield was slight, but *D. glomerata* and *P. pratensis* were also occasionally infected (Channon and Hissett, 1984).

In Japan chocolate spot disease of *L. multiflorum* is caused by *Pseudomonas coronafaciens* var. *atropurpurea*. Symptoms are necrotic spots surrounded by a chlorotic halo. The bacteria produce an extracellular toxin, coronatine, which may control the production of enzymes which increase the plasticity of the cell wall before the cell expands by turgor pressure (Sakai *et al.*, 1979).

8.6 VIRUSES

Many viruses infect grasses, and some grasses are susceptible to many viruses. Catherall (1981) considered 22 viruses to have been identified from grasses in Britain, only eight of which were sufficiently damaging or widespread to be important. However, many viruses are poorly characterized and identification is difficult, especially of those with isometric particles (Paul *et al.*, 1980).

The effects of viruses on their grass hosts range from symptomless infection, through mild mosaics to necrosis and, infrequently, death. All crop infection is systemic, consequently once they are infected plants remain diseased until they die and cutting or grazing infected crops may spread viruses and their vectors. In contrast, cutting removes inoculum of fungal and bacterial disease and often leaves conditions that are rarely as favourable for pathogen attack as before.

Virus damage is more often insidious than spectacular and methods of husbandry and the effects of viruses interact (Carr, 1975). Thus, viruses which stunt their host and increase tiller number are relatively more damaging when the sward is cut for hay than when intensively grazed. Viruses which decrease tiller number but not height are most damaging in closely grazed swards. Thus, losses caused by viruses are made up of many interacting factors some of which,

Viruses

such as the increased content of water-soluble carbohydrate seen in some virus-infected plants, appear to be advantageous. Virus-infected plants also interact with other pathogens. In *Lolium* spp. infection by ryegrass mosaic virus (RMV) suppressed crown rust infection by 75%, whereas infection by barley yellow dwarf virus (BYDV) had no effect. That these interactions are complex is illustrated by the effect of infection by both viruses; in *L. multiflorum* rust infection was intermediate between that on plants infected with either virus alone, but in *L. perenne* dual virus infection decreased rust infection to less than that seen on plants infected by either virus alone (Latch and Potter, 1977). Virus infection also interacts with virus vectors. Gibson (1976) found that the mite vectors of RMV did not thrive on RMV-infected plants and dispersed more rapidly from them than from healthy plants.

Some grass viruses, such as BYDV, seem to be ubiquitous as most members of the Gramineae are susceptible; its aphid vectors are numerous; others, such as oat mosaic virus, are restricted in both host range (*Avena* spp. only), and by having a soil-borne fungal vector which spreads only slowly even in the presence of a susceptible host. Many viruses that infect grasses also infect cereals, and it is from these hosts that much of our knowledge of the effects of the disease derives. While some extrapolation from cereals to grasses is justified the differences between grasses and cereals in management and utilization are such that such comparisons should be treated with caution.

8.6.1 Barley yellow dwarf virus

Barley yellow dwarf virus (BYDV) is probably the most widely-distributed and common virus in the world (Plumb, 1983). It is spread only by aphids, of which at least 23 species are known to be vectors. The virus multiplies only in the phloem and spreads rapidly throughout the plant once introduced; it is not seed-borne. Most species of Gramineae tested have proved susceptible, but few grasses show obvious symptoms: when they do occur they usually appear as pale yellow to purplish red discolorations of leaves and stunting of tillers. Infection sometimes stimulates tiller production, resulting in low, cushiony growth. *Lolium* and *Festuca* spp. most frequently show symptoms.

Infection by BYDV can be very damaging. The yields of single plants of *Lolium multiflorum*, *L. perenne*, *Festuca arundinacea* and *Poa pratensis* were decreased by 45, 63, 71 and 72%, respectively, after infection by BYDV (Catherall, 1971). However, in simulated swards of *L. perenne* and white clover (*Trifolium repens*) the uninoculated sward out-yielded the inoculated by 20%; fewer fertile tillers were produced by infected plants, but the contribution of the clover component to yield increased compared with uninoculated swards. In their second year of growth both inoculated and uninoculated plot yields were initially the same, but when cut for hay infected plots yielded significantly less than the uninfected; in contrast, when frequently cut, infected swards yielded more. In totally infected pure grass swards the yield loss was twice as great when plots were cut infrequently as when cut frequently; in mixed swards the yield loss was unaffected by cutting frequency (Catherall, 1966). These results suggests that, depending on the method of management, BYDV can seriously decrease yield especially in single species swards. In New Zealand Latch (1980) concluded that while infection by BYDV was widespread, with up to 84% of some swards infected (Latch, 1977), it was of no economic significance in well-managed pastures because the presence of white clover counteracted the adverse affects of BYDV.

Infection by BYDV changes the seasonal growth pattern, decreasing autumn growth but stimulating growth in early spring. To account for these changes Wilkins and Catherall (1977) suggested that photosynthate in the autumn was diverted to storage and utilized the following spring. However, there has been little work on the physiology of BYDV-infected plants, and most of our knowledge is derived from investigations of small grain cereals.

The presence of BYDV in phloem (Jensen, 1969) results in degeneration and necrosis of sieve elements, companion cells and neighbouring parenchyma cells (Esau, 1957). Infection results in large increases in the water-soluble carbohydrate content of leaves and possible changes in their type (Moline and Jensen, 1975); up to 15 times the normal soluble carbohydrate content has been detected in infected barley whilst starch also accumulates (Jensen, 1969). These effects have visual expression in the discoloration and stiffer, more upright habit of infected leaves. Stunting, the other characteristic symptom, is caused by a decrease in cell number rather than a change in size. The change in cell number may be due to a decrease in endogenous gibberellins in infected plants (Russell and Kimmins, 1971).

The accumulation of carbohydrates increases dry weight and may itself be the cause of a decrease in other metabolic activities, especially photosynthesis, by an inhibitory feedback mechanism. By contrast with the effects on above-ground tissue, BYDV infection decreases the soluble carbohydrate content of the roots of cereals (Orlob and Arny, 1961).

The rate of photosynthesis is usually decreased by infection, the decrease depending on how the results are expressed, but ranging from 50 to 15% (Jensen, 1968). The method of expressing results has an even bigger effect on respiration rates than it does on photosynthesis. Respiration rates expressed per unit fresh weight and per unit area are more than doubled by infection but, because of the associated large increases in dry matter, respiration per unit dry matter does not differ significantly from that of uninfected plants (Jensen, 1968).

8.6.2 Ryegrass mosaic virus

Ryegrass mosaic virus (RMV) is transmitted via sap and by the eriophyid mite *Abacarus hystrix*. Infection is common in northwestern Europe, and the virus is present in North America. Only members of the Gramineae are susceptible,

Viruses

including species of *Avena, Dactylis, Agrostis, Bromus* and *Lolium*. The principal hosts are *Lolium* spp., *L. multiflorum* being especially susceptible and damaged by infection (Heard *et al.*, 1974). Yield losses can be as much as 30% (A'Brook and Heard, 1975), but depend on the severity of the virus isolate and seasonal growth. The effects of infection are usually greatest when crop growth is most rapid, and some isolates in Britain can cause severe necrosis and death. In North America most isolates cause mild symptoms.

Irrespective of isolate there is a marked interaction between infection and the amount of nitrogen applied to the crop. Where no nitrogen was used RMV had little effect on yield, but where 400 kg nitrogen ha^{-1} was applied yield was decreased by 26%. When nitrogen applied to a sward 70–80% infected with RMV was doubled from 189 to 378 kg nitrogen ha^{-1} the effect of the virus was to eliminate the 15% increase in yield obtained on virus-free plots (Holmes, 1980) (Fig. 8.3).

Decreased yield in field crops is associated with a decrease in tiller number and height. Using simulated swards of *L. multiflorum* and *L. perenne* at least 80% infected with RMV, Jones *et al.* (1977) found that leaf area index was decreased by 32–42% but that light interception differed little between infected and healthy swards. Tiller numbers in both species were decreased by

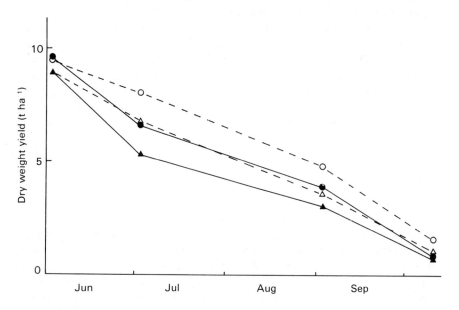

Figure 8.3 Effects of infection by ryegrass mosaic (RMV) on dry weight yield of *Lolium multiflorum* cv. S.22 at different levels of fertilizer application. (Redrawn from Holmes, 1980.) 189 kg nitrogen ha^{-1}: ●——●, uninoculaed (max. RMV 9.5%), ▲——▲, inoculated (max. RMV 72.0%). 378 kg nitrogen ha^{-1}: ○---○, uninoculated (max. RMV 4.5%), △---△, inoculated (max. RMV 83.5%).

infection, but there were no differences between rate of leaf emergence and leaf extension. Net canopy photosynthesis at six levels of irradiance up to full sunlight was less in infected than in healthy swards, and the difference was greatest at the highest levels of irradiance.

At an irradiance of $200\,\mathrm{W\,m^{-2}}$ (PAR) the photosynthetic rates of the youngest fully expanded leaves were consistently less for infected than for healthy leaves, and the difference was again greatest at the highest levels of irradiance. Canopy dark respiration was greater in infected than in healthy swards; the maximum difference was 33%.

It was concluded that the yield losses due to RMV infection were largely attributable to the decrease in net canopy photosynthesis and increased losses through the greater rate of dark respiration. The decrease in canopy net photosynthesis was partly attributed to an increase in respiration but there was also a large decrease in gross photosynthesis (Jones et al., 1977). However, the effects of individual factors such as nutrition, water supply, temperature, radiation and management, which may interact with RMV infection, were not investigated, and interaction between several pests and diseases is a common phenomenon in crops and may change the physiological responses of the host. Cytoplasmic changes caused by infection are seen in mesophyll cells in which large aggregates of the virus occur (Plumb and James, 1973). Infected cells contained large proteinaceous laminar inclusions and 'pinwheels'.

Quality as well as quantity of herbage is affected by virus infection. Organic matter content, organic matter digestibility, water-soluble carbohydrate content and DMD of *L. perenne* were all decreased by RMV infection (Holmes, 1979b). When infected by a virus strain which caused only a mild mottle, the decrease in digestibility was mostly accounted for by the decrease in water-soluble carbohydrate. However, when infected by an isolate causing necrosis, the most important contributory cause was the loss of green tissue.

8.6.3 Brome mosaic virus

Brome mosaic virus (BMV) is widely distributed in Europe, North America and South Africa, infecting many species of Gramineae. In barley infection initially increases respiration rate which, although fluctuating, generally remains greater than in non-infected plants. This respiratory increase appears not to be associated with uncoupling of phosphorylation or increased availability of substrate. Mitochondria appear to be affected, as the activity of succinoxidase from mitochondria was decreased by virus infection (Burroughs et al., 1966). Schmidt (1967) found that infection resulted in chloroplast enlargement, and subsequently they and mitochondria ruptured. Brome mosaic virus reaches very large concentrations (10^6 particles/$75\,\mu\mathrm{m}^3$ cell) and almost all cells are invaded (Paliwal, 1970); mesophyll cells are most severely affected. It is surprising that such a common virus, and one that is readily

Viruses

transmitted by sap (as well as by some nematodes) and reaches such large concentrations in infected hosts, does not cause more damage.

8.6.4 Cocksfoot streak virus

Cocksfoot streak virus (CFSV) is widespread in *D. glomerata* in Europe. Infection decreases tiller production, especially of vegetative tillers, and flowering tillers produce fewer fertile seeds. Infection is most common and damaging in crops grown for hay or seed. In closely grazed or frequently cut swards infection is uncommon because the effect of the virus makes infected plants much less able to compete (Catherall and Griffiths, 1966).

8.6.5 Cocksfoot mottle virus

Cocksfoot mottle virus (CFMV) causes a lethal disease in *D. glomerata*, and infection is widespread in swards more than 2 years old. The severe effects of the virus have probably been a contributory cause of the decline in the area of *D. glomerata* grown in UK.

8.6.6 Cocksfoot mild mosaic virus

The relationships of *cocksfoot mild mosaic virus* (CFMV) to other isometric viruses that infect Gramineae has been investigated by Catherall and Chamberlain (1977) and Paul *et al.* (1980), and it seems that the viruses known as phleum mottle, brome stem leaf mottle, festuca mottle and holcus transitory streak viruses are inter-related. Various of these isolates are probably widespread in northern and western Europe, but there is little information except from Britain and the Federal Republic of Germany. On cocksfoot CMMV usually causes a diffuse or mild mottle, but chlorotic streaking is sometimes seen. Related viruses also cause only mild symptoms on naturally-infected hosts.

8.6.7 Oat sterile dwarf virus

Oat sterile dwarf virus (OSDV) is commonest in north, central and western Europe, and mainly affects *Lolium* and *Arrhenatherum* spp. Infection results in extreme dwarfing and proliferation of tillers, with leaves becoming twisted, distorted and a dark blue-green. Small enations 1–2 mm long develop on the abaxial surface as a result of proliferation of the phloem to which tissue the virus is confined. Serious damage has been reported to *Arrhenatherum elatius* in the German Democratic Republic, and to *A. elatius* and *L. multiflorum* in the Federal German Republic, but its most damaging effects are on cereals. In Britain the virus is transmitted by the planthopper vector *Javesella pellucida*, but symptoms are rarely seen, possibly because infected plants are quickly eliminated by competition from healthy plants.

8.7 CONCLUDING REMARKS

The pests and pathogens briefly considered in this chapter are only a few of those known to affect grasses. That so few justify comment is indicative as much of a lack of knowledge as of the comparative resilience of the grass crop, especially when grown in a mixed sward. It is in intensively managed crops, especially single-species swards, where pests and pathogens are most damaging, but as standards of husbandry improve yield lost to pests and diseases is likely to increase in all grass crops and may then become the limiting factor for increased productivity. A better understanding of the biology, epidemiology and ecology of grass pests and diseases and how they interact with management and have their effects seems to be the best way of ensuring that the losses they cause are minimized.

REFERENCES

A'Brook, J. and Heard, A. J. (1975) The effect of ryegrass mosaic virus on the yield of perennial ryegrass swards. *Ann. Appl. Biol.*, **80**, 163–8.

Anon. (1948) Wireworms and food production. A wireworm survey of England and Wales 1939–42. *MAFF Bull.*, **128**.

Anon. (1977) Amenity grasslands – the needs for research. *Natural Envir. Res. Coun. Publ.*, Ser. C, No. 19.

Årsvoll, K. (1977) Effects of hardening, plant age and development in *Phleum pratense* and *Festuca pratensis* on resistance to snow mould fungi. *Meld. Norg. Landbrukshøgsk.*, **56** (28).

Bacon, C. W., Porter, J. K., Robbins, J. D. and Luttrell, E. S. (1977) *Epichloë typhina* from toxic tall fescue grasses. Appl. Envir. Microbiol., **34**, 576–81.

Bean, G. A. and Wilcoxson, R. D. (1964) Helminthosporium leaf spot of bluegrass. *Phytopathology*, **54**, 1065–70.

Bezooijen, J. van (1979) Nematodes in grasses *Medesletin gen. van de Facultat Landbouwettenschappen Rijksuniversiteit Gent.*, **44**, 339–349.

Blackshaw, R. P. (1983) The annual leatherjacket survey in Northern Ireland, 1965–82, and some factors affecting populations. *Pl. Pathol.*, **32**, 345–9.

Blank, R. H. and Olson, M. H. (1981) The damage potential of the black field cricket *Teleogryllus commodus*. *N.Z. J. Agric. Res.*, **24**, 251–8.

Boag, B. and Trudgill, D. L. (1977) Damage caused by *Longidorus elongatus*. *Rep. Scot. Hort. Res. Inst. for 1976*, pp. 89–90.

Bradshaw, A. D. (1959) Population differentiation in *Agrostis tenuis* Sibth. II. The incidence and significance of infection by *Epichloë typhina*. *New Phytol.*, **58**, 310–5.

Burroughs, R., Goss, J. A. and Sill, W. H. (1966) Alterations in respiration of barley plants infected with bromegrass mosaic virus. *Virology*, **29**, 580–5.

Cagas, B. (1979) Estimation of economic losses due to *Puccinia coronata* Corda var. *coronata*. *Ochrana Rostlin*, **15**, 253–8.

Cagas, B. (1981) The use of Bayleton in the protection of seed cultures of smooth-stalked meadow grass. *Ochrana Rostlin*, **17**, 61–6.

References

Carr, A. J. H. (1971) Grasses, in *Diseases of Crop Plants* (ed. J. H. Western), Wiley, London, pp. 286–307.

Carr, A. J. H. (1975) Diseases of herbage crops – some problems and progress. *Ann. Appl. Biol.*, **81**, 235–9.

Carr, A. J. H. (1979) Causes of sward change – diseases, in *Changes in Sward Composition and Productivity* (eds A. H. Charles and R. J. Haggar), Occasional Symposium No. 10, British Grassland Society, pp. 161–6.

Carr, A. J. H. and Catherall, P. L. (1964) The assessment of disease in herbage crops. *Rep. Welsh Pl. Breeding Stn. for 1963*, pp. 94–100.

Catherall, P. L. (1966) Effects of barley yellow dwarf virus on the growth and yield of single plants and simulated swards of perennial rye-grass. *Ann. Appl. Biol.*, **57**, 155–62.

Catherall, P. L. (1971) Virus diseases of grasses and cereals, in *Diseases of Crop Plants* (ed. J. H. Western), Wiley, London, pp. 308–22.

Catherall, P. L. (1981) Virus diseases of grasses. *MAFF Advisory Leaflet 595*.

Catherall, P. L. and Chamberlain, J. A. (1977) Relationships, host-ranges and symptoms of some isolates of phleum mottle virus. *Ann. Appl. Biol.*, **87**, 147–57.

Catherall, P. L. and Griffiths, E. (1966) Influence of cocksfoot streak virus on the growth of single cocksfoot plants. *Ann. Appl. Biol.*, **57**, 141–8.

Channon, A. G. and Hissett, R. (1984) The incidence of bacterial wilt caused by *Xanthomonas campestris pv graminis* in pasture grasses in the West of Scotland. *Pl. Pathol.*, **33**, 113–21.

Christensen, M. J. (1979) *Rhizoctonia* species associated with diseased turf grasses in New Zealand. *N.Z. J. Agric. Res.*, **22**, 627–9.

Clements, R. O. (1980) Grassland pests – an unseen enemy. *Outlook Agric.*, **10**, 219–23.

Clements, R. O. and Henderson, I. F. (1979) Insects as a cause of botanical change in swards, in *Changes in Sward Composition and Productivity* (eds A. H. Charles and R. J. Haggar), Occasional Symposium No. 10, British Grassland Society, pp. 157–60.

Clements, R. O., Bentley, B. R. and Henderson, I. F. (1980) Pest problems in grass and legume crops. *Rep. Grassland Res. Inst. for 1979*, p. 60.

Clements, R. O., Henderson, I. F. and Bentley, B. R. (1982a) The effects of pesticide application on upland permanent pastures. *Grass, Forage Sci.*, **37**, 123–8.

Clements, R. O., French, N., Guile, C. T., Golightly, W. H., Serfiah Lewis and Savage, M. J. (1982b) The effects of pesticides on establishment of grass swards in England and Wales. *Ann. Appl. Biol.*, **101**, 305–13.

Coghill, K. J. and Gair, R. (1954) The estimation in the field of the damage caused by timothy flies (*Amaurosoma* spp.). *J. Br. Grassland Soc.*, **9**, 329–34.

Cook, F. G. (1975) Production loss estimation in *Drechslera* infection of ryegrass. *Ann. Appl. Biol.*, **81**, 251–6.

Cook, R. and York, P. A. (1980) Nematodes and herbage improvement. *Rep. Welsh Pl. Breeding St. for 1979*, pp. 177–207.

Coulson, J. C. and Butterfield, J. E. L. (1978) The animal communities of upland Britain, in *The Future of Upland Britain* (ed. R. B. Tranter), Centre for Agricultural Strategy, Reading, pp. 417–35.

Cruickshank, I. A. M. (1957) Crown rust of ryegrass. *N.Z. J. Sci. Technol.*, **A38**, 539–43.

Cummins, G. B. (1971) *The Rust Fungi of Cereals, Grasses and Bamboos,* Springer Verlag, Berlin.

Cunningham, I. J. (1958) Non-toxicity to animals of ryegrass endophyte and other endophytic fungi of New Zealand grasses. *N.Z. J. Agric. Res.,* **1,** 489–97.

Davies, H., Williams, A. E. and Morgan, W. A. (1970) The effect of mildew and leaf blotch on yield and quality of cv. Lior Italian ryegrass. *Pl. Pathol.,* **19,** 135–8.

Dean, G. J. W. (1973) Bionomics of aphids reared on cereals and some Gramineae. *Ann. Appl. Biol.,* **73,** 127–35.

Dumbleton, L. J. (1942) The grass grub (*Odontria zealandica* White): A review of the problem in New Zealand. *N.Z. J. Sci. Technol.,* **A23,** 305–21.

East, R., King, P. D. and Watson, R. N. (1981) Population studies of grass grub (*Costelytra zealandica*) and black beetle (*Heteronychus arator*) (Coleoptera: Scarabaeidae). *N.Z. J. Ecol.,* **4,** 56–64.

Egli, T. and Schmidt, D. (1982) Pathogenic variation among the causal agents of bacterial wilt of forage grasses. *Phytopath. Z.,* **104,** 138–50.

Elliott, E. S. (1962) Disease damage in forage grasses. *Phytopathology,* **52,** 448–51.

Eriksson, K. B. (1972) Nematode diseases of pasture legumes and turf grasses, in *Economic Entomology* (ed. J. M. Webster), Academic Press, London, pp. 66–96.

Esau, K. (1957) Phloem degeneration in Gramineae affected by the barley yellow-dwarf virus. *Am. J. Bot.,* **44,** 245–51.

Eshed, N. and Dinoor, A. (1981) Genetics of pathogenicity in *Puccinia coronata*: the host range among grasses. *Phytopathology,* **71,** 156–63.

Falloon, R. E. (1980) Seedling emergence responses in ryegrasses (*Lolium* spp.) to fungicide seed treatments. *N.Z. J. Agric. Res.,* **23,** 385–91.

Fischer, G. W. (1938) Some new grass smut records from the Pacific North West. *Mycologia,* **30,** 385–95.

Fischer, G. W. (1953a) Some of the 125 rusts of grasses, in *Plant Disease,* Yearbook of Agriculture, United States Department of Agriculture, Washington, DC, pp. 276–80.

Fischer, G. W. (1953b) Smuts that parasitize grasses, in *Plant Disease,* Yearbook of Agriculture, United States Department of Agriculture, Washington, DC, pp. 280–4.

Frauenstein, K. (1977) Zur Schadwirkung des Mehltaubefalls an Wiesenrispe. *Arch. Phytopath. Pflanzenschutz,* **13,** 385–90.

French, N. (1969) Assessment of leather jacket damage to grassland and economic aspects of control. *Proc. 5th Br. Insecticide, Fungicide Conf.,* **2,** 511–21.

Gair, R. (1953) Observations on grass aphids in Derbyshire, 1950–52. *Pl. Pathol.,* **2,** 117–21.

Gaskin, T. A. and Britton, M. P. (1962) The effect of powdery mildew on the growth of Kentucky bluegrass. *Pl. Dis. Reporter,* **46,** 724–5.

Gibbs, A. F. and Wilcoxson, R. D. (1972) Effect of sugar content of *Poa pratensis* on *Helminthosporium* leaf spot. *Physiol. Pl. Pathol.,* **2,** 279–87.

Gibeault, V. A., Autio, R., Spaulding, S. and Younger, V. B. (1980) Mixing turfgrasses controls *Fusarium* blight. *Calif. Agric.,* **34,** 11–2.

Gibson, R. W. (1976) Diseases of grass and forage crops. *Rep. Rothamsted Exp. Stn. for 1975,* Part I, p. 258.

Goss, R. L. and Gould, C. J. (1971) Inter-relationships between fertility levels and *Corticium* red thread disease of turf grasses. *J. Sports Turf Res. Inst.,* **47,** 4–53.

References

Gould, C. J., Miller, V. L. and Goss, R. L. (1967) Fungicidal control of red thread disease of turf grasses in western Washington. *Pl. Dis. Reporter,* **51,** 215–9.

Gray, E. G. and Copeman, G. J. F. (1975) The role of snow moulds in winter damage to grassland in northern Scotland. *Ann. Appl. Biol.,* **81,** 245–51.

Gross, D. F., Mankin, C. J. and Ross, J. G. (1975) Effect of diseases on *in vitro* digestibility of smooth bromegrass. *Crop Sci.,* **15,** 273–5.

Hampton, J. G. and Scott, D. J. (1980) Blind seed disease of ryegrass in New Zealand. II Nitrogen fertiliser: effect on incidence and possible mode of action. *N.Z. J. Agric. Res.,* **23,** 149–53.

Hardison, J. R. (1963) Commercial control of *Puccinia striiformis* and other rusts in seed crops of *Poa pratensis* by nickel fungicides. *Phytopathology,* **53,** 209–16.

Hardwick, N. V. and Heard, A. J. (1978) The effect of *Marasmius oreades* in pasture. *Pl. Pathol.,* **27,** 53–7.

Heard, A. J. (1972) The grasses: pests and diseases, in *Grasses and Legumes in British Agriculture* (eds C. R. W. Spedding and E. C. Diekmahns), Bulletin 49, Commonwealth Bureau of Pastures and Field Crops, Commonwealth Agricultural Bureaux, Farnham Royal, England, pp. 85–97.

Heard, A. J. and Roberts, E. T. (1975) Disorders of temporary ryegrass swards in south-east England. *Ann. Appl. Biol.,* **81,** 240–3.

Heard, A. J., A'Brook, J., Roberts, E. T. and Cook, R. J. (1974) The incidence of ryegrass mosaic virus in crops of ryegrass grown for seed in some southern counties of England. *Pl. Pathol.,* **23,** 119–27.

Heath, G. W. (1960) Ley deterioration and soil insects. *J. Br. Grassland Soc.,* **15,** 209–11.

Henderson, I. F. and Clements, R. O. (1974) The effect of pesticides on the yield and botanical composition of a newly-sown ryegrass ley and an old mixed pasture. *J. Br. Grassland Soc.,* **29,** 185–90.

Henderson, I. F. and Clements, R. O. (1977) Grass growth in different parts of England in relation to invertebrate numbers and pesticide treatment. *J. Br. Grassland Soc.,* **32,** 89–98.

Henderson, I. F. and Clements, R. O. (1979) Differential susceptibility to pest damage in agricultural grasses. *J. Agric. Sci., Camb.,* **73,** 465–72.

Henderson, I. F. and Clements, R. O. (1980) The effect of insecticide treatment on the establishment and growth of Italian ryegrass under different sowing conditions. *Grass, Forage Sci.,* **35,** 235–41.

Hewitt, G. B. (1978) Reduction of western wheatgrass by the feeding of two rangeland grasshoppers *Aulocara elliotti* and *Melanoplus infantilis*. *J. Econ. Entomol.,* **71,** 419–21.

Hill, A. R. (1971) The reproductive behaviour of *Metopolophium festucae* (Theobald) at different temperatures and on different host plants. *Ann. Appl. Biol.,* **67,** 289–95.

Hims, M. J., Dickinson, C. H. and Fletcher, J. T. (1984) Control of red thread, a disease of grasses caused by *Laetisaria fuciformis*. *Pl. Pathol.,* **33,** 513–6.

Holmes, S. J. I. (1976) A comparative study of the infection of perennial ryegrass by *Fusarium nivale* and *F. culmorum* in sterilized soil. *Ann. Appl. Biol.,* **84,** 13–19.

Holmes, S. J. I. (1979a) Effects of *Fusarium nivale* and *F. culmorum* on the establishment of four species of pasture grass. *Ann. Appl. Biol.,* **91,** 243–50.

Holmes, S. J. I. (1979b) Effect of ryegrass mosaic virus on the quality of perennial ryegrass. *Ann. Appl. Biol.*, **91**, 75–9.

Holmes, S. J. I. (1980) Field studies on the effects of ryegrass mosaic virus on the yield of Italian ryegrass cv. S.22. *Ann. Appl. Biol.*, **96**, 209–17.

Isawa, K., Abe, A. and Nishihara, N. (1974) Influence of diseases on the chemical composition and nutritive value of forage crops. I. Chemical composition of Italian ryegrass infected with crown rust. *Ann. Phytopath. Soc. Jpn.*, **40**, 86–92.

Janson, H. W. (1959) Aphids on cereals and grasses in 1957. *Pl. Pathol.*, **8**, 38.

Jensen, S. G. (1968) Photosynthesis, respiration and other physiological relationships in barley infected with barley yellow dwarf virus. *Phytopathology*, **58**, 204–8.

Jensen, S. G. (1969) Occurrence of virus particles in the phloem tissue of BYDV-infected barley. *Virology*, **38**, 83–91.

Jones, D. P. (1945) Gall midges and grass seed production. *Agriculture, Lond.*, **52**, 248–51.

Jones, M. B., Heard, A. J., Woledge, J., Leafe, E. L. and Plumb, R. T. (1977) The effect of ryegrass mosaic virus on carbon assimilation and growth of ryegrasses. *Ann. Appl. Biol.*, **87**, 393–405.

Labruyère, R. E. (1977) Contamination of ryegrass seed with *Drechslera* species and its effect on disease incidence in the ensuing crop. *Neth. J. Pl. Pathol.*, **83**, 205–15.

Lacey, J. (1975) Airborne spores in pastures. *Trans. Br. Mycol. Soc.*, **64**, 265–81.

Lam, A. (1982) Presence of *Drechslera* species in certified ryegrass seed-lots. *Grass, Forage Sci.*, **37**, 47–52.

Lam, A. (1985) Effect of fungal pathogens on digestibility and chemical composition of Italian ryegrass (*Lolium multiflorum*) and tall fescue (*Festuca arundinacea*). *Pl. Pathol.*, **34**, 190–9.

Lancashire, J. A. and Latch, G. C. M. (1966) Some effects of crown rust (*Puccinia coronata* Corda) on the growth of two ryegrass varieties in New Zealand. *N.Z. J. Agric. Res.*, **9**, 628–40.

Lancashire, J. A. and Latch, G. C. M. (1969) Some effects of stem rust (*Puccinia graminis* Pers.) on the growth of cocksfoot (*Dactylis glomerata* L. "Grasslands Apanui"). *N.Z. J. Agric. Res.*, **12**, 697–702.

Large, E. C. (1954) Surveys for choke (*Epichloë typhina*) in cocksfoot seed crops, 1951–53. *Pl. Pathol.*, **3**, 6–11.

Latch, G. C. M. (1966a) Fungous diseases of ryegrasses in New Zealand I. Foliage diseases. *N.Z. J. Agric. Res.*, **9**, 394–409.

Latch, G. C. M. (1966b) Fungous diseases of ryegrasses in New Zealand. II Foliage, root and seed diseases. *N.Z. J. Agric. Res.*, **9**, 808–19.

Latch, G. C. M. (1977) Incidence of barley yellow dwarf virus in ryegrass pastures in New Zealand. *N.Z. J. Agric. Res.*, **20**, 87–9.

Latch, G. C. M. (1980) Effects of barley yellow dwarf virus on simulated swards of Nui perennial ryegrass. *N.Z. J. Agric. Res.*, **23**, 373–8.

Latch, G. C. M. and Potter, L. R. (1977) Interaction between crown rust (*Puccinia coronata*) and two viruses of ryegrass. *Ann. Appl. Biol.*, **87**, 139–45.

Lefebvre, C. L., Howard, F. L. and Grau, F. V. (1953) How to keep turf grass healthy, in *Plant Disease,* Yearbook of Agriculture, United States Department of Agriculture, Washington, DC, pp. 285–91.

Lewis, S. and Webley, D. (1966) Observations on two nematodes infesting grasses. *Pl. Pathol.*, **15**, 184–6.

References

Madsen, J. P. and Hodges, C. F. (1980a) Pathogenicity of some select soilborne dematiaceous Hyphomycetes on germinating seed of *Festuca rubra*. *Phytopathology*, **70**, 21–5.

Madsen, J. P. and Hodges, C. F. (1980b) Nitrogen effects on the pathogenicity of *Drechslera sorokiniana* and *Curvularia geniculata* on germinating seed of *Festuca rubra*. *Phytopathology*, **70**, 1033–6.

Makela, K. (1981) Winter damage and low temperature fungi on leys in North Finland in 1976–1979. *Ann. Agric. Fenn.*, **20**, 102–31.

Michail, S. H. and Carr, A. J. H. (1966) Italian rye-grass, a new host for *Ligniera junci*. *Trans. Br. Mycol. Soc.*, **49**, 411–8.

Moline, H. E. and Jensen, S. G. (1975) Histochemical evidence for glycogen-like deposits in barley yellow dwarf virus-infected barley leaf chloroplasts. *J. Ultrastruct. Res.*, **53**, 217–21.

Moore, D. and Clements, R. O. (1984) Stem-boring Diptera in perennial ryegrass in relation to fertiliser I. Nitrogen level and form. *Ann. Appl. Biol.*, **105**, 1–6.

Mowat, D. J. (1974) Factors affecting the abundance of shoot-flies (Diptera) in grassland. *J. Appl. Ecol.*, **11**, 951–62.

Newbold, J. W. (1981) The control of leather jackets in grassland by winter pesticide applications. *Proc. Crop Protection N. Britain*, pp. 207–11.

Noble, M. and Richardson, M. J. (1968) An annotated list of seed-borne diseases. *Phytopath. Papers No. 8*, Commonwealth Mycological Institute, Kew.

Orlob, G. B. and Arny, D. C. (1961) Some metabolic changes accompanying infection by barley yellow dwarf virus. *Phytopathology*, **51**, 768–75.

O'Rourke, C. J. (1975) Common and newly-recorded forage crop diseases in Ireland. *Ann. Appl. Biol.*, **81**, 243–7.

O'Rourke, C. J. (1976) *Diseases of grasses and Forage Legumes in Ireland*, An Foras Taluntais, Dublin.

Paliwal, Y. C. (1970) Electron microscopy of bromegrass mosaic virus in infected leaves. *J. Ultrastruct. Res.*, **30**, 491–502.

Pass, B. C., Buckner, R. C. and Burrus, P. R. (1965) Differential reaction of Kentucky bluegrass strains to sod webworms. *Agron. J.*, **57**, 510–1.

Paul, H. L., Querfurth, G. and Huth, W. (1980) Serological studies on the relationships of some isometric viruses of Gramineae. *J. Gen. Virol.*, **47**, 67–77.

Plumb, R. T. (1983) Barley yellow dwarf virus – a global problem, in *Plant Virus Epidemiology, the Spread and Control of Insect-borne Viruses* (eds R. T. Plumb and J. M. Thresh), Blackwell Scientific, Oxford, pp. 185–98.

Plumb, R. T. and James, M. (1973) Virus aggregates and pinwheels in plants infected with mite-transmitted ryegrass mosaic virus. *J. Gen. Virol.*, **18**, 409–11.

Price, P. C., Fisher, J. M. and Kerr, A. (1979) Annual ryegrass toxicity: parasitism of *Lolium rigidum* by a seed-gall forming nematode (*Anguina* sp.). *Ann. Appl. Biol.*, **91**, 359–69.

Radcliffe, J. E. (1970) Some effects of grass grub (*Costelytra zealandica* (White)) larvae on pasture plants. *N.Z. J. Agric. Res.*, **13**, 87–104.

Raw, F. (1952) The ecology of the garden chafer *Phyllopertha horticola* (L.) with preliminary observations on control measures. *Bull. Entomol. Res.*, **42**, 605–46.

Reinert, J. A. (1976) Control of sod webworms (*Herpetogramma* spp. and *Crambus* spp.) on Bermuda grass. *J. Econ. Entomol.*, **69**, 669–72.

Roberts, D. A., Sherwood, R. T., Fezer, K. D. and Ramamurth, C. S. (1955) Diseases of forage crops in New York, 1954. *Pl. Dis. Reporter,* **39,** 316–7.
Robinson, P. W. and Hodges, C. F. (1977) Effect of nitrogen fertilization on free amino acid and soluble sugar content of *Poa pratensis* and on infection and disease severity by *Drechslera sorokiniana. Phytopathology,* **67,** 1239–44.
Robinson, P. W. and Hodges, C. F. (1981) Nitrogen-induced changes in the sugars and amino acids of sequentially senescing leaves of *Poa pratensis* and pathogenesis by *Drechslera sorokiniana. Phytopath. Z.,* **101,** 348–61.
Russell, S. L. and Kimmins, W. C. (1971) Growth regulators and the effect of barley yellow dwarf virus on barley (*Hordeum vulgare* L.). *Ann. Bot. (Lond.) New Ser.,* **35,** 1037–43.
Sakai, R., Nishiyama, K., Ichihara, A., Shiraishi, K. and Sakamura, S. (1979) Studies on the mechanism of physiological activity of coronatine. Effect of coronatine on cell wall extensibility and expansion of potato tuber tissue. *Ann. Phytopath. Soc. Jpn,* **45,** 645–53.
Sakuma, T. and Narita, T. (1961) Heterosporium leafspot of timothy and its causal fungus. *Heterosporium phlei* Gregory. *Bull. Hokkaido Agric. Exp. Stn,* **7,** 77–90.
Sampson, K. and Western, J. H. (1954) *Diseases of British Grasses and Herbage Legumes,* Cambridge University Press.
Schmidt, H. B. (1967) Histologische Veränderungen in Dünnschnitten Trespen-mosaik – Virus-infizierter Gerstenpflanzen. *Phytopath. Z.,* **62,** 66–78.
Smith, J. D. (1965) *Fungal Diseases of Turf Grasses,* Sports Turf Research Institute, Bingley.
Smith, J. D. (1980) Is biologic control of *Marasmius oreades* fairy rings possible? *Pl. Dis.,* **64,** 348–54.
Spaull, A. M. and Clements, R. O. (1982) The effect of root ectoparasitic nematodes upon grass establishment. *Grass, Forage Sci.,* **37,** 183.
Spaull, A. M., Newton, P. G. and Clements, R. O. (1985) Establishment and yield of three ryegrasses following aldicarb use, and changes in abundance of plant parasitic nematodes. *Ann. Appl. Biol.,* **106,** 313–21.
Stalpers, J. A. and Loerakker, W. M. (1982) *Laetisaria* and *Limonomyces* species (Corticiaceae) causing pink diseases in turf grasses. *Can. J. Bot.,* **60,** 529–37.
Starks, K. J. and Thurston, R. (1962) Control of plant bugs and other insects on Kentucky bluegrass grown for seed. *J. Econ. Entomol.,* **55,** 993–7.
Stegmark, R. (1979) Occurrence of leafspot disease (*Drechslera* sp. = *Helminthosporium* sp.) in different cultivars of perennial ryegrass (*Lolium perenne*) and meadow fescue (*Festuca pratensis*). *Växtskyddsrapporter,* **6,** 21–62.
Stynes, B. A. and Bird, A. F. (1980) *Anguina agrostis,* the vector of annual ryegrass toxicity in Australia. *Nematologica,* **26,** 475–90.
Todd, D. H. (1959) Black beetle, *Heteronychus sanctae-helenae* Blanch., in pastures in New Zealand. *N.Z. J. Agric. Res.,* **2,** 1262.
Troll, J. and Rohde, R. A. (1966) The effects of nematocides on turfgrass growth. *Pl. Dis. Reporter,* **50,** 489–92.
Ullrich, J. (1977) Die mitteleuropäischen Rostpilze der Futter- und Rasengräser. *Mitt. Biol. Bundesanst. Land- und Forstwirtsch.,* **175.**
Wadley, F. M. (1931) Ecology of *Toxoptera graminum,* especially as to factors affecting importance in the northern United States of America. *Ann. Entomol. Soc. Am.,* **24,** 325–95.

References

Webley, D. P. (1960) The effect of dieldrin on a direct reseeded ley. *Pl. Pathol.*, **9**, 92–3.

Wilkins, P. W. (1973a) Infection of *Lolium* and *Festuca* spp. by *Drechslera siccans* and *D. catenaria*. *Euphytica*, **22**, 106–13.

Wilkins, P. W. (1973b) Infection of *Lolium multiflorum* with *Rhynchosporium* species. *Pl. Pathol.*, **22**, 107–11.

Wilkins, P. W. and Catherall, P. L. (1977) Variation in reaction to barley yellow dwarf virus in ryegrass and its inheritance. *Ann. Appl. Biol.*, **85**, 257–63.

Williams, R. D. (1984) *Crop Protection Handbook – Grass and Clover Swards,* British Crop Protection Council.

Wright, C. E. (1967) Blind seed disease of ryegrass. *Euphytica*, **16**, 122–30.

CHAPTER 9
The grass crop in perspective: selection, plant performance and animal production

Alec Lazenby

9.1 INTRODUCTION

The grasses commonly used in improved grasslands owe their importance in agriculture to their ability to withstand repeated defoliation, their potential to supply large quantities of nutritious forage for ruminant feed and their adaptability to a wide range of growing conditions.

The early chapters of this book contain descriptions of the grass plant, with its vegetative and reproductive tillers, its growth and regrowth following defoliation; considerable evidence is presented on the physiological processes involved, and the effect on these of a range of factors, some out of human control and others manipulable. Much of the information relates to perennial ryegrass (*Lolium perenne*), the most important grass in the improved grasslands of the UK, parts of Western Europe and New Zealand.

A major objective of these chapters was to provide a *better understanding of growth,* through the study of plant physiology. Those involved directly in studying performance in the field are usually more concerned to determine, experimentally or by prediction, the *effects of such growth,* e.g. the level of dry matter (DM) production of grasses grown in small plots, or the total nutrients which grassland can contribute to an animal production enterprise. They also seek information on the influence of the factors most significant in determining performance whether of, say, a new grass cultivar or a forage-based system of animal production. The conditions prevailing in a region, or the level of management of an individual farmer, may be all-important in determining the level of such performance.

The grass crop is grown almost entirely to provide feed for the ruminant – for the maintenance of the animal's body condition, its growth and the production of wool, meat and milk. Put simply, making good use of grassland for animal production can be said to be dependent on four things:

(a) having plants which can perform well in the prevailing conditions;

(b) providing, within the limits of the natural environment, good conditions for the plants' growth;
(c) ensuring efficient utilization by the animal of the herbage produced;
(d) having a system of animal production appropriate to the environment.

This chapter therefore contains an overview of the significance of plant selection, growing conditions and management on the performance of improved grasslands. An attempt is made to highlight factors important in determining such performance and to illustrate their significance in different systems of animal production based on improved grasslands. An indication is also given of how factors limiting output from such grasslands may change over time. The examples used are drawn heavily from the author's experience in two contrasting environments – the UK and temperate Australia.

9.2 SELECTION

9.2.1 General considerations

Grassland plants provide the foundation for sustained and high output. However, if swards are not based on species or cultivars which are suited to the growing conditions, the farmer has no chance of reaching and maintaining the level of production from his grassland which is possible in his region. The plants selected for use in improved grasslands thus need to be adapted to the natural environment, being both able to survive, and perform well in, the prevailing conditions. Performance includes not only the potential to produce large quantities of good quality herbage for the ruminant but also other plant characteristics which may be of importance in sward establishment, management and animal output.

The adaptability of a grass plant to any region is determined primarily by the prevailing climate and inherent characteristics of the soil – factors over which there is little or no control. The two most important factors affecting plant survival and growth are temperature and rainfall. The temperature regime has a major influence on adaptation, largely determining whether temperate (C_3) or tropical (C_4) species are best suited to a region, and the potential pattern of their annual growth. However, in many parts of the world, it is rainfall which has the biggest effect on plants; for example, low rainfall sets an absolute limit to the distribution and performance of grassland plants throughout much of Australia, determining not only whether plants can survive, but having the major influence on total and seasonal DM production and its reliability (Fitzpatrick and Nix, 1970).

Unpredictable rainfall and a build-up of soil moisture deficits during the main growing season are characteristic features of most grassland situations in Australia. Seasonal drought is of special significance, as many perennials are unable to survive prolonged periods of moisture stress, a fact which explains the widespread use of annuals in sown grasslands in Australia. There are few

Selection 313

indigenous species suitable for grassland improvement, which is therefore based almost entirely on introduced species. In contrast, the grasses used for such improvement in the UK have been derived, in the main, from populations long adapted to the conditions prevailing; these include a more hospitable natural environment for plant growth than that in Australia, and a long established agriculture (see Chapter 1).

9.2.2 Grass selection in the UK

Stapledon, the first Director of the Welsh Plant Breeding Station (WPBS), established at Aberystwyth in 1919, and Jenkin, in charge of grass breeding there, dominated early grassland research in Britain. Their influence still remains, not only in the UK, but in many other parts of the world. Stapledon determined that the first main objective of the WPBS would be to improve grassland in Wales, and that this required the use of long leys (Stapledon, 1933). The grasses then available commercially in the UK were too short-lived to guarantee their performance over the whole life of the ley, hence the need to select more persistent cultivars. It was Jenkin who largely determined both the criteria adopted for selecting improved cultivars of grasses and the methods and techniques used (Jenkin, 1937).

In addition to persistence, a high potential for DM production, with growth spread as evenly as possible over a long growing season, and rapid recovery after defoliation, were other important criteria used in selecting improved cultivars; nutritious and palatable herbage, general resistance to disease and uniformity were also sought (Jenkin, 1937). The investigations of Fagan and his colleagues (e.g. Fagan, 1927) resulted in the breeder equating nutritive value with percentage protein, and thus selecting plants with a high proportion of leaf (Stapledon, 1933). Trials with sheep indicated a general association between leaf and both palatability and nutritive value (Stapledon and Jones, 1927; Jones and Jones, 1930).

Material was collected as tillers or seeds from various grasslands in Britain, particularly the famous fattening pastures of the English midlands. Performance was assessed, and initial selections made, on widely spaced plants, sometimes supplemented by information collected from tiller rows or simulated swards. Selected plants were then emasculated and crossed in a diallel manner, normally using hand-pollination, to determine their compatibility and the uniformity of their progeny. Plants which performed well were then multiplied clonally and allowed to inter-pollinate in glasshouses. The seed so produced provided material for comparative small-plot sward tests; seed multiplication followed for those cultivars judged to be superior (summarized from Lazenby and Rogers, 1963).

Since World War II, the grass breeder has concentrated on more specific objectives. For example, much effort has been put into attempts to select for improved quality (e.g. Hughes, 1971); plants have been sought with increased winter production (e.g. Rogers, 1961) and winter hardiness (e.g. Eagles *et al.*, 1984), better resistance to drought and diseases (e.g. Wright, 1957) and for less fertile areas (Beddows, 1958). The quality advantages of some polyploids has resulted in the release of several such cultivars, e.g. Sabrina, a tetraploid Italian × perennial hybrid (Breeze, 1971), while the potential of some intergeneric hybrids has been explored (e.g. Lewis, 1971).

Plant collections made both in Britain (e.g. Rogers, 1961) and overseas (e.g. Hughes *et al.*, 1962) have increased the gene pool available to the breeder. Direct selections of improved cultivars have been made from some introduced grasses, e.g. populations of tall fescue from North Africa; other introductions have been included in breeding programmes. Laboratory techniques have been developed to improve the precision of selection; for example, for quality (Tilley and Terry, 1963), winter hardiness and winter growth (e.g. Eagles *et al.*, 1984). More basic work has also proved valuable; for example, results on studies of the effects of light, temperature and available moisture on plant growth and development have provided a better basis for predicting where plants with desirable characteristics might be found. They have also been instrumental in providing a more optimistic prognosis of the possibility of combining winter hardiness with some winter growth.

Some challenges remain. For example, it seems both desirable and possible to select for increased resistance to specific pests such as frit-fly (Clements, 1980), and diseases such as ryegrass mosaic virus (A'Brook and Heard, 1975); it would also be worthwhile to breed for better winter growth combined with frost hardiness, and for improved drought resistance. In the longer term it might be possible to increase yields by selecting for reduced respiration loss (Robson, 1982; Wilson and Jones, 1982), and to increase the efficiency of nitrogen use (Lazenby and Rogers, 1965). However, prospects of any early improvement in quality – certainly in ryegrass – seem less bright, in spite of the recorded 11% difference in live weight gain over 4 years between animals grazing two cultivars (Conolly *et al.*, 1977).

The early grass breeders can claim many real achievements. Their stated objectives to select cultivars which were leafy, persistent and uniform, and which retained their integrity from one generation to the next, have been attained. Sward management has been made easier and the farmer can now purchase cultivars and have confidence in their performance; for instance, a number of ryegrass cultivars which are available have the potential both to persist indefinitely and to produce high yields of good quality herbage under grazing, whilst the range of cultivars with different times of inflorescence emergence enables a spread of the workload for conservation without loss of yield or quality.

A number of other effects have flowed from the efforts of the breeders. For example, grass has now become recognized as a crop, and its potential for animal production has appreciated increasingly. Seed mixtures have become perceptibly simpler, with the ryegrasses – which are adaptable, easy to establish, respond well to nitrogen and produce good quality forage –

becoming increasingly popular; by 1978 ryegrass comprised more than 90% of the total grass seed sown for agricultural purposes in the UK (Hopkins, 1979). Further, once a range of good grasses became commercially available there was little reason for the choice of cultivar to limit grassland output; making best use of grassland for animal production therefore became more dependent on providing good conditions for grass growth and properly utilizing the herbage for ruminant feed.

9.2.3 Improved grasses for temperate Australia

Some of the many differences between conditions in Australia and those prevailing in the UK bear on the selection of improved cultivars. Australia's harsh climate and inherently low soil fertility not only restrict the area over which grassland improvement can occur, but also ensure that extensive systems of animal production for wool and beef are the norm; this contrasts with the potential for intensive grassland output for fat lamb, beef and milk production possible in many parts of the UK.

Grassland improvement, more recent than in Britain, has involved the replacement of the indigenous plants with introduced species, with their greater potential for animal production. Early introductions of such plants, sometimes made by chance and sometimes because of their value in UK agriculture, were effected without any appreciation of their performance in the Australian environment. There was often insufficient testing of species before their use in grassland improvement. For example, for some two decades perennial ryegrass was the main grass used in the pasture improvement programme on the Northern Tablelands of New South Wales; it was only its inability to survive moisture stress in severe drought which prompted testing of a wider range of introduced grasses and the consequent demonstration that both phalaris (*Phalaris aquatica*) and tall fescue (*Festuca arundinacea*) were better adapted than perennial ryegrass to much of the region.

Whilst some selections of improved cultivars were made from introduced populations at varying stages of their adaptation to the environment, organized collections of plants, thought to have potential, soon became an integral part of grassland improvement. Hartley's (1963) studies of the relationship between the centres of diversity of valuable grasses and their prevailing climatic factors did much to indicate where species with desirable characteristics might be found, and thus provide a more logical basis for plant collecting expeditions. The organization and methods used for plant introductions, outlined by Hartley and Neal-Smith (1963), included tests designed to measure field performance in different ecological and agronomic situations, whilst a list of criteria designed to enable improved objectivity in assessing potentially improved cultivars before their release was suggested by the NSW Herbage Plant Liaison Committee (1967). Some of these criteria e.g. yield and quality characters, reaction to grazing and an indication of their potential for animal production, are common to those used by the breeder in the UK; others, such as seed production, compatability with other species and toxin levels, are more significant requirements in Australia. In contrast with the UK, the performance of grasses in improved pastures in Australia is dependent on the presence of suitable legumes to fix the N required for plant growth. It is not surprising, therefore, that much of the effort in selecting better pasture plants has been put into legumes rather than grasses. However, some improved cultivars of temperate grasses have been bred and released. One such grass, phalaris, provides a good illustration of the challenges faced by breeders, and their achievements in producing improved cultivars.

Commercial phalaris, derived from a Mediterranean population, originally introduced into Australia in 1884, is a winter growing and summer dormant grass, able to withstand climatic extremes from drought to flood; it can also survive fire and overgrazing, and is resistant to many common pests and diseases. The plant is widely adapted to the conditions prevailing in south-east Australia and has been used extensively for grassland improvement, not only contributing considerably to increased output and weed control, but also reducing erosion. Nevertheless, commercial phalaris has a number of deficiencies. Two of these – lack of tolerance to summer rain and to acid soils – reduce its range of adaptability, whilst its toxicity to animals, seed shedding at maturity, low seedling vigour (which makes it difficult to establish) and relatively poor winter growth all limit the plant's productivity.

The problem of seed shedding was lessened following the selection, by McWilliam (1963), of Seedmaster from an Argentinian cultivar. Seedmaster is similar in most characteristics to, though with considerably greater seed yields than, common phalaris which it gradually replaced. Whilst it has not been possible to transfer the Seedmaster mechanism of seed retention into other cultivars, a seed-retaining mutant was discovered in commercial phalaris. Hybridization of this mutant with normal plants of commercial resulted in a cultivar with seed yields both at least double those of commercial phalaris and of better quality. This cultivar, Uneta, is likely to replace both commercial and Seedmaster (Oram, 1984).

Sirocco, with large seedlings and better winter growth than commercial phalaris, was selected from a Moroccan population. It is the most summer dormant and earliest flowering cultivar of phalaris, being particularly adapted to a short growing season and a dry summer. However, two superior cultivars have subsequently been released following extensive breeding programmes. One, Sirolan, selected following crosses involving Sirocco and a number of other accessions from Morocco, not only has seedling vigour and good winter growth, but is the most adaptable of all phalaris cultivars presently available. Sirolan, which is now replacing Sirocco, is drought hardy, and thus suited to low rainfall areas, and also performs well in the main phalaris-growing belt. In addition, it is the only cultivar virtually free of tryptamine alkaloids (Oram, 1984) that are so potentially poisonous to sheep. The other newly released cultivar, Sirosa, which was developed following hybridization between commercial phalaris and Moroccan plants, combines seedling vigour and winter growth; it is relatively late flowering and is suited to conditions found in the main phalaris-growing regions.

Considerable progress has thus been made in improving the adaptability of phalaris; in addition, selection of cultivars with a number of other valuable economic characters seems likely. Incorporation of the seed-retaining mechanism from commercial into an improved Sirolan appears both possible and desirable, thus enabling the selection of one or more cultivars containing good seed retention and better seed quality with seedling vigour, winter growth, summer dormancy and low alkaloid content (Oram *et al.*, 1985). Further, genetic studies suggest the possibility of developing cultivars with increased tolerance to soil acidity (Culvenor and Oram, 1985).

The major objective in selecting pasture plants for temperate Australia remains that of seeking better legumes. However, in contrast with the UK, the grasses presently available do not always have the potential to enable the achievement of the level of grassland output possible. Thus, greater opportunities exist for the grass breeder to make substantial and rapid improvement in grass cultivars than appears likely in Britain. It is clear that breeding for increased adaptability is an important requirement, with drought resistance an integral component in this objective. It is also possible that winter growth could be increased by making more use of grass populations from the Mediterranean and North Africa, e.g. tall fescues on the Northern Tablelands of NSW. Further, improvement in herbage quality is another important objective (Hutchinson, 1987), even though the potential for enhancing nutritive value is unlikely to be as spectacular as the improvement recorded in Bermuda grass (Burton *et al.*, 1967; Utley *et al.*, 1974).

9.3 GROWTH AND DM YIELDS

9.3.1 Effects of temperature and rainfall

The classic controlled environment studies of Mitchell (1956) showed a broad division of grasses into two groups – temperate (C_3) and tropical (C_4) species – according to their growth response to temperature (Fig. 9.1). Mitchell's investigations suggested an optimum growth temperature of some 30 °C for the typical C_4 species *Paspalum dilatatum* (paspalum), about 10 °C higher than for the temperate species, perennial ryegrass; whilst paspalum had a considerably higher maximum growth rate, the growth of ryegrass was nearer its potential for a much greater temperature range.

Colman *et al.* (1974) showed the growth rate in the field of the temperate grasses perennial ryegrass and phalaris, to increase by almost 13 kg ha^{-1} day^{-1} for each degree rise in temperature between 8 and 15 °C, and that for paspalum

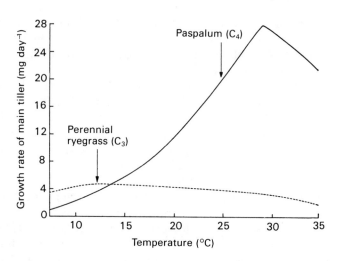

Figure 9.1 Effect of temperature on tiller growth of C_3 perennial ryegrass and C_4 paspalum. (Adapted from Mitchell, 1956.)

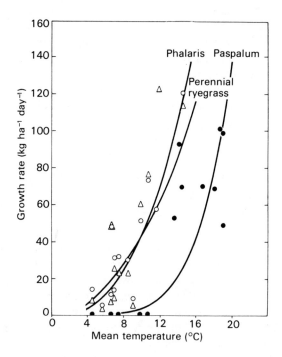

Figure 9.2 Relations between mean temperature and growth rate of (○) perennial ryegrass and (△) phalaris in winter–spring and (●) paspalum in spring–summer on irrigated high N plots. (From Colman et al., 1974.)

Growth and DM yields

by some $21\,\text{kg}\,\text{ha}^{-1}\,\text{day}^{-1}\,°\text{C}^{-1}$, within the range 14–20 °C. Little growth occurred below 4 and 12 °C for the temperate species and paspalum, respectively (Fig. 9.2).

Dry matter yields of more than $40\,\text{t}\,\text{ha}^{-1}\,\text{year}^{-1}$ have been recorded commonly for C_4 tropical grasses, with $85\,\text{t}\,\text{ha}^{-1}\,\text{year}^{-1}$ reported for Napier grass (*Pennisetum purpureum*) (Vicente-Chandler *et al.*, 1959); such data compare with yields of C_3 temperate species which rarely exceed about $20\,\text{t}\,\text{DM}\,\text{ha}^{-1}\,\text{year}^{-1}$ (Wright, 1978). Further, tropical grasses are both better able to withstand moisture stress and are more efficient in converting water into dry matter production than temperate grasses (e.g. Downes, 1969; see Chapter 1).

The interrelationship of prevailing temperature and rainfall on the growth of tropical legumes at Bileola, Queensland (Fig. 9.3) was illustrated by Fitzpatrick and Nix (1970) using a modelling technique. Whilst there was little

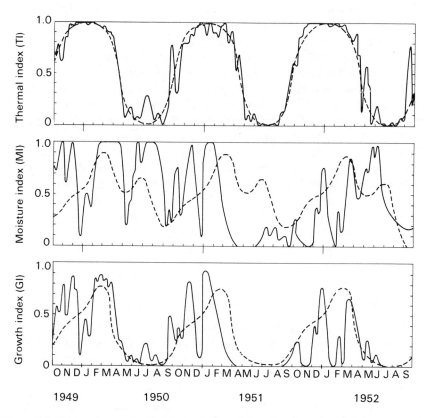

Figure 9.3 Comparisons of estimated actual weekly values (——) with long-term mean weekly values (–––) of thermal, moisture and growth indices for tropical legumes at Bileola, central Queensland. (From Fitzpatrick and Nix, 1970.)

or no growth in winter, when temperatures were too low, the level of available moisture was responsible for much of the within- and between-year variation in production. Low and unpredictable rainfall is the main cause of the unreliable grass growth so typical of Australia.

Colman *et al.* (1974) investigated the growth patterns of the temperate grasses *Lolium perenne* and *Phalaris aquatica* and the tropical species *Paspalum dilatatum* at Armidale; the centre is typical of the Northern Tablelands of New South Wales with summer temperatures suitable for good growth of C_4 grasses and winter temperatures low enough to greatly reduce growth of C_3 species (Fig. 9.4). Both temperate grasses grew most rapidly in spring, with peak rates in October coinciding with flowering. In contrast, paspalum grew actively from October to February, with daily growth rates varying from 50 to 100 kg ha^{-1}; growth ceased completely in April, the swards then remaining dormant until October. Rainfall had a significant effect on the growth patterns, a matter discussed later.

In Britain, nationwide trials undertaken by the Agricultural Development Advisory Service (ADAS) and the Grassland Research Institute (GRI) (now part of the Institute for Grassland and Animal Production) have produced a clearer picture both of the likely maximum DM yields of grass and their year-to-year and seasonal variation. Annual yields from cut non-irrigated plots of S23 ryegrass sown at 28 sites in England and Wales ranged from less than 5 to almost 18 t ha^{-1}. Whilst both light and temperature restricted growth at the beginning and end of the season (Corrall and Chapman, pers. comm.), it was available soil moisture which had the greatest influence on yield – characteristically limiting grass growth in summer, especially in the drier south and east of England. Differences in available soil moisture, affected by both rainfall and soil water holding capacity, accounted for more than 60% of the annual yield variation between sites (Morrison *et al.*, 1980). Average grass growing days in the year, determined largely by the prevailing temperature and available moisture, vary from more than 300 in the extreme south-west of England to less than 200 in East Anglia and the uplands of England and Scotland (Fig. 9.5).

9.3.2 Effect of nitrogen, defoliation and the grazing animal

In the absence of irrigation – too expensive to justify in most grassland situations – human influence on grass growth is effected largely through changes in soil fertility, in defoliation and in his management of the grazing animal. Of the nutrients required by grass, N is the most significant overall in

---

Figure 9.4 Weekly rainfall irrigation and potential evapotranspiration and 4-weekly moisture deficits, mean temperatures and growth rate of grasses on high N plots. (From Colman *et al.*, 1974.)

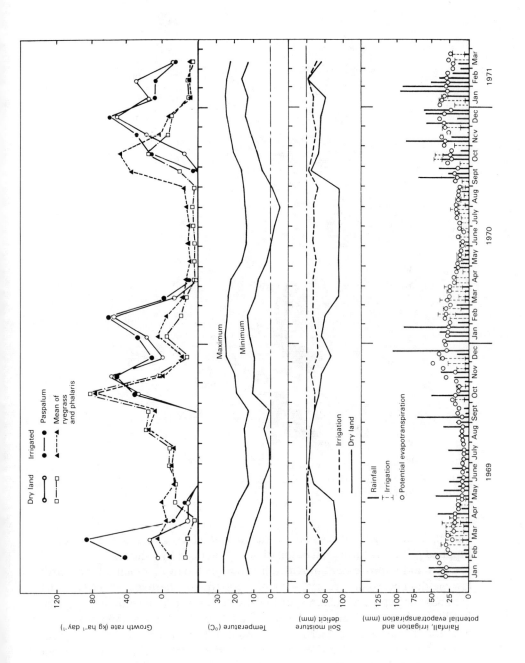

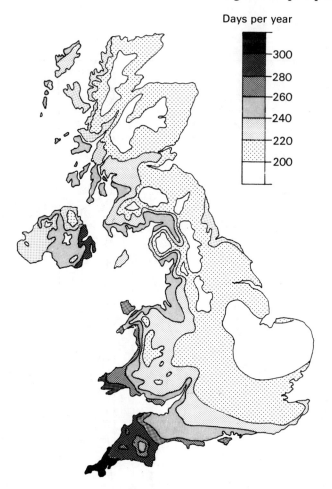

Figure 9.5 Grass growing days – soil temperature adjusted for drought and altitude. (From Down *et al.*, 1981.)

limiting yields. In countries such as Australia and New Zealand, the nutrient is supplied primarily through fixation of atmospheric N by legumes.

Grassland improvement in Australia involves the replacement of the native low-yielding perennial grasses, with their poor quality herbage, which typically dominate unimproved pastures, by improved plants, and the building up of the fertility of the soil. Very low levels of available N, P, sometimes S and, on occasion, the absence of micronutrients such as Mo, are characteristic features of most Australian soils. Three predictable stages in grassland improvement follow the application of superphosphate and the sowing of improved pasture plants; these are associated with dramatic changes both in botanical

Growth and DM yields

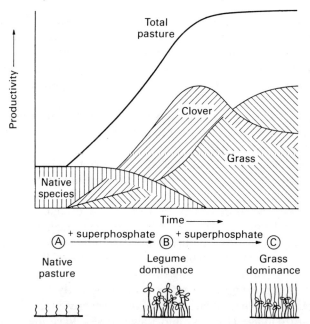

Figure 9.6 Stages in pasture improvement. (From Wolfe, 1972.)

composition and productivity (Fig. 9.6). Within 2 or 3 years the pasture becomes dominated by the sown legume, e.g. subterranean clover (*Trifolium subterraneum*) or white clover (*Trifolium repens*). Some of the N fixed by the legume and released into the soil becomes available for the growth of improved grasses sown either with the legume or subsequently. Given good growing conditions the grasses, which are effective competitors for nutrients, moisture and light, become the dominant plants in 3–5 years.

Changes in producitvity parallel those in botanical composition. The legume-dominant (Stage B) yields typically two to five times that of the unimproved pasture (Stage A), as well as producing much better quality herbage. The grass-dominant stage (C) normally yields slightly more than Stage B and, in addition, provides a more stable sward than one dominated by legumes (Wolfe, 1972).

The effect of N on grass growth is most easily investigated in monoculture swards, e.g. of perennial ryegrass, typical of Britain. Data collected from plots at some 30 sites in England and Wales indicate that, irrespective of the availability of soil N, a linear increase of 22–25 kg DM can be expected for every kilogram of N applied to temperate grasses – whether cut or grazed – at least up to an annual application of 300 kg nitrogen ha^{-1} (Richards, 1977; Morrison et al., 1980). In contrast, with tropical grasses, a linear response as high as 70 kg DM can be expected up to 400 kg nitrogen ha^{-1}year^{-1}, assuming good temperature and moisture conditions (Vicente-Chandler et al., 1959).

Herbage yields can also be influenced by frequency of defoliation. In general, the less often the herbage is cut, the higher the annual DM production. For example, trials in Britain have shown mean DM yields of S23 ryegrass to be 11.1 t ha^{-1} year^{-1} for a four-cut conservation-type regime, compared with 9.6 t ha^{-1} year^{-1} for a six- to seven-cut simulated grazing regime (Morrison, 1980).

Most of the data on grass yields have been collected from cut plots – hence legitimate questions arise as to their applicability to 'real' grassland situations involving the grazing animal. Even in the UK, where conservation plays such an improtant role in intensive grassland enterprises, more than 70% of the grass consumed by the ruminant is eaten by the grazing animal (Wade, 1979). Parsons (see Chapter 4) demonstrates clearly both how, and why, the classical growth curve of ryegrass in the UK, derived from data collected from cut plots, can be modified by grazing – particularly continuous grazing.

Grazing animals can also modify grass growth through the effects of their excreta, treading and differential grazing (see Chapter 3). The concentration of nutrients, caused through camping of sheep, is a particular problem of real significance in some improved grazings on inherently low fertility soils in Australia (Hilder, 1964). Poaching can have severe effects in wet regions, causing sward damage (Newton and Laws, 1983), lowering DM yields and reducing utilization of the herbage by the grazing animal (Garwood *et al.*, 1986; Peel *et al.*, 1986); data being collected from permanent and newly established grassland can be expected to shed further light on the effects of poaching (Wilkins, 1986).

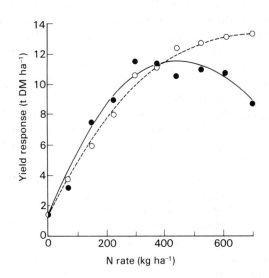

Figure 9.7 Yield response to N on cut (○) and grazed (●) swards. (From Richards, 1977.)

Growth and DM yields

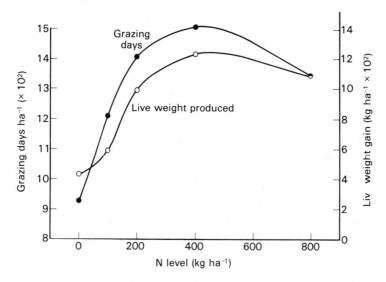

Figure 9.8 Animal production response to N on old unimproved permanent pasture. (From Tallowin *et al.*, 1986.)

There are few data comparing directly the effects of N on grass growth under cutting and grazing. In one such comparison (Richards, 1977), grass grazed rotationally by sheep was somewhat higher yielding than that of cut swards, at N levels up to 350 kg ha^{-1} year^{-1} – the N supply presumably being effectively increased by recycling, primarily through urine. However, at greater levels of N the higher stocking rate needed to utilize the additional DM produced led to sward damage through treading and urine scorch, a poorer response to N, and hence DM yields both lower in total and less than those from cut plots (Fig. 9.7).

Relatively few data exist on the effects of N on animal performance. However, in one recent investigation involving a study of the influence of N on permanent grassland grazed by young beef cattle (Tallowin *et al.*, 1986), significant increases in grazing days per hectare, overall rate of live weight gain and live weight produced per hectare were detected as the N level was increased to 400 kg ha^{-1}; performance was reduced at the highest N level (Fig. 9.8), a result associated with sward damage through scorching in hot, dry conditions.

In Britain, therefore, where most intensively used grasslands consist largely or entirely of grass, it is now possible to make reasonable predictions of DM yields following applications of N – at least up to 300 kg ha^{-1} year^{-1}. Growth in summer is somewhat uncertain – particularly in the drier areas of the east and south of the country – and variations in radiation and temperature add a further element of uncertainty to yield predictions at the beginning and end of the year.

9.4 QUALITY

Raymond and his colleagues at the GRI were responsible for laying the foundations of our present knowledge of herbage quality (Terry, 1974), through measurements of its digestibility (the ratio of digestible to total matter) and, subsequently, its D-value (the proportion of digestible organic matter in dry matter); both are essentially measures of energy content. The development of an *in vitro* method for the determination of digestibility and D-value, simulating ruminant digestion, has become integral to the assessment of herbage quality; it is now widely used by plant breeders in selection programmes, as well as in the measurement both of quality changes in cut or grazed pastures and of conserved forage.

The higher the D-value of grass, the better the animal performance. In part this is a consequence of the inherent nutritive advantages of the more digestible material, and in part because of the more rapid passage of high D-value herbage through the ruminant's alimentary tract, and thus its greater intake, A reduction from 70 to 65 in the D-value of forage, used as the sole feed for young cattle weighing 200–400 kg, reduced daily live weight gains from 800 to 500 g (Kilkenny *et al.*, 1978). Further, 75-D forage fed *ad libitum* to dairy cows provided sufficient nutrients both to maintain the body weight of the animals and produce 23 kg milk day^{-1}; in contrast, dairy cattle fed only 55-D forage not only barely maintained their body weight, but gave hardly any milk (ADAS, 1977).

A generally negative correlation exists between yield and quality (Fig. 9.9);

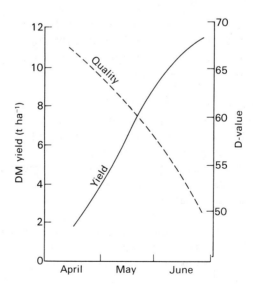

Figure 9.9 Yield and quality relationship in perennial ryegrass cv. S24. (From Green *et al.*, 1971.)

Quality

hence maximizing one inevitably means some sacrifice in the other. Very high grass yields, produced at the expense of quality, or high D-material with its inevitable yield reduction, both carry a penalty for animal production.

The quality of C_3 grasses is significantly higher than that of C_4 plants, comparable mean figures from a survey undertaken by Minson and McLeod (1970) being 68.2 and 55.4% dry matter digestibility, respectively (Fig. 9.10). Growing temperate and tropical grasses at their respective optimal temperatures results in biggest quality differences; the digestibility of the C_4 plants is increased when grown in temperatures optimal for temperate grasses, and that of C_3 species reduced at temperatures optimal for tropical plants. Quality differences between C_3 and C_4 grasses go a long way towards explaining why dairy animals eating only high-quality temperate grass can both maintain their body condition and yield more than $20\,kg\,milk\,day^{-1}$; whilst those grazing tropical pastures rarely produce more than $6–8\,kg\,milk\,day^{-1}$; they are also mainly responsible for the higher growth rate of ruminants fed on temperate compared with tropical grasses.

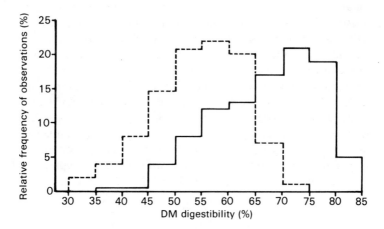

Figure 9.10 The frequency distribution of digestibility in (---) tropical and (——) temperate grasses. (From Minson and McLeod, 1970.)

Whilst digestibility (or D-value) is the single most important indicator of quality, other differences in the nutritive value of digested herbage are apparent on occasion. Examples range from insufficient P-levels to meet even the modest demands of wether sheep (Christie, 1979), higher growth rates and milk production from white clover than from all grass forage (both with D-values of 75–80) (Thomson, 1979), and indications of differences in milk yield as high as $15\,kg\,day^{-1}$ when dairy cattle at comparable stages in their lactation are fed spring compared with autumn grass of comparable D-values (Le Du and Hutchinson, 1982).

9.5 UTILIZATION

9.5.1 Grazing

Efficient use of a grazed sward requires the maintenance of good-quality herbage; thus tillers must be defoliated whilst they are still highly digestible. Typical D-values for a well-grazed ryegrass pasture in Britain during the growing season are: April, 75; May, 73; June, 69; July, 69; August, 70; September, 71; and October, 71 (from GRI data). There are also indications that the digestibility of herbage selected by sheep grazing some well-managed improved pastures in temperate Australia seldom falls below 70% (see Fig. 9.16(b), below). Full exploitation of grass also requires that herbage should be present in sufficient quantity for the animal to eat its fill; when intake falls, so does milk production or animal growth rate. It also demands the minimum of wastage through undergrazing. Utilization of 75–80% of grown herbage by the grazing animal is a target which should be achievable in a good grass growing environment typical of much of the UK. However, a lower utilization of the grass is good management practice in the less-favourable growing conditions characteristic of much of Australia; conservative stocking rates are needed there both to ensure the long-term stability of the typically continuously grazed pastures, and to provide some buffer against feed shortage and thus losses in animal live weight.

Most quantitative studies on utilization have been undertaken on intensively used grassland. Contrary to earlier beliefs, research has shown little difference in animal production from continuously and rotationally grazed all-grass swards. However, grazing studies have shown a relationship between sward canopy height and intake varying somewhat with the grazing system. Cattle on rotationally grazed ryegrass pastures find it difficult to eat to appetite when the stubble falls below 9–10 cm (Ernst *et al.*, 1980) (Fig. 9.11); this compares with a

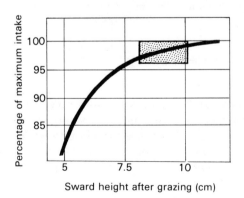

Figure 9.11 Herbage intake under rotational grazing. Critical height range is shaded. (From Le Du and Hutchinson, 1982.)

Utilization

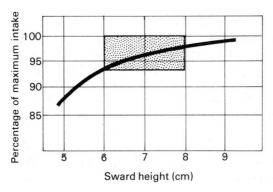

Figure 9.12 Herbage intake under continuous grazing. Critical height range is shaded. (From Le Du and Hutchinson, 1982.)

critical canopy height of 6–8 cm on continuously grazed ryegrass swards (Ernst *et al.*, 1980; Wilkins, 1986) (Fig. 9.12). Height of sward canopy thus emerges as a basis for management decisions designed to maximize output from grazed grass.

9.5.2 Conservation

Whilst conservation is of doubtful importance in the more extensive grassland enterprises such as wool or beef production in Australia, it is vital to the achievement of maximum animal output on intensive grass-based systems for milk or meat production in the UK and Western Europe.

Making good-quality hay is difficult in the vagaries of the British weather and though investigations have been undertaken to improve its quality and predictability (summarized in Lazenby, 1981), silage is now more commonly used in Britain to conserve grass – especially in intensive systems. Improved understanding of the processes of fermentation (Wilkins and Wilson, 1974; Woolford, 1984) and the means of manipulating and controlling them (McDonald, 1976; Woolford, 1984) have resulted in the identification of the causes of deterioration both within the silo and on exposure at feeding (Crawshaw and Woolford, 1979; Woolford, 1984). As a result technology is now available enabling silage of predictable quality to be made (Thomas, 1980), thus allowing exploitation of its inherent advantages over hay making because it is less weather dependent, less labour intensive and more suited to mechanization. Energy losses during silage making are inevitable; nevertheless, the aim should be to reduce these to as near 20% as possible, rather than the 30% which is widely accepted.

Some controversies remain on silage production and utilisation (Wilkins, 1980), particularly the value of wilting (important in reducing effluent and associated losses, but with little evidence on any direct effect on animal production) and the use of additives. These are a valuable insurance to assist desirable fermentation processes, especially in making silage from grass low in sugar and available data suggest an average increase in daily milk production of 1–2 kg by animals fed on silage with an additive. Good conditions for making and storing silage result in little effect on the digestibility of the original grass and the efficiency with which the animal uses the energy content in the forage; this contrasts with the poor utilization of the N in the silage.

9.6 MEETING THE NUTRITIONAL REQUIREMENTS OF THE ANIMAL

Whilst DM yields and their quality, as measured by digestibility or D-value, are very valuable in describing grassland production, they go only part way towards measuring grassland output in meaningful terms, i.e. as utilized animal feed and animal products. The development of the concept of Metabolizable Energy (ME) and the so-called ME system (ARC, 1965) was a major step towards achieving this objective. ME, like D-value, is essentially a measure of energy, and a fairly close relationship exists between them; for example, forage with a D-value of 70 can be predicted to have an ME value of some 11.5 MJ kg^{-1} (Fig. 9.13). However, the ME system enables both the feed value of forage and the energy requirements of ruminants to be expressed in common terms; it is therefore possible to calculate the proportion of such energy needed in different animal enterprises which can be supplied by forage.

For example, a production curve, typical of an autumn calving dairy cow weighing 600 kg and producing 6000 kg of milk during the course of one lactation, is shown in Fig. 9.14(a). During the first 10 weeks or so after calving, the dairy cow obtains some of its energy requirements from its own body

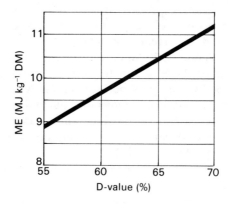

Figure 9.13 Relationship between metabolizable energy value of silage (ME, MJ kg^{-1} DM) and D-value. (Adapted from Thomas and Golightly, 1982.)

Meeting nutritional requirements

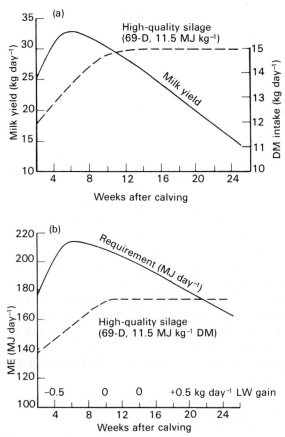

Figure 9.14 (a) Lactation curve and silage DM intake of a 600 kg autumn-calving cow (from GRI data). (After Lazenby, 1981.) (b) Contribution of high-quality silage to energy requirement of a 600 kg autumn-calving cow (from GRI data). (After Lazenby, 1981.)

reserves, losing an average of about 0.5 kg day^{-1}, which is made up later in the lactation (Fig. 9.14(b)).

The contribution which good-quality silage can make to the animal's ME requirements is dictated by the level of intake of such bulky forage. Peak daily intake of some 2.5% of body weight is reached about 12 weeks after calving (Fig. 9.14(a)); at 2 and 6 weeks after calving maximum intake is some 20 and 10%, respectively, less than the peak (Thomas, pers. comm.).

It is clear (Fig. 9.14(b)) that a large proportion, but not all, of the ME requirements in early lactation can be supplied by good-quality silage; the balance must be met by concentrates (in later lactation all the ME requirements can be met from grazed grass). However, each kilogram of concentrate eaten by the animal reduces the silage intake by 0.7 kg in the first 12 weeks or so in the lactation, and by some 0.5 kg thereafter.

In addition to good-quality silage fed *ad libitum,* approximately 1 t of concentrate (with an ME value of 12.5 MJ kg^{-1}) appears necessary for an autumn-calving dairy cow, weighing 600 kg, to produce 6000 kg of milk during her lactation. At the GRI, flat-rate feeding of just over 1 t of such concentrates to autumn calvers during the first 120 days of their lactation (together with *ad libitum* feeding of good-quality silage) resulted in an increase of their mean milk yields from 5100 to 5800 kg between 1973/74 and 1977/78 (Thomas, 1980). Over 75% of their energy requirements were thus met by forage, compared with less than 50% from equivalent yielding herds from the Milk Marketing Board (MMB/LCP Scheme) (MMB, 1978).

The ME system is now the standard method used to provide rations for dairy cows, whilst the ME concept is also widely used for developing production targets for a range of intensive grassland systems. It has been demonstrated, both experimentally and in practice, that production targets for systems such as spring-calving dairy cows, intensive beef and lamb production, can also be met largely or even entirely from forage.

It is thus clear, in biological terms at least, that grass can make a greater contribution than at present to the feeding of high-producing and rapidly growing ruminants. In the UK this is especially true for the high-yielding dairy cow. However, using grass as the main energy source for the achievement of high animal production targets is dependent on a reliable supply of known and good-quality material.

9.7 SYSTEMS BASED ON GRASS

Because grassland output, and the factors important in affecting it, vary greatly from one situation to another, two examples have been chosen to illustrate systems adapted to different environments; considerable differences exist between them. In one environment selected, a specific region of Australia, growing conditions are suited to relatively extensive systems for wool or beef; the other, typified by many parts of the UK, is more favourable for grass growth, thus enabling the development of intensive animal production systems for milk, beef or fat lamb.

9.7.1 Grass/legume system for wool production on the Northern Tablelands of New South Wales

The Northern Tablelands comprise a discrete and essentially pastoral area of some 3.6 million ha, lying between approximately 29 and 32°S and 151 and 153° E, and 800–1300 m above sea level. As already indicated, in common with most of Australia, the main climatic factors affecting growth are temperature and rainfall. Armidale, typical of the region, has mean max/min temperatures

Systems based on grass

of 26.5/13 °C in January and 12.5/1 °C in July. Rainfall averages some 760 mm year^{-1} with a coefficient of variation of 25%, more reliable than in most parts of Australia; approximately 60% of the rain falls in the 6 months including summer (mid-October to mid-April), with rainfall exceeding potential evaporation only in the winter period from about mid-May until early August (Fig. 9.4).

The native species dominating unimproved pastures are typically *Bothriochloa macra* (red grass), *Sporobolus elongatus* (Parramatta grass) and *Chloris truncata* (umbrella or windmill grass), all C$_4$ plants intolerant of cold and dying back in winter. The low-quality feed, particularly in winter, was not only responsible for low stocking rates of such pastures, it also resulted in cattle having to be sold off as stores rather than as fat animals, and made sheep breeding so difficult that, before pasture improvement, the grazier's normal practice was to buy wethers from the Western Plains for his wool crop. The impetus for pasture improvement came largely from high prices for quality wool, coupled with the availability of a suitable technology, involving aerial seeding and application of superphosphate, which had been highly successful in

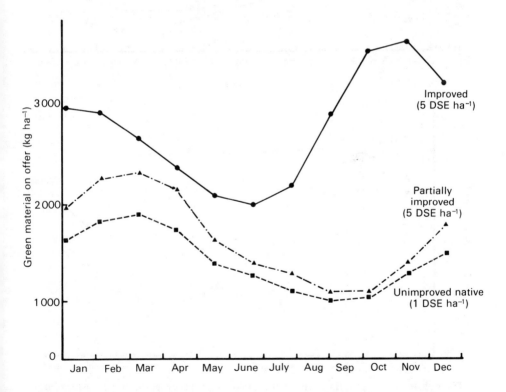

Figure 9.15 Green material DM on offer on three pasture types on the Northern Tablelands of NSW (Smith and Lazenby, unpublished.)

more-southerly latitudes of Australia. A series of good rainfall years was further encouragement for rapid pasture improvement.

Between 1950 and 1965 the area of improved pastures increased from less than 35 000 ha to more than 600 000 ha, with an associated increase in annual superphosphate use from 2000 to 125 000 t. During that period the pattern of pasture development followed the stages previously described. The increased DM yields on the improved area, with a longer seasonal spread of growth (Fig. 9.15) coupled with higher-quality herbage, especially in winter, enabled sheep to select more-nutritious herbage than they could on unimproved pastures. Their intake appeared fairly constant throughout the year, at approximately 7.5 kg DM week^{-1} (Fig. 9.16). Wool growth was more uniform (Fig. 9.17) and,

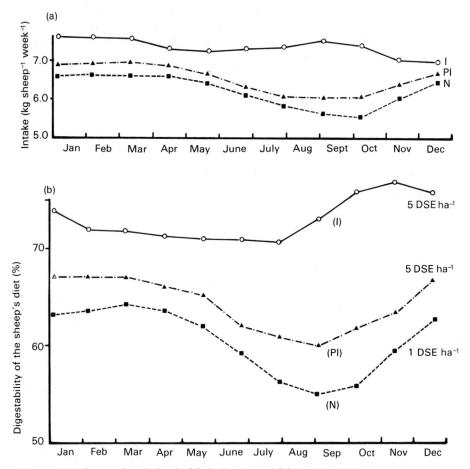

Figure 9.16 Seasonal variation in (a) the intake and (b) the digestibility of pasture eaten by sheep grazing native (N), partially improved (PI) and improved (I) pasture on the Northern Tablelands of NSW. (Smith and Lazenby, unpublished.)

Systems based on grass

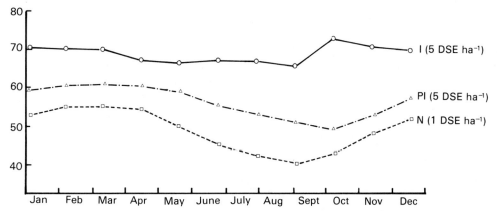

Figure 9.17 Wool growth rate for sheep grazing native (N), partially improved (PI) and improved (I) pasture on the Northern Tablelands of NSW. (Smith and Lazenby, unpublished.)

unlike sheep on native pastures, there was no fall in body weight in winter. Whilst in practice the increase in stocking rate (S/R) was less than the level predicted as possible (Fig. 9.18), it was nevertheless considerable. Overall, stocking rates averaged 7.5 dry sheep equivalents (DSE) ha^{-1}, three times greater than that on the better unimproved pastures. Wool production on the tablelands more than doubled to over 20 million kg in 1965. Almost 70% of this improvement was a result of the bigger S/R, the rest because of heavier fleeces, the average weight of which increased from 3.5 to 4.2 kg (McDonald, 1968).

It was the 1965/66 drought which exposed the limitations of the technology used in pasture improvement. Whilst perennial ryegrass and white clover, the main species used in the programme, had yielded well when soil moisture was relatively plentiful, neither was able to survive severe moisture stress. The pastures, thus deprived of both improved species and nitrogen, deteriorated quickly with both broad-leaved weeds and native grasses becoming increasingly common constituents of the swards. Failures in pasture establishment became more common and pasture development did not proceed from the clover-dominant to the grass-dominant stage. Further, bloat in cattle was widespread on the clover-dominated swards.

A quite different set of factors thus limited output from the improved grassland compared with those of the unimproved pastures. In addition to the limitations associated with species, establishment, pasture development and deterioration, the available soil moisture was a more important limiting factor on the improved grasslands. A broadly-based experimental programme – involving co-ordinated controlled environment, glasshouse and field experiments, with associated trials on properties – was thus developed to investigate these problems (Hartridge, 1979).

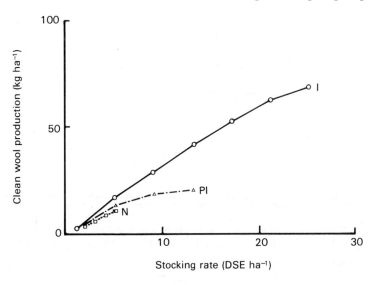

Figure 9.18 Relationship between stocking rate and clean wool production for native (N), partially improved (PI) and improved (I) pasture on the Northern Tablelands of NSW. (Smith and Lazenby, unpublished.)

The growth of a range of pasture species was studied in an attempt to select plants better able to withstand moisture stress than perennial ryegrass and white clover. Both phalaris and tall fescue were shown to be less affected by soil moisture deficits than perennial ryegrass, with tall fescue having the more desirable seasonal distribution of growth (Lazenby and Swain, 1972). Lucerne was superior to white clover in both its ability to survive and its DM production under drought conditions. With the exception of lucerne, no differences were apparent in the water use of the plants; the deeper-rooting system of lucerne enabled water to be moved from the soil down to 150 cm, some 30 cm deeper than under the other plants (Johns and Lazenby, 1973a, b). Whilst lucerne presents some management problems for the grazier, it appears well adapted to the natural environment on the tablelands where soils which dry out are recharged periodically by rain.

In addition, models were developed to investigate the potential DM production and the relative importance of soil moisture and temperature on the growth of improved pastures (Smith and Stephens, 1976) (Fig. 9.19). Predicted annual DM yields ranged from $5\,t\,ha^{-1}$ in the drought year of 1965, to a maximum of $14\,t\,ha^{-1}$ on good moisture retentive soils in 1971. It appeared that soil moisture was the main factor limiting growth from the beginning of October to the end of May, and temperature from the end of May to early October. However, there were periods during both May and October when both factors had a similar probability of being most limiting. Soil moisture

Systems based on grass

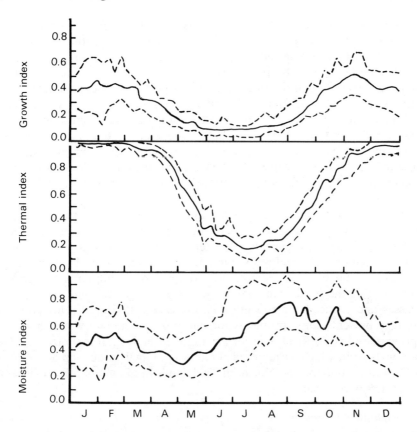

Figure 9.19 Indices of growth, temperature and soil moisture for Armidale NSW (1950–73). Median (solid line) with upper and lower quartiles. The growth index is a dimensionless ratio of actual growth rate/potential growth rate. The growth index is the product of the climatic indices. (From Smith and Stephens, 1976.)

variability seems to restrict pasture growth in most years to some 50% of its potential; such variations cause more within-year variability in growth than does temperature, except in April and September when the two factors appear equal in their effect.

Studies on pasture establishemt (Campbell, 1972; Campbell and Swain, 1973) showed that whilst available soil water, physical conditions on the soil surface and competition from existing vegetation were all significant in influencing pasture establishment from aerially sown seeds, the first of these was the most important. Most favourable soil moisture conditions were shown to occur from June to August (Smith and Johns, 1975), a period when there is also least competition from native pasture plants. Convincing field data (Smith and Stephens, 1976) subsequently demonstrated sowing seeds in early winter to be more likely to result in successful pasture establishment than sowing in autumn or spring, the times previously recommended (Grantham, 1961).

The rate of pasture development was shown to be very strongly dependent on the level of superphosphate applied, with P being the most important element affecting such development and S significant to a lesser extent. The other variables studied, namely N application, grass species sown and sowing rate had little or no effect. Increasing the normal rate of application of superphosphate from 100 to 300 kg ha^{-1} both hastened the rate of pasture development and made it more certain to progress to the grass-dominant stage. In addition, grasses sown with the clover were able to establish in the low fertility conditions and survive until the fertility increased (Wolfe and Lazenby, 1973a, b).

Pasture deterioration occurred almost immediately superphosphate application was discontinued; dying out of improved species resulted in reduced production, especially in winter, as well as an associated increase of native species such as red grass. Only under continuing levels of application of superphosphate could the botanical composition be influenced by grazing; then, resting the pasture in winter and early spring increased the improved grasses and reduced the red grass (Cook *et al.*, 1978a, b). However, this strategy presents some problems for the grazier. The late winter and early spring period is a difficult one for pasture feed; indeed, it is shortage of feed during this time which is a major factor limiting animal production on improved pastures, not only on the Northern Tablelands but in many other temperate grasslands in Australia.

(a) Changes and challenges

The research results, which have been used to draw up guidelines for successful pasture development, have been responsible for significant changes in the practice of grassland improvement on the Tablelands. Wider use is now made of phalaris and tall fescue, and early winter is the most common time for sowing. Grass is now generally sown with white clover at the beginning of the programme, rather than 2 or 3 years afterwards. Many graziers have increased the levels of superphosphate applied in their pasture improvement. A widely used soil-testing programme has been developed to assist graziers to decide when the heavier rates of superphosphate, needed to ensure full pasture development, could be reduced to maintenance requirements (of approximately 100 kg ha^{-1} year^{-1}) and still enable the pasture to remain grass dominant. Graziers' confidence in their pastures has improved greatly, with many beef producers feeling they can control bloat within their pasture-improvement programme (Griffith and Blair, 1981). However, there remain many challenges for the researcher. As in Australia as a whole, grassland improvement on the Tablelands is dependent on a legume able to both survive in the conditions prevailing and provide sufficient N for continuing pasture growth. The N needed for such growth is considerably greater than that on

Systems based on grass

unimproved pastures. The DM yields of the latter, say $2\,t\,ha^{-1}\,year^{-1}$ require an annual N input of $40\,kg\,ha^{-1}$, assuming a 2% N content of the herbage; the low S/R, coupled with likely little loss of nutrient, probably ensures that the N required is sustainable through natural inputs (Smith and Lazenby, unpublished). In contrast, improved pastures producing say $9\,t\,DM\,ha^{-1}\,year^{-1}$, with a 3% N content, would need an annual requirement of the nutrient of $270\,kg\,ha^{-1}$ (Frizzel, 1978). High N fixation and release is therefore needed to prevent the nutrient limiting growth, even assuming that only half the herbage is eaten and there is rapid recycling through the grazing animal. As already indicated, white clover, which is the legume normally used in pasture improvement on the Tablelands, leaves something to be desired in this respect. Whilst there appear possibilities for wider use of lucerne, its persistence is limited on continuously grazed swards. Improved cultivars of white clover are needed, particularly drought resistant forms, and a search for other legumes adapted to the region may be warranted.

Output from pastures is still limited by their growth in winter, the one time of year when moisture is plentiful. The demonstration (Schiller and Lazenby, 1975) that Mediterranean populations of tall fescue are capable of increasing winter production by some $10\,kg\,DM\,ha^{-1}\,day^{-1}$, provides both a real challenge and interesting alternatives for the grazier and breeder to introduce such Mediterranean plants into grazing practice. Drought-tolerant grasses, able to recover rapidly after summer drought, are also very desirable; it would be worthwhile exploring further whether C_4 grasses–the dominant plants in unimproved pastures–have any role in improved grasslands. Their drought resistance and superior ability to convert available moisture into DM are clear advantages compared with C_3 plants (see Chapter 6). Further, *Paspalum dilatatum* has been shown capable of producing total annual DM yields similar to phalaris, perennial ryegrass and tall fescue, though with growth concentrated in summer (Colman *et al.*, 1974; Harris and Lazenby, 1974).

There is little doubt that improved grasses have a better water use efficiency (WUE) than the species found in unimproved pastures (Smith and Lazenby, unpublished), growing and using available water in periods when native plants are dormant. However, this characteristic increases the likelihood of improved pastures suffering more frequent and severe water stress during summer (Begg, 1959). Too high a S/R also increases the risk of feed shortage. In this context, modelling has an important role in future research. Climatic models should enable more accurate definition of optimum long-term S/Rs being related to year-to-year variation in DM production (Smith and Stephens, 1976). They are also relevant to the optimum use of superphosphate. Because superphosphate is becoming increasingly expensive–it now costs some $A150 per tonne spread–it is important that it be used efficiently; hence the value of soil tests to provide advice on optimum levels of superphosphate use.

Experiments during the past decade have shown that the perennial native grass *Danthonia pallida* is resistant to grazing, remains green in winter (Lodge

340   The grass crop in perspective

and Whalley, 1983) and responds to fertility (Lodge, 1981). Further, quite high animal production levels have been achieved from improved native pastures at a lower fertility input than that required in the traditional method of pasture improvement (Wyndham, pers. comm.). Two options therefore exist for the further improvement of pasture on the Northern Tablelands. The first, which can be considered a development of the more traditional method, involves a grass + legume pasture more persistent and capable of better growth in winter, and in summer following water stress, than the present system; this option requires major economic inputs. The second possibility is to make greater use of native species together with an introduced legume to provide more-stable, lower cost input pastures. The work on *Danthonia* has given impetus to the exploration of this second option (Smith and Lazenby, unpublished).

9.7.2 Intensive grass – nitrogen system of milk production in the UK

The potential contribution which grass can make to animal output is well illustrated in a perennial ryegrass – N system intensively used for milk production, which is adapted to much of the UK. Factors affecting grass growth and its utilization have already been discussed. Nevertheless, they warrant some further mention before examining their integration in the system.

(a) DM production

The DM response of grass to N, described previously, coupled with the rainfall between April and September and the moisture-holding capacity of the soil, provide the basis for predicting seasonal and total DM production and its reliability. Deep loam or clay soils with an average summer rainfall of more than 400 mm offer the best grass-growing conditions and thus the highest potential yields; they have been designated by Corrall *et al.* (1982) as Class 1 sites, whilst shallow, gravelly, chalky or sandy soils with less than 300 mm of rain, which can be expected to produce lowest DM production, are classified as Class 5 sites. Yield differences between sites become most apparent from June onwards (Fig. 9.20), following the build up of soil moisture deficits which are most severe on poor moisture-retaining soils receiving low summer rainfall. Table 9.1 indicates the yields which can be expected on the different sites, with an N application at target levels, defined by Corrall *et al.* (1982) as the N needed to achieve 90% potential DM production (Fig. 9.21). The predictions are conservative, being based on the lowest summer rainfall, and thus the lowest yields, expected in seven years out of ten. Good agronomic practice includes a regime of N application which results in both a good DM response from the grass and a minimum of loss of the nutrient. The timing of the first application is now being made increasingly on the so-called T-sum (i.e. the dates when accumulated mean daily air temperatures from 1 January reach 200), coupled with the ground conditions; it may be as early as mid-February in parts of the

Systems based on grass

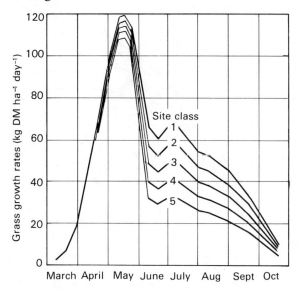

Figure 9.20 Seasonal pattern of dry matter production from a perennial ryegrass (cv. 23) sward at five site classes (see text for explanation of site classes). (From Corrall et al., 1982.)

TABLE 9.1 *Target levels of nitrogen fertilizer and yields of grass cut or grazed monthly at different site classes (from Corrall et al., 1982)*

Site class	Target N (kg ha^{-1})	Target dry matter yield (t ha^{-1})
1	450	12.7
2	410	11.6
3	370	10.5
4	330	9.5
5	300	8.4

south of England and 2–3 weeks later in the more northerly parts of Britain. The last N dressing normally occurs in August.

Seasonal production can be manipulated to an extent, without detriment to annual DM yields, by altering the pattern of N applications. For example, application of a large proportion of the annual dressing in spring would increase the percentage of the total yield produced early in the season; such growth could be exploited for conservation. Applying a large proportion of N in summer will result in the more even distribution of herbage available for grazing (Morrison et al., 1980).

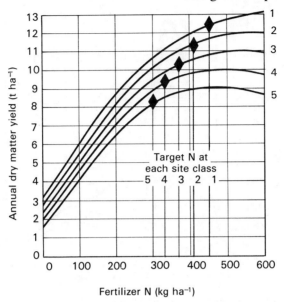

Figure 9.21 Yield and response of grass at grazing stage (68+ D) to fertilizer N at different site classes, showing target yield and target N use. (From Corrall et al., 1982.)

Frequency of defoliation not only affects herbage yields, but also herbage quality (Fig. 9.22). This provides the basis for determining a cutting regime to provide sufficient conserved feed of the quality needed for winter feeding.

(b) Utilization

The objective of good grazing management in a grass-based system used for milk production is to achieve both efficient utilization of the herbage and maximum milk yields per unit area. Stocking rate is the main factor used to achieve a supply of good quality grass for the herd. However, it has to be matched with a number of other factors, including the grass growing conditions, the level of N applied, the pattern of grass growth during the grazing season, the cutting strategy adopted for conservation and the milk yield of the dairy cow, which in itself is related to the stage in lactation (Fig. 9.23).

Decisions on when to change the S/R can be made on the basis of sward height. This enables an element of flexibility to be introduced in the area set aside for grazing and conservation, allowing day-to-day decisions to be made to help efficient herbage utilization.

The relation between S/R and milk yield per cow (Fig. 9.24) shows that a reduction in the potential milk yield per animal is necessary to achieve the maximum milk production per unit area. Data from one investigation in Northern Ireland (taken from Le Du and Hutchinson, 1982) show that a 7% reduction in potential milk yield per animal resulted in a 22% increase in milk

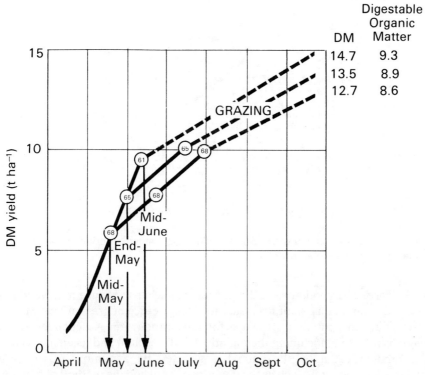

Figure 9.22 The effect of cutting at various digestibilities on the yield of perennial ryegrass (cv. S23) at site class 1. (From Corrall *et al.*, 1982.)

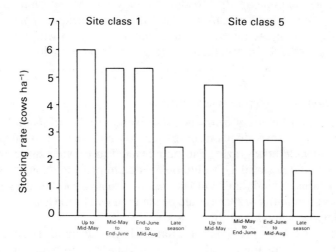

Figure 9.23 Effect of season and class site on grazing stocking rates (autumn calvers). (From Le Du and Hutchinson, 1982.)

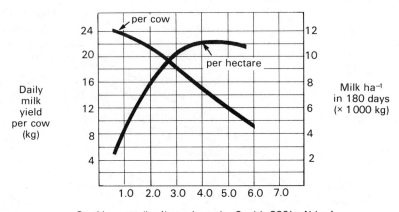

Figure 9.24 An example of the effect of stocking rate on milk production per cow and per hectare. (From Holmes, 1980.)

per hectare. The good grassland manager thus seeks to reduce milk yield per cow to the minimum extent necessary to achieve efficient grass utilization.

Attaining the potential of silage for milk production is dependent on a number of factors, including the quantity of silage eaten and its energy value. The amount eaten by the animal is affected by a number of practical management decisions such as the quantity offered and the method of feeding, as well as the fermentation, digestibility and level of concentrates fed. The best use of grassland for animal production can only be achieved if efficient conservation is combined with a feeding strategy with predetermined roles for grazed grass, silage and concentrates. This involves a calculation of the amount and quantity of silage required for winter feeding together with the minimum concentrates needed; an accurate estimate can then be made of the area of grass needed to produce the silage (Fig. 9.25) (Thomas and Golightly, 1982).

(c) Integration of conservation with grazing

So far the relationship of grazing and conservation to the needs of the animal have been discussed as individual components of the system. Typical questions considered include how grazing can be managed to provide sufficient good-quality feed; how much silage would be needed to provide enough for winter feeding; the area of land that would be required to achieve this level; and how the quality and quantity of silage could be manipulated by the level of N applied and decisions on when to cut the grass.

However, the real problem in grassland management is to integrate conservation with grazing in order to match grass growth with herd requirements (Fig. 9.26). Effective management therefore involves changing the S/R during the season to provide enough good-quality grass for the animals

Systems based on grass

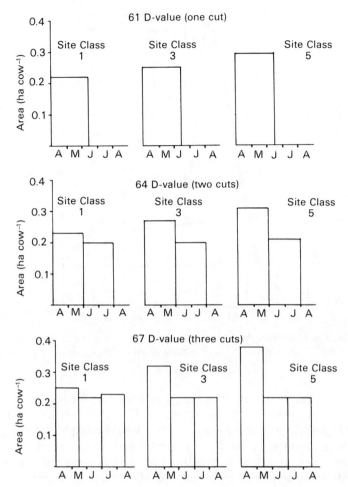

Figure 9.25 Areas required per cow for autumn calvers yielding 6000 kg. (From Thomas and Golightly, 1982.)

throughout the growing season, and the conservation of surplus material. Fresh grass must be grazed when young if its quality is to be maintained, whereas silage can be fed when needed (Doyle *et al.*, 1982).

One objective of conservation, namely to produce winter feed with a minimum loss of yield and quality, is well understood; the other, to assist in the management of grass by adjusting the area available for grazing, is much less widely appreciated. The timing of conservation cuts not only influences the quality and quantity of the herbage to be conserved, it also affects the regrowth of grass available for grazing. It is therefore a key element in providing enough quality grass throughout the grazing season.

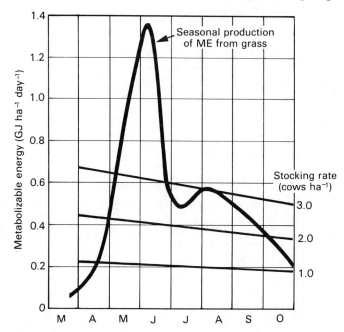

Figure 9.26 The ME requirements of spring-calving cows at three stocking rates compared with the seasonal production of metabolizable energy from grass. (From Doyle et al., 1982.)

In practice the quality of herbage required for winter feeding determines the number of cuts needed for conservation. For example, a low-quality silage – say of 61D with an ME of 9.8 MJ kg^{-1} DM – would probably be produced by cutting the grass once, in mid-June; in contrast, three cuts would be needed – in mid-May, the end of June and mid-August – to produce a high-quality silage of 67 or 68D with an ME of 10.7 MJ kg^{-1} DM. The cutting regime also affects the efficiency of grassland management. A one-cut system leads not only to a lower UME (see below) than do two or three cuts, but also to more wastage of grass; typically less than 60% of the herbage grown is utilized by the grazing animal in such a one-cut system. In contrast, the overall grazing efficiency of a three-cut system approaches 80%, with better utilization than the one-cut system particularly evident in early May, July and August (Fig. 9.27).

Doyle et al. (1982) suggest four reasons for the low utilization of one-cut systems, namely: (a) reduction in management flexibility following the closing up of a fixed area of land for a long period, which makes it difficult to balance grazing and conservation needs early in the season; (b) slower regrowth as a result of the heavier crop; (c) a reduced land area available for grazing in mid season; and (d) the release of a larger area for grazing when the requirements of the animals are lower than the grass available.

A calculation can be made of the energy derived from grazed and conserved

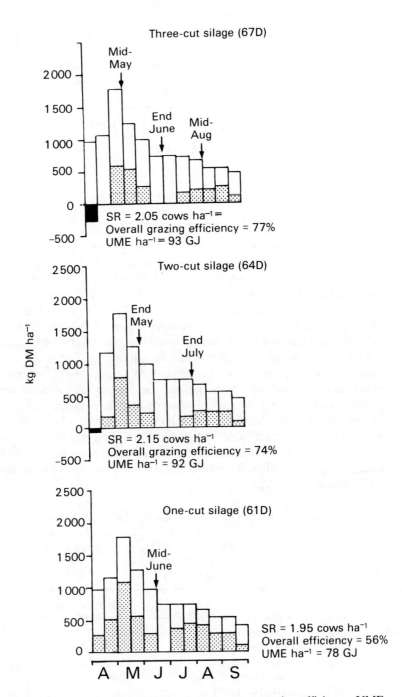

Figure 9.27 The effect of cutting strategy on grass wastage, grazing efficiency, UME output and stocking rate (autumn calvers, site Class 3). ☐, Grass utilized; ▦, grass wasted; ■, supplementary feeding. Arrows show cutting dates. (From Doyle et al., 1982.)

TABLE 9.2 *Target utilized metabolizable energy (UME) values at different site classes (from Doyle et al., 1982)*

Site class	Target UME (GJ ha^{-1})
1	115
2	105
3	95
4	85
5	75

grass which is eaten by the animal (Baker, 1984). The energy thus derived from the grassland is calculated as utilized metabolizable energy (UME) and generally expressed as Gigajoules (GJ) of ME ha^{-1}. It is also possible to estimate the UME which should be attainable from well-managed grassland on the different site classes for a particular level of N applied (Table 9.2); this provides a target for the farmer. A comparison of the calculated and target UME values thus indicates how efficiently the grassland is being managed.

Making the maximum use of grass to achieve specific levels of milk production thus requires N to be applied at target levels, efficient grazing management, good quality silage and an integration of conservation and grazing. Many more decisions are required to manage such a system efficiently for milk production than in one based on concentrates. Nevertheless, a high potential exists for profitable milk production based on maximum use of grass (Table 9.3), whilst there are many data showing greater profitability of grass-based compared with concentrate-based systems of milk production (Doyle and Richardson, 1982).

TABLE 9.3 *Potential for profitable milk production from feeding minimum concentrates and making maximum use of grass: projected gross margins at early 1982 prices (site class 3) (from Doyle and Richardson, 1982)*

	Autumn calvers
Milk yield (l cow^{-1})	6000
Concentrates (kg cow^{-1})	925
Fertilizer (kg ha^{-1})	370
Stocking rate (cows ha^{-1})	2.05
UME (GJ ha^{-1})	92.9
Grass utilized (%)	77
Gross margin per cow (£)	586
Gross margin per ha (£)	1201

Systems based on grass

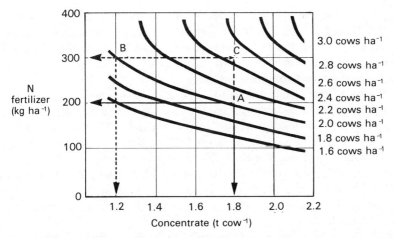

Figure 9.28 The relationship between level of N fertilizer and concentrate for different target stocking rates (site Class 3, autumn calvers). (From Doyle and Richardson, 1982.)

In seeking to achieve specific milk yield targets, the N used, S/R and level of concentrates fed should be properly balanced (Walsh, 1979; Doyle and Richardson, 1982). If they are not, then problems are likely; for example, if the S/R is too low, grass will be wasted, whilst if it is too high milk yields will fall. Figure 9.28 illustrates the relationship between the N level and concentrate usage for different stocking rates in an autumn-calving herd with a target milk yield of 6000 kg, fed on 64D silage. Animals on average (Class 3 site) grassland, stocked at 2 cows ha^{-1}, fed 1.8 t concentrate (of 12.5 MJ kg^{-1}), should achieve the milk yield with 200 kg nitrogen ha^{-1} (A). Lower N levels are likely to lead to shortage of grass, and thus a need for more concentrates to achieve the target. Increasing the N application from 200 to 300 kg ha^{-1} should enable the milk yield to be achieved by either reducing concentrate usage to 1.2 t cow^{-1} (B), or increasing S/R to 2.4 cow ha^{-1} (C).

(d) Changes and challenges

Developments and opportunities for improving the grass by N system revolve around its intensification and more efficient use; many are typical of the changes and challenges facing British grasslands as a whole. Reference has already been made to several trends in grassland practice arising from the efforts of grass breeders. There have been other changes. For example, the increase in land under leys in the period embracing World War II and the years immediately following, has now given way to a marked reduction in such temporary grassland (Green and Lazenby, 1981); more land is permanently under grass, which has become concentrated in the west and north of the UK. The length of life of swards has increased, as has their ryegrass content, a development associated with their more intensive use. Greater use of fertilizer is apparent, and the average application of N on specialist dairy farms now exceeds 200 kg ha^{-1} year^{-1} (Archer, 1986). Stocking rate has also increased, especially up to the mid-1970s, whilst more grass is now conserved as silage than hay – an indication both of farmers' confidence in the new technology and the increased intensity of grassland use.

Yet output from British grasslands generally remains considerably below its potential. Annual DM yields average 5–6 t ha^{-1} (Green and Baker, 1981), with UME output of lowland grassland currently some 47 GJ ha^{-1} (Wilkins and Down, 1984), compared with the 80–90 GJ ha^{-1} achieved by some permanent grassland farmers (Forbes *et al.*, 1980) and up to 140 GJ ha^{-1} from farm-scale experiments in Northern Ireland (Gordon, 1981). It is pertinent to enquire why grassland is so under used, especially when it is both such a cheap source of ruminant feed and capable of making a considerably increased contribution to animal production.

One factor is the relatively low mean S/R; this results in lower than possible utilization of grass, with figures of 50–60% often quoted (e.g. Forbes *et al.*, 1980). Increasing the N applied on dairy farms to 300–350 kg ha^{-1} year^{-1} should result in mean UME output of at least 80 GJ ha^{-1}, providing the grass was utilized efficiently. However, it is clear that high output is dependent on more than simply applying greater dressings of N and increasing the S/R; as already indicated, full exploitation of grass requires proper integration of grazing with conservation in a system geared to a specific animal production enterprise. The UME levels achievable under different grass growing conditions make it possible to set targets appropriate to the local environment.

Yet a key factor affecting output is the differing ability of individual farmers to manage their grass. For example, whilst DM production on farms with good grass-growing conditions and high N use may be twice that of less favourable environments, differences in the standard of management can increase the range of utilised output to three- or even four-fold (Table 9.4). Some farmers waste their grass, relying unnecessarily on concentrates for animal production; others not only use a higher proportion of the herbage grown, but put it to a specific use in the animal's diet. For example, Forbes *et al.* (1980) showed that

TABLE 9.4 *Range of outputs of utilized metabolizable energy (UME) on farms with high or low grass yields and with different levels of utilization (from GRI/ADAS Permanent Pasture Group data)*

	High		Low	
DM yield (kg ha^{-1})	12 000		6000	
Utilization (%)	75	50	75	50
Utilized DM (kg ha^{-1})	9000	6000	4500	3000
ME kg^{-1} DM (MJ)	10.5	9.5	10.0	8.0
UME output (GJ ha^{-1})	94.5	57.0	45.0	24.0

Systems based on grass

the 10% of permanent grassland farmers making most of their grass as a production feed achieved an average of 4700 kg milk ha^{-1} from grass, which supplied about 75% of the total ME requirements of their animals. In contrast, considerably less than the maintenance requirements of animals were met by grass on farms where least use was made of it in the animal diet.

Grass can play somewhat different roles in the feeding of animals in intensive forage-based systems and it is important that the practice adopted fits the circumstances of the individual farm. Table 9.5 illustrates this principle. Whilst all three farmers rely considerably on grass for their milk production, Farm A is limited in size and maximum output is achieved by combining very high milk yields per cow with a high S/R and relatively high concentrate usage. Farm B, with a lower rainfall than Farm A, has less pressure on land and a combination of moderate milk yields and a low use of concentrates provides good profits. Farm C has growing conditions similar to those found at the GRI. The results from this farm appear achievable by many dairy farmers in England and Wales. Thus, whilst grass is used intensively on all three farms, its proportional contribution to the energy requirements of dairy animals varies from 72 to 86%.

Nationally, however, grass is still generally regarded as a maintenance feed, with many farmers lacking confidence in its ability to provide the major part of the diet needed for high animal production. Such lack of confidence is an important factor preventing full exploitation of the potential of grassland in Britain. Economic considerations are also significant, whilst lack of incentive was cited by a surprisingly large number of farmers in a national survey as the reason for not intensifying their grassland use (Forbes *et al.*, 1980). There is thus no single cause responsible for the present underuse of British grasslands; a mixture of technical, economic and sociological factors are involved, and they vary in importance from one farm to another.

TABLE 9.5 *Sample farms from ICI Dairymaid scheme*

Location	Herd A NW	Herd B East Yorks	Herd C W. Midlands
Herd size	109	130	76
Milk yield (l cow^{-1})	7103	5220	6064
Concentrates (t cow^{-1})	1.66	0.58	1.06
Concentrates (kg l^{-1})	0.23	0.11	0.17
Stocking density (cows ha^{-1})	2.92	2.30	1.90
N (kg ha^{-1})	395	450	331
Milk from forage (l cow^{-1})	3413	3935	3713
ME from forage (%)	72	86	78
Margin (£ cow^{-1})	528	412	489
Milk feed and forage costs (£ ha^{-1})	1541	948	929

Future R & D on grassland output needs to be concentrated on the possibility of growing grass more cheaply and converting it more effectively into animal products. Work on N warrants more effort in this context. Fertilizer N is the single most costly input in producing high yields of grass in Britain, and it is therefore important that it be used efficiently. Yet there are many gaps in our knowledge of N in grassland (Lazenby, 1982; Ryden, 1983); for instance, there is remarkably little information on its effects on DM yields of grazed pastures, whilst its relationship with animal production is also far from clear. We need to know much more than we do about the pathways of N loss from grassland, and of the factors involved in reducing such losses, especially those occurring as a result of volatilization and leaching (Ryden 1983) (see Chapter 5). Such information would not only enable better use of the nutrient by the grass, but would also reduce the risk of pollution of the atmosphere and of ground water.

Large quantities of soil N exist under some intensively used permanent grassland. Increasing the rate of turnover of such N, making more available to the grass without damaging the soil structure or losing the nutrient from the system, would be a major research achievement and could greatly reduce the need for fertilizer N.

Nitrogen can also be supplied through fixation by legumes; if sufficient N could be supplied via white clover it could have a considerable effect on the yield and quality of British grasslands, particularly permanent pasture. There are problems associated with using white clover, including difficulties in its establishment, its susceptibility to moisture stress and severe grazing, and its association with various animal disorders. Nevertheless, the encouraging performance of some cultivars released in the past 10 years, has raised hopes that it may be possible to select white clovers both more persistent and able to perform better than those presently available. Such a development, with its effects both in saving the cost of N fertilizer and improving the efficiency of the whole system, would represent one of the most important advances in grassland R & D for many years.

There are many other problems which need investigation. For example, grazing and conservation are key factors in the economic realization of the potential for animal production; they need to be understood fully. We know less than we should about the sward/animal interface in the pasture (see Chapters 3 and 4). Whilst there have been major developments in the methodology and techniques available for measuring sward growth (Hodgson et al., 1981) and herbage intake (Leaver, 1982), further investigations could provide better insight into the interaction between the sward and the ruminant. Taken together with the effects of fertilizers (especially N) on the grazed sward and on animal production, they should provide the basis for more-rational fertilizer use, improved management practice and better prediction of animal output.

Opportunities also exist for reducing DM losses in conserved forage (Wilkinson, 1981) and improving N utilization by the animal. There appears

Postscript 353

a real chance to break the negative correlation between yield and quality. Improvement of the digestibility, and thus the intake, of poor-quality conserved forage by treatment with alkalis is a possibility; the development of a suitable on-farm method for applying alkalis, thereby increasing the nutritive value of forage without sacrificing yield, would have considerable and obvious advantages.

It is clear that models have a major role in future R & D. They are particularly suited to a comparison of various farming systems (e.g. forage-based and concentrate-based systems for milk and beef production) and feeding strategies (e.g. different methods of feeding and rates of concentrate use). Models should also indicate the likely biological input and economic returns attainable following various inputs and management decisions; they are also helpful in suggesting the factors most needing research. Predictions should be compared with results obtained from systems experiments, preferably conducted over several years. Though lengthy and expensive, such experiments would provide credible and practical information, particularly if they could be extended to commercial farms.

9.8 POSTSCRIPT

Many of the changes which have occurred in world agriculture during the past two decades are directly relevant to grassland enterprises. The food shortages predicted for the early-1970s have been replaced by overproduction, at least in many of the advanced, and some of the developing, countries. Agricultural surpluses in dairy products and meat characterize the European Economic Community, whilst Australia, which has traditionally exported two-thirds of its agricultural products, faces increasing competition to retain its world markets. The fall in commodity prices has had profound effects on farmers, not least those relying on grass-based systems for their livelihood.

R & D has not escaped the consequences of these changes. The very success of research on factors limiting production has prompted questions on both the type of investigations which should now be undertaken and the extent to which they should be publicly funded. Perceptible changes have become apparent in grassland R & D. Whilst some important projects, e.g. the search for better pasture plants for some Australian grasslands, and studies on the effects of poaching on DM yields and utilization of grass on poorly drained soils in Britain, fall into the category of improving production, grassland R & D is no longer directed primarily towards increased or potential output. Increasing effort is being put into improving the efficiency of herbage and animal production, as typified by some of the N work in the UK and attempts to determine minimum maintenance dressings of superphosphate for improved pastures in Australia.

One of the other major changes in grassland R & D has been the increased use of models for prediction purposes. Whilst some models have been developed for studying components of systems, more effort is now possible on modelling whole systems. Rather than studying an effect of an individual variable or a specific management decision in isolation, it became possible to investigate their likely effects on production and productivity of the system as a whole. It also became possible to quantify the likely effect of individual decisions or whole systems on profit and risk. Options were able to be developed, making it possible for the farmer to be better informed when making decisions. Further, targets became used increasingly in assessing grassland output.

The continuing cost-price squeeze can only stimulate grassland farmers to make better use of their grass. Indeed, it appears the only course open for many farmers to maintain their standard of living and for some even to retain their livelihood (Lazenby and Doyle, 1981). Economic pressures have also been the major catalyst in increasing the study of alternative, lower-input forage-based systems for animal production, e.g. the possible greater use of native grasses in improving some Australian pastures, or introducing legumes as an alternative to fertilizer N in British grasslands. However, the potential for environmental pollution following high use of N fertilizer has also played a role in influencing recent research. The quantification of the polluting effects of such fertilizer in both the atmosphere and ground water should provide the basis for better methods of management to reduce the problem in shorter term. The danger of environmental pollution should also ensure adequate resources to help in the development of a more reliable grass/legume system, which, hopefully, will become available in the longer term.

REFERENCES

A'Brook, J. and Heard, A. J. (1975) The effect of ryegrass mosaic virus on the yield of perennial ryegrass swards. *Ann. Appl. Biol.*, **80**, 163–8.
ADAS (1977) Higher yields from dairy herds. *Profitable Farm Enterprises Series, No. 12,* Agricultural Development and Advisory Services. Ministry of Agriculture, Fisheries and Food, London.
ARC (1965) *Nutrient Requirements of Farm Livestock No. 2,* Ruminants, Technical Review and Summaries, Agricultural Research Council, HMSO, London.
Archer, J. R. (1986) Grassland manuring past and present in *Grassland Manuring* (eds J. P. Cooper and W. F. Raymond), Occasional Symposium No. 20, British Grassland Society, pp. 5–14.
Baker, R. D. (1984) UME – a way to assess the efficiency of utilisation of grass, in *Grassland Research Today* (ed. J. Hardcastle), Agricultural and Food Research Council, London, pp. 16–7.
Beddows, A. R. (1958) Grass breeding. *Rep. Welsh Pl. Breeding Stn 1950–56,* pp. 11–35, University College of Wales, Aberystwyth.
Begg, J. E. (1959) Annual pattern of soil moisture stress under sown and native pastures. *Austral. J. Agric. Res.*, **10**, 518–29.

References

Breeze, E. L. (1971) Herbage breeding. *Rep. Welsh Pl. Breeding Stn 1970*, pp. 13–24, University College of Wales, Aberystwyth.

Burton, G. W., Hart, R. H. and Lowrey, R. S. (1967) Improving forage quality in Bermudagrass by breeding. *Crop Sci.*, **7**, 329–32.

Campbell, M. H. (1972) Factors responsible for losses during germination, radicle entry, establishment and survival of surface sown pasture species. Ph.D. Thesis, University of New England.

Campbell, M. H. and Swain, F. G. (1973) Factors causing surface losses during the establishment of surface sown pastures. *J. Range Mgmt*, **26**, 355–9.

Christie, E. K. (1979) Ecosystem processes in semi-arid grasslands. II. Litter production, decomposition and nutrient dynamics. *Austral. J. Agric. Res.*, **30**, 29–42.

Clements, R. O. (1980) Pests – the unseen enemy. *Outlook Agric.*, **10**, 219–23.

Colman, R. L., Lazenby, A. and Grierson, J. (1974) Nitrogen fertilizer responses and seasonal production of temperate and warm climate grasses on the Northern Tablelands of New South Wales. *Austral. J. Exp. Agric. Animal Husbandry*, **14**, 362–72.

Connolly, V., Ribeiro do Valle, M. and Crowley, J. G. (1977) Potential of grass and legume cultivars under Irish conditions, in *An Foras Taluntais* (ed. B. Gilsenan). *Proc. Int. Mtg Animal Production from Temperate Grassland*, Dublin, pp. 23–8.

Cook, S. J., Lazenby, A. and Blair, G. J. (1978a) Pasture degeneration I. Effect on total and seasonal pasture production. *Austral. J. Agric. Res.*, **29**, 9–18.

Cook, S. J., Blair, G. J. and Lazenby, A. (1978b) Pasture degeneration II. The importance of superphosphate, nitrogen and grazing management. *Austral. J. Agric. Res.*, **29**, 19–29.

Corrall, A. J., Morrison, J. and Young, J. W. O. (1982) Grass production, in *Milk from Grass* (eds C. Thomas and J. W. O. Young), ICI and GRI, pp. 1–19.

Crawshaw, R. and Woolford, M.K. (1979) Aerobic deterioration of silage in and out of the soil. *ADAS Q. Rev.*, **34**, 151–78.

Culvenor, R. A. and Oram, R. N. (1985) Breeding phalaris for greater tolerance to acid soils. *Proc. 3rd Austral. Agron. Conf.*, 231, Hobart.

Down, K. M., Jollans, L. L., Lazenby, A. and Wilkins, R. J. (1981) The distribution of grassland and grassland usage in the U.K. in *Grassland in the British Economy* (ed. J. L. Jollans), CAS Paper 10, Centre for Agricultural Strategy, Reading, pp. 580–3.

Downes, R. W. (1969) Differences in transpiration rates between tropical and temperate grasses. *Planta*, **88**, 261–73.

Doyle, C. J., Corrall, A. J., Le Du, Y. L. P. and Thomas, C. (1982) The integration of conservation with grazing, in *Milk from Grass* (eds C. Thomas and J. W. O. Young), ICI and GRI, pp. 59–74.

Doyle, C. J. and Richardson, E. (1982) Economics of production in *Milk from Grass* (eds C. Thomas and J. W. O. Young), ICI and GRI, pp. 75–85.

Eagles, C., Pollock, C., Thomas, H. and Wilson, D. (1984) Breeding for better winter-hardiness and drought resistance, in *Grassland Research Today* (ed. J. Hardcastle), pp. 6–8.

Ernst, E. P., Le Du, Y. L. P. and Carlier, L. (1980) Animal and sward production under rotational and continuous grazing managements – a critical appraisal, in *The Role of Nitrogen in Intensive Grassland Production* (eds W. H. Prins and G. H. Arnold), Pudoc, Wageningen, pp. 119–26.

Fagan, T. W. (1927) The nutritive value of grasses as pasture, hay and aftermath as shown by their chemical composition. *Advisory Bull., Agric. Dep, Univ. Coll. Wales*, **2**, 23.

Fitzpatrick, E. A. and Nix, H. A. (1970) The climatic factor in Australian grassland ecology, in *Australian Grasslands* (ed. R. M. Moore), ANU Press, Canberra, pp. 3–26.

Forbes, T. J., Dibb, C., Green, J. O., Hopkins, A. and Peel, S. (1980) *Factors Affecting the Productivity of Permanent Grassland, a National Farm Study*, Grassland Research Institute and Agricultural Development and Advisory Service Joint Permanent Pasture Group, Hurley.

Frizzel, M. J. (ed.) (1978) *Cycling of Mineral Nutrients in Agricultural Ecosystems*, Elsevier Scientific, Amsterdam.

Garwood, E. A., Tyson, K. C., Denehy, H. L., Stone, A. C., Reid, T. C., Ryden, J. C., Armstrong, A., Atkinson, J. L. and Hallard, M. (1986) The effects of field drainage on sward productivity and utilisation, soil physical condition and nutrient movement. *Final Annual Report 1984–85*, Grassland Research Institute, Hurley, pp. 79–82.

Gordon, F. J. (1981) Potential for change in the output of milk from grassland, in *Grassland in the British Economy* (ed. J. L. Jollans), CAS Paper 10, Centre for Agricultural Strategy, Reading, pp. 429–43.

Grantham, H. A. (1961) *Pastures for the Armidale District, New South Wales*, Department of Agriculture, Division of Plant Industry.

Green J. O. and Baker, R. D. (1981) Classification, distribution and productivity of U.K. grasslands, in *Grassland in the British Economy* (ed. J. L. Jollans), CAS Paper 10, Centre for Agricultural Strategy, pp. 237–47.

Green, J. O. and Lazenby, A. (1981) Improvement of permanent grassland: is reseeding necessary? *Span*, **24**, 66–9.

Green, J. O., Corrall, A. J. and Terry, R. A. (1971) Grass species and varieties. Relationships between stage of growth, yield and forage quality. *Tech. Rep. No. 8*, Grassland Research Institute, Hurley.

Griffith, J. L. P. and Blair, G. J. (1981) *An Evaluation of the Productive Pastures (Bloat Safe) Project*, Agricultural Business Research Institute, Armidale.

Harris, W. and Lazenby, A. (1974) Competitive interaction of grasses with contrasting temperature responses and water stress tolerances. *Austral. J. Agric. Res.*, **25**, 227–46.

Hartley, W. (1963) The phytogeographic basis of pasture plant introduction. *Genet. Agraria*, **17**, 135–60.

Hartley, W. and Neal-Smith, C. A. (1963) Plant introduction and exploration in Australia. *Genet. Agraria*, **17**, 483–500.

Hartridge, F. (1979) *Pastoral Research on the Northern Tablelands, New South Wales*, Department of Agriculture, New South Wales.

Hilder, E. J. (1964) The distribution of plant nutrients by sheep at pasture. *Proc. Austral. Soc. Animal Production*, **5**, 241–8.

Hodgson, J., Baker, R. D., Davies, A., Laidlaw, A. S. and Leaver, J. D. (eds) (1981) *Sward Measurement Handbook*, British Grassland Society.

Holmes, W. (1980) *Grass: Its Production and Utilization*. Blackwell Scientific, Oxford.

References

Hopkins, A. (1979) The botanical composition of grassland in England and Wales; an appraisal of the role of species and varieties. *J. R. Agric. Soc. Engl.*, **140**, 140–50.
Hughes, R. (1971) *Grassland Agronomy. Rep. Welsh Pl. Breeding Stn 1970*, pp. 20–24, University College of Wales, Aberystwyth.
Hughes, R., Borrill, M., Davies, I. and Hayes, J. D. (1962) Plant collection in Portugal, June 1961. *Rep. Welsh Pl. Breeding Stn 1961*, pp. 137–144, University College of Wales, Aberystwyth.
Hutchinson, K. J. (ed.) (1987) *Improving the Nutritive Value of Forage*, CSIRO Working Party Report.
Jenkin, T. J. (1937) Some aspects of strain-building in the herbage grasses. *Rep. IVth Int. Grassland Congr.*, pp. 54–60.
Johns, G. G. and Lazenby, A. (1973a) Defoliation, leaf area index, and the water use of four temperate pasture species under irrigated and dryland conditions. *Austral. J. Agric. Res.*, **24**, 783–95.
Johns, G. G. and Lazenby, A. (1973b) Effect of irrigation and defoliation on the herbage production and water use efficiency of four temperatre pasture species. *Austral. J. Agric. Res.*, **24**, 797–808.
Jones, M. G. and Jones, Ll. I. (1930) The effect of varying the periods of rest in rotational grazing. *Bull. 11, Ser. H, Welsh Pl. Breeding Stn*, pp. 38–59, University College of Wales, Aberystwyth.
Kilkenny, J. B., Holmes, W., Baker, R. D., Walsh, A. and Shaw, P. G. (1978) Grazing management. *Beef Production Handbook No. 4*, Beef Improvement Service, Meat and Livestock Commission, Bletchley.
Lazenby, A. (1981) British grasslands; past, present and future. *Grass, Forage Sci.*, **36**, 243–66.
Lazenby, A. (1982) Nitrogen in grassland ecosystems. *Proc. XIV Int. Grassland Congr.*, Lexington, Kentucky pp. 56–63.
Lazenby, A. and Doyle, C. J. (1981) Grassland in the British economy – some problems, possibilities and speculations, in *Grassland in the British Economy* (ed. J. L. Jollans), CAS Paper 10, Centre for Agricultural Strategy, Reading, pp. 14–50.
Lazenby, A. and Rogers, H. H. (1963) Grass breeding in the United Kingdom. *Herbage Abstr.*, **3**, 73–80.
Lazenby, A. and Rogers, H. H. (1965) Selection criteria in grass breeding; V. Performance of *Lolium perenne* genotypes grown at different nitrogen levels and spacings. *J. Agric. Sci., Camb.*, **65**, 79–89.
Lazenby, A. and Swain, F. G. (1972) Pasture species, in *Intensive Pasture Production* (eds A. Lazenby and F. G. Swain), Angus and Robertson, pp. 67–96.
Leaver, J. D. (ed.) (1982) *Herbage Intake Handbook*, British Grassland Society.
Le Du, Y. L. P. and Hutchinson, M. E. (1982) Grazing, in *Milk from Grass* (eds C. Thomas and J. W. O. Young), ICI and GRI, pp. 43–57.
Lewis, E. J. (1971) Intergeneric hybrids. *Rep. Welsh Pl. Breeding Stn 1970*, p. 14, University College of Wales, Aberystwyth.
Lodge, G. M. (1981) The role of plant mass, basal area and density in assessing the herbage mass response to fertility of some native perennial grasses. *Austral. Rangeland J.*, **3**, 92–8.

Lodge, G. M. and Whalley, R. D. B. (1983) Seasonal variations in the herbage mass, crude protein and in-vitro digestibility of native perennial grasses on the north west slopes of New South Wales. *Austral. Rangeland J.*, **5**, 20–7.

McDonald, G. T. (1968) Recent pasture development on the Northern Tablelands of New South Wales. *Austral. Geogr.*, **X**, 382–91.

McDonald, P. (1976) Trends in silage making, in *Microbiology in Agriculture, Fisheries and Food, Proceedings of Fourth Symposium of Society of Applied Bacteriology* (eds F. A. Skinner and J. G. Carr), Academic Press, London, pp. 109–23.

McWilliam, J. R. (1963) Selection for seed retention in *Phalaris tuberosa* L. *Austral. J. Agric. Res.*, **14**, 755–64.

Minson, D. J. and McLeod, M. N. (1970) The digestibility of temperate and tropical grasses. *Proc. XI Int. Grassland Congr.*, Surfers Paradise, Queensland, pp. 719–22.

Mitchell, K. J. (1956) The influence of light and temperature on the growth of pasture species. *Proc. VII Int. Grassland Congr.*, Palmerston North, New Zealand, pp. 58–69.

MMB (1978) An analysis of LCP costed farms, 1977/78. *LCP Info. Unit Rep. No. 16*, Milk Marketing Board, Thames Ditton.

Morrison, J. (1980) The influence of climate and soil on the yield of grass and its response to fertilizer nitrogen, in *The Role of Nitrogen in Intensive Grassland Production* (eds W. H. Prins and G. H. Arnold), Pudoc, Wageningen, pp. 51–7.

Morrison, J., Jackson, M. V. and Sparrow, P. E. (1980) The response of perennial ryegrass to fertilizer nitrogen in relation to climate and soil. Report of the Joint GRI/ADAS Grassland Manuring Trial, GM20. *Tech. Rep. No. 27*, Grassland Research Institute, Hurley.

Newton, J. E. and Laws, J. A. (1983) Effect of different winter stocking rates on animal and sward performance. *Annual Report, 1982*, Grassland Research Institute, Hurley, pp. 80–1.

NSW Herbage Plant Liaison Committee (1967) Sub-committee Interim Report on the Evaluation of New Herbage Cultivars.

Oram, R. N. (1984) Australian cultivars of phalaris. *Austral. Seed Ind. Mag.*, **2**, 29–30.

Oram, R. N., Schroeder, H. E. and Culvenor, R. A. (1985) Domestication of *Phalaris aquatica* as a pasture grass. *Proc. XV Int. Grassland Congr.*, Kyoto, Japan, pp. 220–1.

Peel, S., Matkin, E. A. and Huckle, C. A. (1986) Grassland production and utilisation on dairy farms in Devon. *Final Annual Report 1984–85*, Grassland Research Institute, Hurley, pp. 84–6.

Richards, I. R. (1977) Influence of sward and sward characteristics on the response to nitrogen, in *An Foras Taluntais* (ed. B. Gilsenan), *Proc. Int. Mtg, Animal Production from Temperate Grassland*, Dublin, June 1977, pp. 45–9.

Robson, M. J. (1982) The growth and carbon economy of selection lines of *Lolium perenne* cv. S23 with 'fast' and 'slow' rates of dark respiration. I. Grown as simulated swards during a regrowth period. *Ann. Bot.*, **49**, 321–9.

Rogers, H. H. (1961) Grasses. *Rep. Pl. Breeding Inst., Cambridge, 1960–61*, pp. 70–4.

Ryden, J. C. (1983) The nitrogen cycle in grassland – a case for studies in grazed pastures. *Essay Paper No. 2, Annual Report (1982)*, Grassland Research Institute, Hurley, pp. 150–66.

References

Schiller, J. M. A. and Lazenby, A. (1975) Yield performance of tall fescue (*Festuca arundinacea*) populations on the Northern Tablelands of New South Wales. *Austral. J. Exp. Agric. Animal Husbandry*, **15**, 391–9.

Smith, R. C. G. and Johns, G. G. (1975) Seasonal trends and variability of soil moisture under temperate pasture on the Northern Tablelands of New South Wales. *Austral. J. Exp. Agric. Animal Husbandry*, **15**, 250–5.

Smith, R. C. G. and Stephens, M. J. (1976) Importance of soil moisture and temperature on the growth of improved pasture on the Northern Tablelands of New South Wales. *Austral. J. Agric. Res.*, **27**, 63–70.

Stapledon, R. G. (1933) *An Account of the Organisation and Work of the Station from its Foundation in April 1919 to July 1933*, p. 167, Welsh Plant Breeding Station, Aberystwyth.

Stapledon, R. G. and Jones, M. G. (1927) The sheep as a grazing animal and as an instrument for estimating the productivity of pastures. *Bull. 5, Ser. H, Welsh Pl. Breeding Stn, Aberystwyth*, pp. 42–54.

Tallowin, J. R. B., Kirkham, F. W. and Brookman, S. K. E. (1986) The response of grazed permanent grassland to N fertilizer. *Final Annual Report 1984–85*, Grassland Research Institute, Hurley, pp. 71–2.

Terry, R. A. (1974) Digestibility of forage. *Silver Jubilee Report 1949–1974*, (eds C. R. W. Spedding and R. D. Williams) pp. 42–8.

Thomas, C. (1980) Factors influencing the potential of silage for milk production. *Br. Grassland Soc. Natn. Silage Competition Conf.*, National Agricultural Centre, Stoneleigh, April.

Thomas, C. and Golightly, A. (1982) Winter feeding, in *Milk from Grass* (eds C. Thomas and J. W. O. Young), ICI and GRI pp. 21–42.

Thomson, D. J. (1979) Effect of proportion of legumes in swards on animal output. *Occas. Symp. 10, Br. Grassland Soc.*, pp. 101–9.

Tilley, J. M. A. and Terry, R. A. (1963) A two-stage technique for the *in-vitro* digestion of forage crops. *J. Br. Grassland Soc.*, **18**, 104–11.

Utley, P. R., Chapman, H. D., Monson, W. G., Marchant, W. H. and McCormick, W. G. (1974) Coastcross–I bermudagrass, coastal bermudagrass and Pensacola bahiagrass as summer pasture for steers. *J. Animal Sci.*, **38**, 490–5.

Vicente-Chandler, J., Silva, S. and Figarella, J. (1959) The effect of nitrogen fertilization and frequency of cutting on the yield and composition of three tropical grasses. *Agron. J.*, **51**, 202–6.

Wade, M. H. (1979) The effect of severity of grazing by dairy cows given three levels of herbage allowance on the dynamics of leaves and tillers in swards of *Lolium perenne*. M.Phil. Thesis, University of Reading.

Walsh, A. (1979) Bigger acres mean more or bigger cows, in *ICI Suppl. to Farmers Weekly*, 23 March.

Wilkins, R. J. (1980) Progress in silage production and utilization. *J. R. Agric. Soc. Engl.*, **141**, 127–41.

Wilkins, R. J. (1986) Evaluation of grass by animal production. *Essay Paper No. 2, Final Annual Report 1984–85*, Grassland Research Institute, Hurley, pp. 155–67.

Wilkins, R. J. and Down, K. M. (1984) Realising the potential of British grassland, in *Grassland Research Today* (ed. J. Hardcastle), Agricultural and Food Research Council, London, pp. 2–3.

Wilkins, R. J. and Wilson, R. F. (1974) Silage, in *Silver Jubilee Report 1949–1974 (eds C. R. W. Spedding and* R. D. Williams), Grassland Research Institute, Hurley, pp. 96–106.

Wilkinson, J. M. (1981) Losses in the conservation and utilisation of grass and forage crops. *Ann. Appl. Biol.,* **98,** 365–75.

Wilson, D. and Jones, J. G. (1982) Effects of selection for dark respiration rate of mature leaves on crop yields of *Lolium perenne* cv. S23. *Ann. Bot.,* **49,** 313–20.

Wolfe, E. C. (1972) Pasture management and its effects on the botanical composition and productivity of pasture, in *Intensive Pasture Production* (eds A. Lazenby and F. G. Swain), Angus and Robertson, pp. 199–220.

Wolfe, E. C. and Lazenby, A. (1973a) Grass-white clover relationships during pasture development. 1. Effect of superphosphate. *Austral. J. Exp. Agric. Animal Husbandry,* **13,** 567–74.

Wolfe, E. C. and Lazenby, A. (1973b) Grass-white clover relationships during pasture development. 2. Effect of nitrogen fertilizer with superphosphate. *Austral. J. Exp. Agric. Animal Husbandry,* **13,** 575–80.

Woolford, M. K. (1984) *The Silage Fermentation,* Microbiology Series, Vol. 14, Marcel Dekker.

Wright, C. E. (1957) *Research Experiments Records, Ministry of Agriculture, Northern Ireland, 1956,* **6,** 1–18.

Wright, C. E. (1978) Maximising herbage production. *Joint Br. Grassland Soc./Br. Vet. Assoc. Conf. Intensive Grassland Use and Livestock Health,* Berkshire College of Agriculture, February, pp. 11–23.

Index

Illustrations are indicated by italic page numbers, tables by bold numbers.

Abbreviations used xiii–xiv
Abscisic acid (ABA) 229
Acclimation
 anaerobic soils 198–9
 metal toxicity in soils 199–200
 salinity 197–8
Aerodynamic method (for gas-exchange measurements) 16
Agrostis spp., *see* Bents
Aldicarb, nematodes controlled by 283
Allelochemic mechanism 191
Aluminium
 content of plants *185*
 essentiality of 186
 seasonal variation *189*
 toxicity in soil 199
Animals
 excreta effects 114–15
 growth rate effects 324–5
 hoof pressures 113
 selective grazing by 116
 sward regrowth affected 101, 113–16
 treading effects 115–16
Annual ryegrass toxicity 284
Aphids 278, 281–2, 283
Argentine stem weevil (*Hyperodes griseus*) larvae 282
Aristidia oligantha 182
Assimilate partition
 parallel system model *256*, 258
 physiological model 254–60
 seasonal changes 145–8
 series system model *256*, 258
Assimilate transformation, physiological model 266–8
Assimilate utilization and partitioning
 defoliation-severity effects 159–61
 dynamics of 53–5
 environmental effects 52, 53
 management effects 52–3
 and meristematic activity 50–5
 physiological model 260–6
 seasonal effects 145–8
 water-stress effects 234, *235*, 264–5, *264*
Australia
 grassland improvement 322–5, 332–40
 insect pests implicated 282
 nematode parasitism in 284
 North Tablelands (NSW) 332–3
  climate 332–3, *337*

Australia *cont.*
 drought conditions 335, 336
 future improvements 338–40
 native grass species 333
 pasture improvement 334–6
 selection methods 315–17
Autumn, leaf extension rates 139, *139*

Bacteria 296
Barium *185*
Barley (*Hordeum* spp.), respiration 58
Barley yellow dwarf virus (BYDV) 297–8
Beer–Lambert law 246, 267
Bents (*Agrostis* spp.)
 browntop (in New Zealand) 1
 seasonal variation in nutrients 186, *188–9*
Bio-clock concept 268
Biochemical pathways, C3 and C4 species 4–6
Biomass production 39–42
 irradiance effects 39–42
 temperature effects 39
Black beetle (*Heteronychus arator* (=*sanctae-helenae*)) larvae 282
Black field cricket (*Teleogryllus commodus*) 282
Black rust (*Puccinia graminis*) 288
Blackman limiting response (of photosynthesis) 249
Blind seed, *Gloeotinia temulenta*-caused 295
Blue grama (*Bouteloua gracilis*), hydraulic resistance in roots 217
Boron *185*, 186
Breeding 17
 see also Selection
Britain, *see* Great Britain; *and* United Kingdom
Brome mosaic virus (BMV) 300–1
Bunts (*Tilletia* spp.) 290
Burning
 management of savannah by 186–7
 nitrogen loss/mineralization by 193
 phosphate returned to soil by *194*

C3 grasses 4, 6–7
 potential yields 18–19, 168
C4 grasses 4–7, 312
 potential yields 19, 317–20, *318*
 quality/digestibility 327, *327*

C4 syndrome 5
Calcium *185*, 186
Calvin cycle 4, 6
Canopy architecture
  concept first used 16
  photosynthesis affected 142–3
Canopy interception (of precipitation) 209–10
Canopy photosynthesis 65–6
  continuously grazing swards 151–2
  contribution of sheath 90
  factors affecting 68–71
  intermittently defoliated swards 157–9
  irradiance effects 250–1, *251*
Capacitance, water content 208
Carbohydrates
  regrowth affected 90–1
  seasonal changes in metabolism 136–7
  water stress 222
Carbon balance
  input/utilization model 260–3
  model seedling sward 62–8
  seasonal changes 143–5
Caryopsis 26–7
  see also Seeds
Cattle/cows
  lactation curve 330, *331*
  nutritional requirements 330–2
  rejection of grazing due to dung 113, 114–5
  tillers in swards affected 101
Ceiling yield
  Donald's concept 71–2
  harvesting at 164
  seedling sward *64*, 65, 72–3
Cell division, water stress effects 225–6
Cell growth, water stress effects 219–25
Celtic period, improvement of grassland in Britain 8–9
Cenchrus ciliaris, water-stress response **224**, 225
Chlorine *185*, 186
Choke (*Epichloë typhina*) 295–6
Chromium *185*, 186
Click beetle (*Agriotes* spp.) larvae, *see* Wireworms
Clover
  possible use in UK pastures 352
  use in Australian pastures 339
  use in New Zealand pastures 7
Cobalt 184, *185*, 186
Cocksfoot (*Dactylis glomerata*)
  first imported to UK 11
  irradiance effects 32
  regrowth response 93, **94**
  root depth effects on yields **216**
  seed size 29
  water-stress responses 225, *227*, **228**
Cocksfoot mild mosaic virus (CFMMV) 301
Cocksfoot mottle virus (CFMV) 301

Cocksfoot streak virus (CFSV) 301
Community, grass sward 62–73
Competition (between species)
  clover and grass *323*
  nutrient effects 195–6
Conservation (of herbage) 329–30
  integrated with grazing 344–9
  objectives of 345
Continuous defoliation 149–56
  defoliation pattern and severity 150–1
  intensity effects on sward growth 153–6
  leaf/canopy photosynthesis characteristics 151–2, **152**, *153*
Continuous grazing
  compared with intermittent grazing 168–9
  critical sward height 329, *329*
  see also Grazing
Controlled-environment cabinet, development of 15
Copper *185*, 186
Cranefly (*Tipula* spp.) larvae, *see* Leatherjackets
Crested dogstail (*Cynosurus cristatus*) 11
Crown rust (*Puccinia coronata*) 286–7
Curvularia geniculata 285
Cylindrocarpon radicicola 284, 285

D-value, *see* Digestibility
Dactylis glomerata, *see* Cocksfoot
Damping off, pathogens causing 284–5
Danthonia pallida 339–40
de Wit model 16, 18
Defoliation
  assimilate utilization affected 52–3
  continuously grazed swards 104–5, 150–1
  long-term effects on stubble composition 105–7
  optimum heights 102–4, 328–9
  regrowth affected 101–7
    height of defoliation 85, 104–5
    time of defoliation 108–10
  severity effects on
    assimilates supply 159–61
    herbage production 102–3, **103**
    losses by death 161–3
  transfer experiments 105
  yield affected 324
  see also Continuous defoliation; Intermittent defoliation
Dew 208–9
  water relations affected 209, *209*
Digestibility of herbage (D-value) 326
  relationship with yield 326–7, *326*
  water-stress effects 234
Dollar spot (*Sclerotinia homoeocarpa*) 293
Drechslera spp. 284, 285, 289–90
Dung pats
  herbage rejection by animals due to 114–15

Index

Dung pats *cont.*
  long-term effects 114

Economic considerations 354
Enclosure (of fields) 10
Enclosure methods (for gas-exchange measurements) 16
Environmental factors
  assimilate partitioning controlled by 52, 53, 254–5
  effects on biomass production 39–42
  photosynthesis affected 43–5, 46–50
  reproductive growth 36–7
  vegetative growth 31–5
Ergot (*Claviceps purpurea*) 294–5
Eriophid mite (*Abacarus hystrix*) 282
Evapotranspiration, soil–plant system 210–11

Fairy rings 294
Fertilizers
  Australian pasture improvement 338
  ecosystem–process responses 191–5
  herbage yield relationship 340–1, **341**, *342*
  use in UK 13, *13*
Fescues (*Festuca* spp.)
  biomass production 40–2, *40*, *41*
  leaf growth affected by temperature 33, *34*
  leaf rolling 233
  regrowth response **94**
  root depth effects on yields **216**
  seed 29
  water-stress response 225
Fire, *see* Burning
Fischer–Turner equation 213
Flooding tolerance 198–9
Flowering
  effect on tillers 26, 113
  see also Inflorescence
Foxtail midge (*Dasineura alopecuri*) 283
Frequent lenient cutting system 166–8
Frit fly (*Oscinella frit*) larvae 278, 279
Fructosans 90
  accumulation of *54*, 136–7
  mobilization of *54*, 137
Fungi
  established sward 285–92
  newly sown grassland 284–5
  seed crops 294–6
  turf grasses 292–4
Fusarium patch (*Micronectriella nivalis*) 292–3
Fusarium spp. 284, 285

Garden chafer (*Phyllopertha horticola*) larvae 279
Georgian period, improvement of grassland in Britain 10–11

Germination processes 28–9
Graminae
  classification **3**
  description 3–4
Grass grubs (*Costelytra* (=*Odontria*) *Zealandica*) 282
Grass seed midges 283
Grass systems
  in Australia 332–40
  in UK 340–53
Grasshoppers 282
Grassland Research Institute (GRI) 320, 326, 332
Grazing
  critical sward height 328–9, *328*, *329*
  growth rate (of grass) affected 324–5
  integration with conservation 344–9
  intensity, effect on sward growth 116, 153–6
  tillers in swards affected 101
  utilization of herbage 328–9
  see also Continuous grazing; Intermittent grazing
Great Britain
  grassland improvement, historical account 8–14
  see also United Kingdom
Greenbug (*Toxoptera* (=*Schizaphis*) *graminum*) 281
Growth analysis, first used 15
Growth rate
  nitrogen nutrient effects 323
  rainfall effects 319–20, *319*, *321*
  temperature effects 317–20, *321*

Halo spot (*Selenophoma donacis*) 291–2
Halophytic grasses 197–8
Harvest timing 72–3, *72*
Haymaking 329
Herbage quality 326–7
Heterodera spp. (nematodes) 283–4
Heteropogon contortus, water-stress responses **224**, 225, 228, *229*
Historical introduction 1–20
Holcus lanatus, seed 27
Humidity, stomatal response to change 229–31
Hydrological cycle 206
  canopy interception 209–10
  inputs of water 208–9
  losses of water 210–11
  transpiration control 212
  uptake of water by plant 215–17
  water use efficiency 212–15
Hyparrhenia rufa, seasonal variation in nutrients 186–7, *188*

ICI Dairymaid scheme **351**
Inflorescence 26, *27*

Infrared gas analyser (IRGA)
  development of 15
  use in transpiration studies 16
Infrequent severe cutting system 165–6
Insect pests 278–83
  established swards 280–2
  newly sown grassland 279
  seed crops 283
  yield loss in Britain due to *280*
Intensive grass systems 340–53
Intermittent defoliation 156–64
  assimilates supply affected by severity 159–61
  losses by death affected by severity 161–3
  photosynthesis affected 157–9
  severity affected by timing of harvest 163–4
Intermittent grazing, compared with continuous grazing 168–9
Ireland, fungal pathogens in 287, 291
Iron
  composition of plants *185*
  essentiality of 186
  seasonal variation *189*
  toxicity in soil 199
Irradiance
  assimilate utilization affected 52, *52*
  biomass production affected 39–42
  leaf affected 31–2
  photosynthesis affected 46–9
  physiological model of canopy interception 246–51
Italian ryegrass *see* Ryegrasses

Japan, pathogens in grass 290, 296
Jenkin, T. J. 12, 313

Kranz anatomy 5

Leaching 191, *192*, 193
Lead
  composition of plants *185*
  essentiality of 186
  seasonal variation *189*
  tolerant species 200
Leaf area
  modelling of expansion 266–7
  regrowth affected 88–90, *89*
Leaf area index (LAI)
  concept first developed 15
  management effects on 150–64
  relation with canopy photosynthesis *158*
  see also Optimum leaf area index
Leaf blotch/scald (*Rhynchosporium* spp.) 291
Leaf extension
  temperature effects 137–41
  water stress effects 219–22, *220*, *221*
Leaf fleck (*Mastigosporium* spp.) 291

Leaf growth 25, 30
  environmental effects 31–5
  irradiance effects 31–2
  morphology 30–1
  temperature effects 32–5
Leaf morphology, water stress effects *219*
Leaf photosynthesis 42–50
  age effects 45–6
  environmental effects 43–5
  intermittently defoliated swards 157–9
  irradiance effects 46–9, 249–50, *250*
  nitrogen deficiency effects 50
  temperature effects 49, *50*
Leaf rolling *219*. 232–3
Leaf water potential, stomatal response to change 227–9
Leaf-mining larvae 280
Leafspot fungi 289–90
Leatherjackets (*Tipula* larvae) 278, 279, 281
Legumes
  competition with grasses 196, *196*
  nitrogen fixation by 322–3
  rainfall effects on growth 319–20, *319*
  temperature effects 319–20, *319*
Ley farming 12–13, 14, 313
Light interception, physiological model 246–51
Ligniera pilorum 285
Lodging, effect on canopy photosynthesis 142, 143
Lolium spp., *see* Ryegrasses
Loose smut (*Ustilago avenae*) 290
Lucerne 336, 339

McCree's (respiration) model 56, 251, 252
Macronutrients
  nitrogen cycle 191–3
  phosphorus cycle 193–5
  supply effect on competitive status of grasses 195–6
Magnesium
  composition of plants *185*
  essentiality of 186
  seasonal variation *188*
Maize (*Zea mays*), water-stress response *220*, *221*
Management techniques
  continuously grazed swards 149–56
  effects of 17, 148–64
  future considerations 169
  intermittently grazed swards 156–64
Manganese *185*, 186
  toxicity in soil 199
Marasmius oreades 294
Meadow foxtail (*Alopecurus pratensis*) 11
Meadow grasses (*Poa* spp.) 11
  germination conditions 28–9
  root depth effects on yields **216**

Index

Mechanization, effect on grassland management 14
Medieval period, improvement of grassland in Britain 9–10
Meloidogyne naasi 283–4
Meristematic activity
  assimilates utilized in 50–5
  defoliation effects 52–3
  environmental effects 52, 53
Metabolizable Energy (ME) concept 330, 332
Metal toxicity, tolerant species 199–200
Metallothionein proteins 200
Meteorological Office Rainfall and Evaporation Calculation System (MORECS) 211
Michaelis–Menten equation *257*, 258
Midlands 'fattening' pastures (UK)
  origin of 10
  plants selected from 12, 313
Mildew infection 288–9
Milk Marketing Board herds 332
Mine wastes, grass–legume swards grown 196
Mineral nutrients
  availability defined 179, 183–4
  composition of plants *185*
  economic aspects 200–1
  essential elements 184–6
  flow in grassland ecosystem *180*, 190–5
  seasonal variation 186–7, *188–9*
  water-stress effects on uptake 233–4
  see also Nitrogen; Phosphorus
Minus 3/2 law 97
Mixed farming 14
Mixed grazing, advantage of 115
Model seedling sward 62–8
  canopy photosynthesis rates 65–6
    factors affecting 68–71
  ceiling yield *64*, 65, 71–3
  leaf growth/death 62–5, *63*
  respiration processes 66, 67, 68
Models
  benefits of 243
  predictive 353–4
  requirements for 244–6
  types of 244
  see also Physiological models
Molybdenum *185*, 186
Monteith's (light attenuation) model 247
Morphology, importance of development 16
Münch hypothesis (for assimilate transport) 244, 258–9
Mycorrhizal fungi, grass plants affected 181–2

Napier grass (*Pennisetum purpureum*)
  potential yields 19
  yields 319
Natural grasslands 1
  actual production data 20

Natural grasslands *cont.*
  comparison with cultivated grasslands 20
  improvement of 7–8
Nature conservation management 201
Nematodes 283–4
Net assimilation rate, concept first developed 15
New Zealand
  fungal pathogens in 285, 286, 288, 289, 294–5
  insect pests implicated 282, 283
  loss of phosphate in soils 193
  mixed ryegrass/clover pastures 7, 8
  nitrogen fixation by legumes 322
  viruses in 298
Nickel
  composition of plants *185*
  seasonal variation *189*
Nitrification inhibitors 182, 191
Nitrogen
  composition of plants *185*, 186
  essentiality of 186
  growth rate effects 320, 322–5
  input/utilization model for 260–3
  macronutrient cycle 191–3
  water-stress effect on uptake of 233–4
  yields affected 323
Nitrogen deficiency, photosynthesis affected 50
Nitrogen fixation 182, 193
  genetic improvement possibilities 182–3, 201
  by legumes 322–3
North America
  fungal pathogens in 285, 288, 290
  insect pests implicated 282
Nutrients, *see* Mineral nutrients
Nutritional requirements (of cows) 330–2

Oat sterile dwarf virus (OSDV) 301
Oats (*Avena* spp.), seasonal variation in nutrients *187*
Ophiobolus patch (*Gaeumannomyces graminis*) 293–4
Optimized production
  frequent lenient cutting 166–8
  grazing methods 168–9
  infrequent severe cutting 165–6
  physiological basis for 164–9
Optimum leaf area index 71
Osmotic adjustment *219*, 222–5

Panicum maximum
  leaf senescence 233
  water-stress responses *220*, 224, 225, 228, 231, *231*, 232
Park Grass experiments 11

Paspalum
  effects of dew 209, *209*
  leaf water potential data 209, *209*
  rainfall effects 320, *321*
  temperature effects 317, *318*
  yield 339
Pathogens 277, 284–301
  bacterial 296
  effects of 278
  fungal 284–96
  virus 296–301
Penman–Monteith equation 210, 211
Perennial ryegrasses, *see* Ryegrasses
Pests 278–83
  definition of 278–9
  different species 277
  effects of 278
  established swards 280–2
  newly sown grassland 279
  seed crops 283
  yield loss in Britain due to 280
Phalaris
  cultivars 316–7
  first introduced into Australia 315, 316
  rainfall effects 320, *321*
  temperature effects *318*
Phleum pratense, *see* Timothy
Phosphatase activity 181
Phosphate cycle 193–5
  legume/grass ratio affected 196, *323*
Phosphorus
  composition of plants *185*
  essentiality of 186
  macronutrient cycle 193–5
  water-stress effect on uptake of 233
  see also Superphosphate
Photoperiod, effects on
  leaf size 32
  reproductive growth 36
Photosynthesis
  continuously grazed swards 151–2, *153*
  intermittently defoliated swards 157–9
  physiological model 246–51
  temperature effects 49, *50*
  water-stress effects 231–2
  see also Canopy photosynthesis; Leaf photosynthesis
Photosynthetic potential, seasonal changes 141–3.
Photosynthetically active radiation (PAR), physiological model of canopy interception 246–51
Physiological models
  assimilate partition 254–60
  assimilate utilization 260–6
  light interception 246–51
  photosynthesis 246–51
  regrowth 268

Physiological models *cont.*
  requirements for 244–6
  respiration 251–4
  symbols used 272–5
  terminology used 244
  tissue death 268–9
  transformation of assimilate into
    leaf area 266–7
    roots 268
    stem tissue 267
  usefulness of 243
Physiology
  optimized production 164–9
  recent developments 14–17
  reproductive vs vegetative growth 25–6
Poa spp., *see* Meadow grasses
Poaching 324
  see also Animals, treading effects
Potassium *185*, 186
Potential production concept 17–20
Powdery mildew (*Erysiphe graminis*) 288–9
Proline accumulation 222
Puccinia spp. (rusts) 286–8
Pythium spp. 284, 285.

Quality
  considerations 326–7
  water stress effects 234

Rainfall, growth rate affected by 319–20, *319*, *321*
Red thread (*Laetisaria fuciformis*) 293
Reductive pentose phosphate (RPP) pathway 5
  see also Calvin cycle
Regrowth
  defoliation systems, effect on 101–7
  physiological model 268
  reproductive swards 107–13
    species differences 112–13
    timing of defoliation, effects on 108–10
  sward structural changes 94–101
    leaf and sheath length 94–6
    leaves per tiller 96–7
    tiller density and turnover 97–101
  vegetative plants 70–1, 87–94
    carbohydrate effects 90–1
    leaf area effects 88–90
    relative importance of leaf area and carbon reserves 91–2
    root effects 92–3
    species/variety differences 93–4, **94**
Rejection (of herbage by animals) 113, 114–5, 156
Reproductive growth 35–7
  environmental effects 36–7
  morphology 35

Index

Reproductive swards
  head numbers 111–12
  regrowth characteristics 107–13
    species differences 112–13
  tiller production 110–11
Respiration processes 55–62
  McCree's model 56, 251, 252
  maintenance component 57–9, 66, 67, 68
  manipulation of 59–62
  physiological model 251–4
  residual problems 59
  synthesis component 56–7, 66, 67, 68
  Thornley's model 252–4, *253*
  yield affected 59–62
Rhizoctonia spp. 284, 285
Rhizosphere bacteria, grass plants affected 181–2
Rhodes grass (*Chloris gayana*) 112
Roman times, improvement of grassland in Britain 9
Root systems
  depth effects on yields **216**
  hydraulic resistance in 216–17
  regrowth affected 92–3
  seasonality patterns 146–7, *146*
  water potential gradients in 215–16
Rotational grazing 168
  critical sward height 328, *328*
Rusts 277, 286–8
Ryegrass mosaic virus (RMV) 297, 298–300
Ryegrasses (*Lolium* spp.)
  assimilate utilization *54*
  canopy photosynthesis in sward *65, 67, 69*
  ceiling yield in sward *64*
  dry-matter production *341*
  inflorescence 26–7, *27*
  irradiance effects *32*
  leaf growth/death in sward *63*
  leaf photosynthesis *43, 45, 47*
  mixed (with clover) pastures, in New Zealand 7, 8
  potential yields 18
  rainfall effects 320, *321*
  regrowth response 70–1, *70*, **94**, *96*
  reproductive tillers 85, *86*
  respiration data *60, 66*
  root depth effects on yields **216**
  seasonal variation
    in growth **138**, 139–40, *139, 140*
    in nutrients 186, *188*
  seeds 27, *27, 28,* 29
  tiller flowering 113
  tiller production 37–9, *38*
  tillers in swards *98, 99, 100*
  value first recognized 10, *11*
  vegetative tillers 85, *86*
  water storage capacity 209

Ryegrasses *cont.*
  water-stress responses *220, 222, 223, 225, 227,* **228**, *230*
  yield data *61,* 62

Salinity-tolerant species 197–8
Savannah grasslands 1
  burning of 186–7, 193
  seasonal variation of typical species 186, *188*
Saxon period, improvement of grassland in Britain 9
Seasonality
  air temperatures *130,* 131, *133*
  assimilate partition 145–8
  carbohydrate metabolism changes 136–7
  carbon balance of sward 143–5
  light environment 130–1
  nutrient ion deployment 186–7, *188–9*
  photosynthetic potential 141–3
  production processes 134–48
    seasonal patterns of production 134–6
  root systems 146–7, *146*
  soil temperatures 131–2, *132, 133*
  temperature effect on leaf extension 137–41
  temperature environment 131–4
Seed crops
  fungi in 294–6
  insect pests in 283
Seedling emergence 29–30
Seeds 26–7
  germination of 28–9
Selection
  in Australia 315–7
  general considerations 312–13
  in United Kingdom 313–15
Selective grazing 116
Selenium *185,* 186
Sheep
  Australian improved pastures 334–5
  facial eczema 286
  Great Britain in Middle Ages 9
  rejection of grazing due to dung 113, 115
  tillers in swards affected 101
Silage 329–30
  additives used 330
  energy-loss target 329
  nutritional requirements supplied by 330–2
  peak daily intake by cow 331
Silicon 184, *185,* 186
Slugs 279
Smuts 277, 290
Snow mould 292
Sodium *185,* 186
Soil
  evaporation from 211
  interdisciplinary approach to study 179

Soil–plant–air continuum *180*, 183–4
   water movement equation 207
Solute potential, lowering 222
Sorghum spp., water-stress response **224**
Specialization (in farming), effect on grassland management 14
Species differences
   reproductive sward regrowth 112–13
   vegative sward regrowth 93–4, **94**
Spring growth 25, 34–5, 139, *139*
Stapledon, R. G. 12, 313
Stem-boring larvae 279, 280, 281
Stocking rate
   effects on milk yield 342, *344*
   effects of season and class site 342, *343*
   in intensive grass–N system 340–9, *349*
   grass–legume systems, NSW 335, *336*, 339
Stomata, role in controlling transpiration 207–8, 212, 226
Stomatal conductances
   response to humidity changes 229–31
   response to leaf water potential changes 227–9
Stone Age, improvement of grassland in Britain 8
Stress
   definitions 197, 206
   see also Water stress
Stripe rust (*Puccinia striiformis*) 288
Stripe smut (*Ustilago striiformis*) 290
Strontium *185*
Stubble composition
   defoliation systems, effect on 105–7
   reproductive swards 107–8
   vegetative swards 105–7
Sulphur *185*, 186
Superphosphate
   cost 339
   effect on Australian pastures *323*, 322–3, 338
   first used 11
   see also Phosphorus
Sward height 328, *328, 329*, 342
   see also Grazing
Sweet vernal (*Anthoxanthum odaratum*) 11
Swift moth (*Hepialus lupulinus*) larvae 279

T-sum 340
Temperate grasses 4, 6
   potential yields 18–19
Temperature
   assimilate utilization affected 53
   biomass production affected 39
   growth rate affected 317–19, *321*
   leaf growth affected 32–5, 137–41
   photosynthesis affected 49, *50*
   seasonal variation 131–4

Thornley's respiration model 252–4, *253*
Tiller production 37–9
   assimilates consumed 101, 255
   reproductive swards 110–11
   vegetative swards 97–101
   water-stress effects 226
Tillers
   importance of 16
   leaf numbers 96–7, 265, 269
   life-lengths 25
Timing (of cut/harvest)
   regrowth affected by 108–10
   severity of defoliation affected by 163–4
   utilization affected by 346, *347*
Timothy (*Phleum pratenis*)
   imported to UK 11
   regrowth response **94**
   root depth effects on yields **216**
   seed 27, *27*, 29
   timing-of-cut effects 108, *109*
   water-stress response 225
Tissue death, physiological model 268–9
Titanium *185*
Transpiration, control by plants 207–8, 212
Treading (by animals), herbage production affected by 115–16
Tropical grasses 4, 6
   germination conditions 29
   potential yields 19
   water-stress responses **224**, 225, 228, **228**, 231, *231*, 232
   see also Cenchrus ciliaris; Heteropogon contortus; Panicum maximum
Tudor period, improvement of grassland in Britain 10
Turf grasses
   fungi in 292–4
   insect pests in 282

United Kingdom
   grass growth days [map] 322
   grassland improvement in 340–53
      historical account 8–14
   intensive grass–nitrogen system (of milk production) 340–53
      conservation integrated with grazing 344–9
      dry-matter production 340–2
      utilization of herbage 342–4
   selection methods 11–12, 313–15
   see also Great Britain
Units, abbreviations used xiii–xiv
Ustilago spp. 290
Utilization
   conservation 329–30
   grazing 328–9
   intensive UK system 342–4

Index

Van den Honert equation (for water movement) 207, 208
Vanadium *185*
Vegetative plants
  leaf extension in 265
  leaves per tiller 96–7, 265, 269
  regrowth in 70–1, 87–94
  shoot morphology 30–5
Viruses 296–301

War, effect on grassland research in UK 12
Water inputs *206, 208–9*
Water losses, soil–plant system 210–11
Water potential gradients 215–16
Water storage capacity, parts of plant 208, 209
Water stress
  assimilate utilization affected 264–5, *264*
  cell division affected 225–6
  cell growth affected 219–25
  development of 217–19
  herbage quality affected 234
  leaf morphology affected *219*
  leaf senescence affected *219*, 233
  morphological changes due to 232–3
  nutrient uptake affected 233–4

Water stress *cont.*
  photosynthesis affected 231–2
  and stomatal conductance 226–31
  yield affected 234–5
Water uptake, grass plants 215–17
Water use efficiency (WUE) 212–15
  factors affecting 213–14
  Fischer–Turner equation 213
  measurement of 213
Waterlogged soils
  lack of information 205
  tolerant species 198–9, 205
Welsh Plant Breeding Station (WPBS) 12, 313
Wilting percentage 215
Wireworms (*Agriotes* larvae) 279, 281
World distribution (of grass species) 1–3, *2*

Yields
  nitrogen nutrient effects 320, 322–5
  temperature effects 320
  water-stress effects 234–5, 320

Zinc *185*, 186
  tolerant species 199–200